S. VAN DER SPOEL · R. P. HEYMAN

A Comparative Atlas of Zooplankton

Biological Patterns in the Oceans

WITH 50 FIGURES AND 143 MAPS

Springer-Verlag Berlin Heidelberg GmbH 1983

Siebrecht van der Spoel
Robert Pieter Heyman

Institute of Taxonomic Zoology
Postbus 20125
NL – 1000 HC Amsterdam

ISBN 978-3-662-02368-6 ISBN 978-3-662-02366-2 (eBook)
DOI 10.1007/978-3-662-02366-2

Library of Congress Cataloging in Publication Data
Spoel, S. van der.
An atlas of comparative zooplankton
Bibliography: p.
Includes index.
1. Marine zooplankton—Geographical distribution.
I. Heyman, R. P. (Robert Pieter), 1952- . II. Title.
QL123.75 1983 592 83-10440

Originally published by Wetenschappelijke uitgeverij Bunge in 1983.
Softcover reprint of the hardcover 1st edition 1983

Contents

I Introduction

It is a well-known fact, that both the geographic distribution and the speciation of plankton are products of the geological history of the oceans, the continental barriers, the current patterns and the limitation of survival of individual species and populations by both biotic and abiotic environmental conditions.

The geographic distribution is also the product of the mobility of populations, the selective pressure and the time during which a taxon has existed. A taxon consists of a population, or a group of populations, sufficiently distinct to be provided with a name, to be ranked in a definite category and to establish a geographic distribution. Each taxon forms part of biological evolution, and is therefore never sharply delimited in time. Ancestors always influence the dispersal of their descendants. Environmental conditons exert a natural selective influence on each population, which results in the survival of the most adequately adapted pheno- and genotypes at a particular place and during a particular period. Consequently, selection determines the presence of a phenotype, or taxon, in an area at a certain moment. In this way, the environment determines the type of taxa, their abundances, their seasonal changes and survival in time. In a broad sense, mobility of populations does not merely depend on migration and transport; reproduction, mortality and population dynamics also influence the mobility. In that sense, mobility comprises all phenomena of movement, including the changes in number of specimens present at a locality. Increase of the number of individuals is a phenomenon of expansion, decrease one of regression; the total area covered by a taxon, and the areal variation in abundance are zoogeographical parameters. As mobility is largely a behavioural process, it is genetically determined, as the population dynamics are, and therefore genetics also contribute to the zoogeography.

Speciation can be defined as the interaction of the factors time, selective pressure and distribution, resulting in a certain genetic configuration of the genome. This means, that the variability or stability of a genotype is directly linked with the variability or stability of distribution patterns. This interaction is pictured in diagram I.

Diagram I

The interaction between speciation and dispersal processes

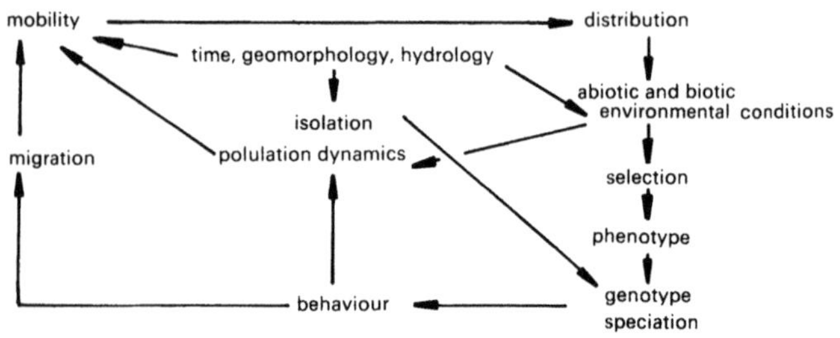

Hence, zoogeography is more than a mere description of presence and absence of taxa.

To conclude from the numerous monographs, faunistic and ecological works and taxonomic papers, the distribution of planktonic species has been treated extensively and with many different approaches. However, a survey on general distribution patterns has not been published before. The large-scale biogeographic trends in marine biota form the topic of this book. To explain the complex system of interacting factors determining the details of distribution, much more information is required than provided in these few pages. There are, however, some influences, usually overruling minor and local or temporary forces, which shaped the general types of distribution.

The major types of distribution have been chosen as a basis to elucidate the general forces in marine biogeography. Hence, this atlas does not give the distributions of all known planktonic species. The intention is to provide material for fundamental comparisons between zoogeographic problems and ecological, oceanographical and taxonomical ones. As the types of distribution here distinguished are each represented by many species, only the most representative or most accurately studied taxa have been selected for this atlas. The boundaries of ranges considered to belong to one and the same distributional type will never coincide fully, yet, this phenomenon does not figure in the maps as it makes the distinguishing of types more difficult, without adding much theoretical information. For the Pacific Ocean, McGowan (1971) published numerous maps showing the variability of ranges of the same distributional type, but it does not seem appropriate to follow this method for the present atlas.

The types of distribution selected are given in the third chapter. They are distinguished on their geomorphological features, and consequently treated in a descriptive manner. It will undoubtedly become possible to discriminate types on the basis of causal zoogeography, using rather the mechanisms behind the patterns than only their shapes, but in our opinion, science has not yet proceeded far enough for this method, though preferable, to be applied.

Distinguishing types of distribution implies giving them names. Concepts such as 'tropical species', 'antiboreal species' etc. are well-known zoogeographic terms but it was not possible to use such geographic-climatological terms for the patterns discussed here, as the complexity is far too great to be covered by such terms. Therefore, the taxonomic name of the taxon which is thus far the best-known example of a particular type of distribution is used to name that type. For example, *Stylocheiron maximum* is the taxon most characteristic of the STYLOCHEIRON MAXIMOIDES type of distribution.

It is questionable whether epi-, meso- and bathypelagic distributions should be separated, especially when they show the same horizontal range. Factors determining bathypelagic patterns usually differ from those responsible for the epi- and mesopelagic ones, so that bathypelagic types may be considered separately. It should not be excluded, however, that interaction exists between meso- and epiplankton on the one hand, and bathyplankton on the other. The dispersal of both epi- and mesoplankton is influenced by partly the same factors, though mesoplankton tends to be more widely distributed. For this reason, patterns of meso- and epiplankton have frequently been taken together into one type and only separated when there were obvious distinct differences.

Even a rigid distinction between plankton and nekton has not been made whenever there was a great similarity in distribution between the two groups. This similarity is caused by the identical shapes of the patterns and less by the influences responsible for the distribution, which may strongly differ in the passively floating plankton and the actively swimming nekton. The high degree of similarity, however, also shows that there must be, probably hydrological, factors which affect plankton in the same way as nekton, and, moreover, both may influence each other's population dynamics.

Although in this atlas the theoretical discussion has been kept to a minimum, the second chapter deals with the species concept and related problems to give an impression of the taxa whose distributions created the patterns represented. For a discussion on speciation and zoogeographical theories the reader is also referred to Van der Spoel & Pierrot-Bults (1979) and Van der Spoel (1983).

The distribution patterns as compared here, have been taken from various authors. For the basic scientific data used in drawing these patterns, the reader is referred to the original papers as listed with the alphabetic index to the maps.

It is a pleasure to mention here the numerous discussions on zoogeography we had with Dr. A. C. Pierrot-Bults, which improved many parts of this book. For linguistic advice and the preparation of the manuscript the authors are very much indebted to Els van der Zee.

THE AUTHORS

2 The taxon and its range

The biotopes of the ocean—Hydrological conditions—Ranges and their interpretations—The plankton species—Boundaries and barriers

In Chapter 3 the types of distribution involve different kinds of species (variable and non-variable, asexually and sexually reproducing ones etc.) and different kinds of ranges (reproductive, non-sterile expatriation, seasonal ranges etc.). In the present chapter, these dissimilarities are discussed to enable further comparisons; a rough description of the environmental parameters is given first.

The biotopes of the ocean

As invisible as the equator is on the earth's surface are the discontinuities in the open ocean. The man-made equator is of no biological importance, the discontinuities are natural and of great significance. To indicate the equator is easy, to draw with any accuracy the lines marking discontinuities on the same map is impossible, and still the zoogeographer needs them more than latitudes and longitudes. Many ocean maps with areal subdivisions are published, but it is hard to discover two which are identical or nearly so; the subdivisions presented are based on different parameters of physical, climatological, chemical or biological nature. Though several geographic terms naming the areas have a well-defined meaning in the different disciplines, it is, due to the different starting-points above-mentioned, impossible to give an exact geographic outline of the areas indicated by the terms. It is interesting, for example, to compare areas indicated in maps by various oceanogeographers with the name Atlantic Temperate.

To pinpoint the exact geographic localities of biotopes, and to provide these with proper names, is as a consequence very difficult. Yet, the student is bound to start with concepts such as 'Subarctic' and 'Transitional' areas as these have a generally accepted meaning, and for that purpose the figs 2-15 have been included, representing different biogeographic interpretations of the ocean.

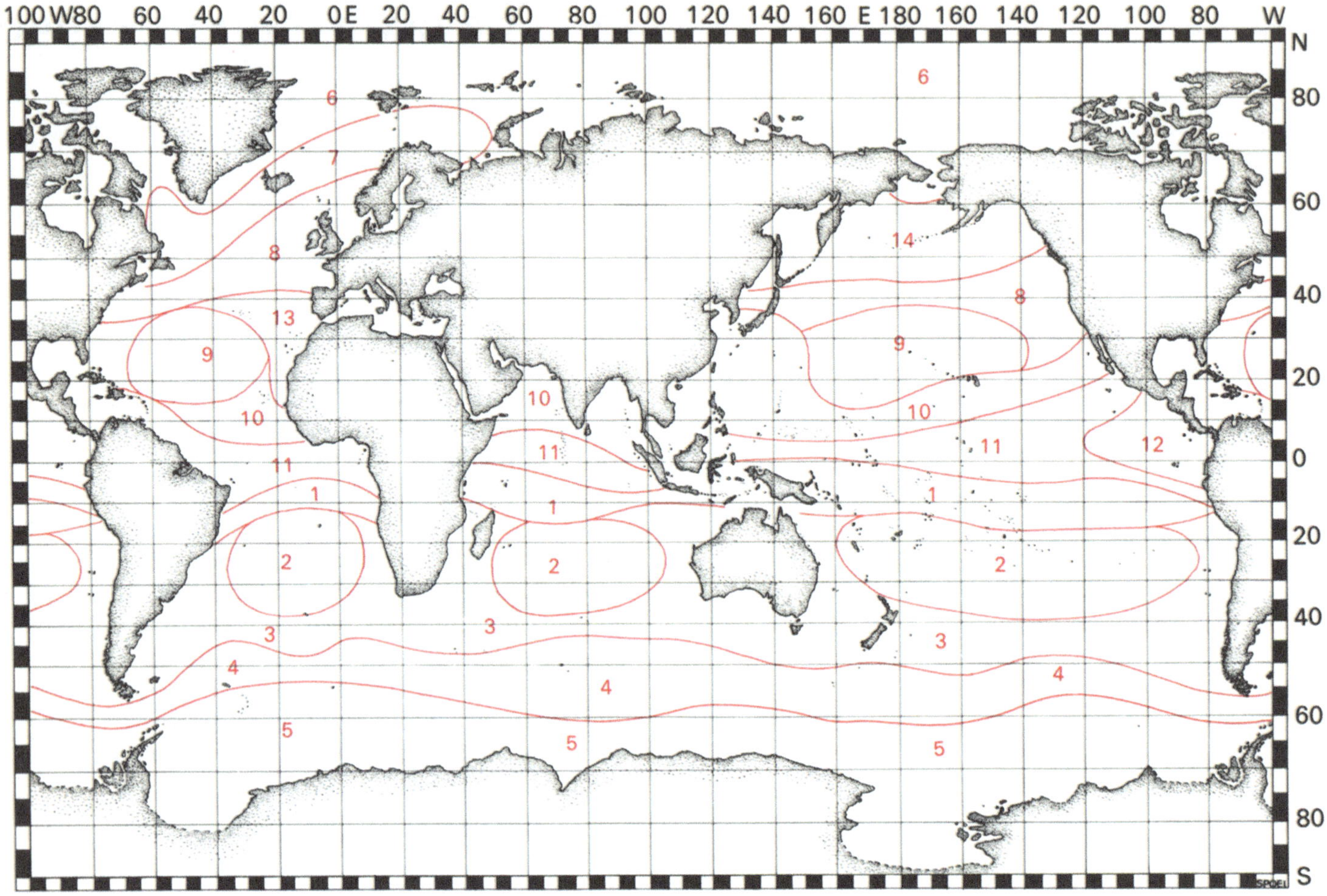

The diagrammatic areal subdivision of the world ocean, given in fig. 1, is used as a general reference for this book (the transparent overlay gives the subdivision of fig. 1 with continuous lines). Only the most important areas are represented here; the boundaries are based on the averages found in several maps (see Literature). Surface waters forming the epipelagic realm are considered to stretch from 0 to about 100 m. It is then assumed, that surface waters of one category are all roughly of the same kind. The mesopelagic realm is considered to extend roughly from 200 to 1000 m depth, below which the bathypelagic realm is found. For a more detailed discussion see p. 137 and fig. 169.

Subdivisions and concepts presented in the following fourteen maps are not proposed as a basis for discussion; they are given because of their great interest for the understanding of zoogeographic terms and related problems.

Definitions of some of the terms frequently used: Central waters are the gyrally flowing waters enclosed by eastern and western boundary currents, equatorial currents and the westward flowing currents at ca. 40° latitude (fig. 1 nos 2 & 9).

Boundary currents are divided into western boundary currents, which flow from equatorial to temperate areas, and eastern boundary currents, which flow from temperate to equatorial areas. Examples of the former are the Gulf Stream and the Kuroshio and Brazil Currents. Examples of eastern boundary currents are the Canary, Benguela and Humboldt Currents.

Neritic areas are the areas over the continental shelf, theoretically bordered by the 200 m depth line, with great fluctuations in temperature and salinity, usually with high production. For a discussion on distant-neritic areas the reader is referred to p. 8 (fig. 7).

Terminal water is a term frequently used also in Soviet papers for the mixed waters of a latitudinal current in between pairs of large gyres. The term 'cyclical' is then applied to the large gyres themselves (fig. 3 nos 1 & 2).

Transitional waters are the waters of a strongly mixed character between Subpolar waters and Subtropical waters (fig. 10 nos 3 & 7; fig. 14 no 7).

The latitudinal physical zonation of water masses (fig. 2), representing zones with stable conditions, without seasonal fluctuations, shows no current influences and neritic conditions or differences in oxygen concentrations and so on, are not traceable. Although this map resembles the one based only on temperatures, it principally gives the more general oceanic climatic differences.

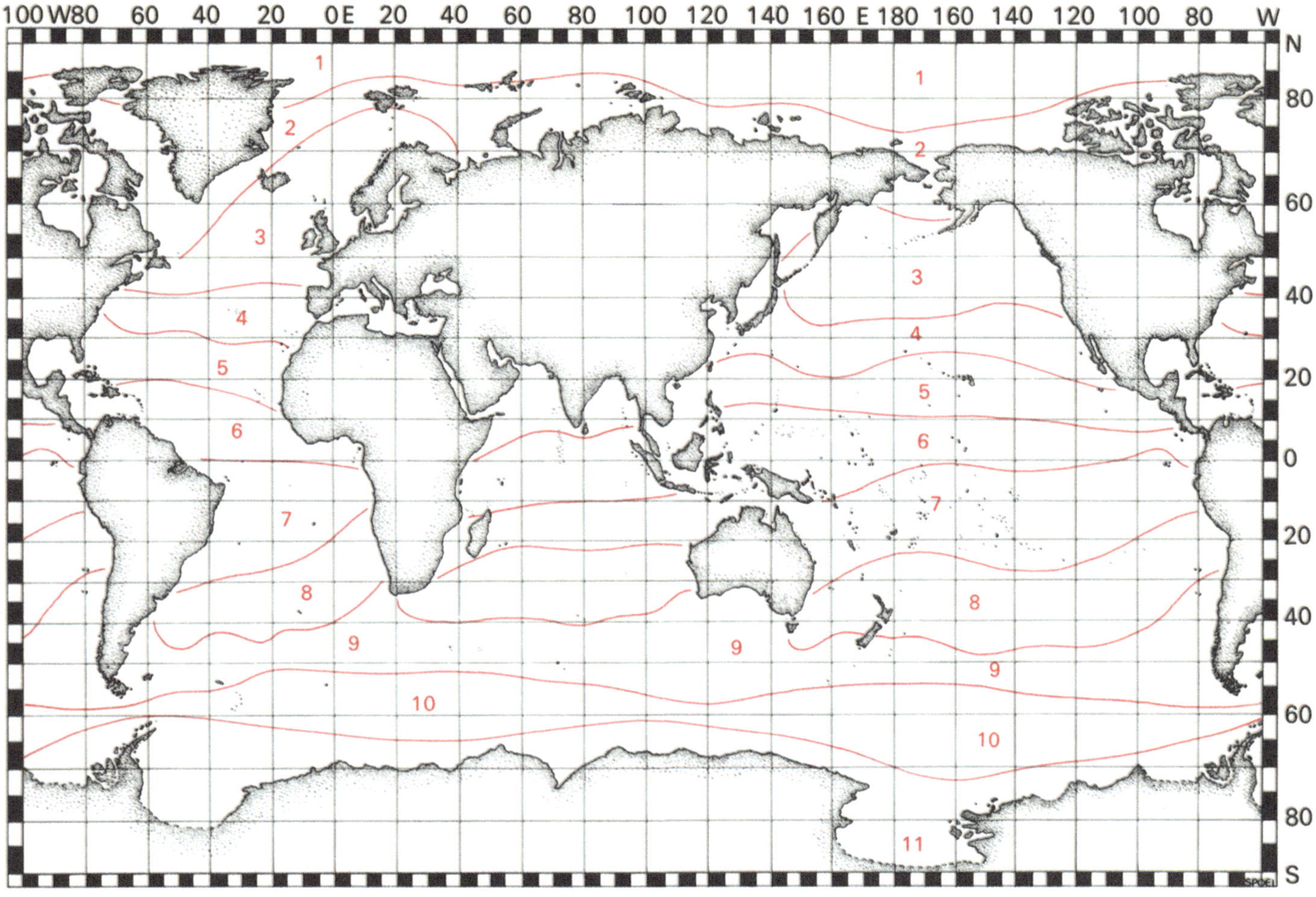

FIG. 1 Diagrammatic areal subdivision of the world ocean. 1 = S. Tropical-Terminal water, 2 = S. Tropical-Central water, 3 = Transitional water, 4 = Subantarctic water, 5 = Antarctic water, 6 = Arctic water, 7 = Subarctic water, 8 = N. Transitional water, 9 = N. Subtropical-Central water, 10 = N. Tropical-Terminal water, 11 = Equatorial water, 12 = E. Pacific water, 13 = Atlanto-Mediterranean water, 14 = Cool-Temperate water; all Transitional, Subtropical and Tropical waters bordering continents can be considered Boundary Current waters.

FIG. 2 Latitudinal-physical geographic zonation of water masses giving zones with stable conditions, without seasonal fluctuations. 1 = Arctic water, 2 = Subarctic water, 3 = N. Temperate water, 4 = N. Subtropical water, 5 = N. Tropical (Trade) water, 6 = Equatorial water, 7 = S. Tropical (Trade) water, 8 = S. Subtropical water, 9 = S. Temperate water, 10 = Subantarctic water, 11 = Antarctic water.

Influences of the currents are well reflected by the areal subdivision of the pelagic communities of the world ocean (fig. 3). In the regions along the coasts no clear areal separations are found and there are no latitudinal influences. A problem of this subdivision into areas occupied by ecological (pelagic) communities is, that they are not representative of individual species. Voronina (1978) stated that ecologically different but adjacent areas may be more similar in species composition than areas of one and the same ecosystem. Especially the cyclical and central areas (no 1) are frequently penetrated by species from adjacent ecosystems. Biotopes, however, are largely characterized by the communities living in them so that the present map is useful for zoogeography.

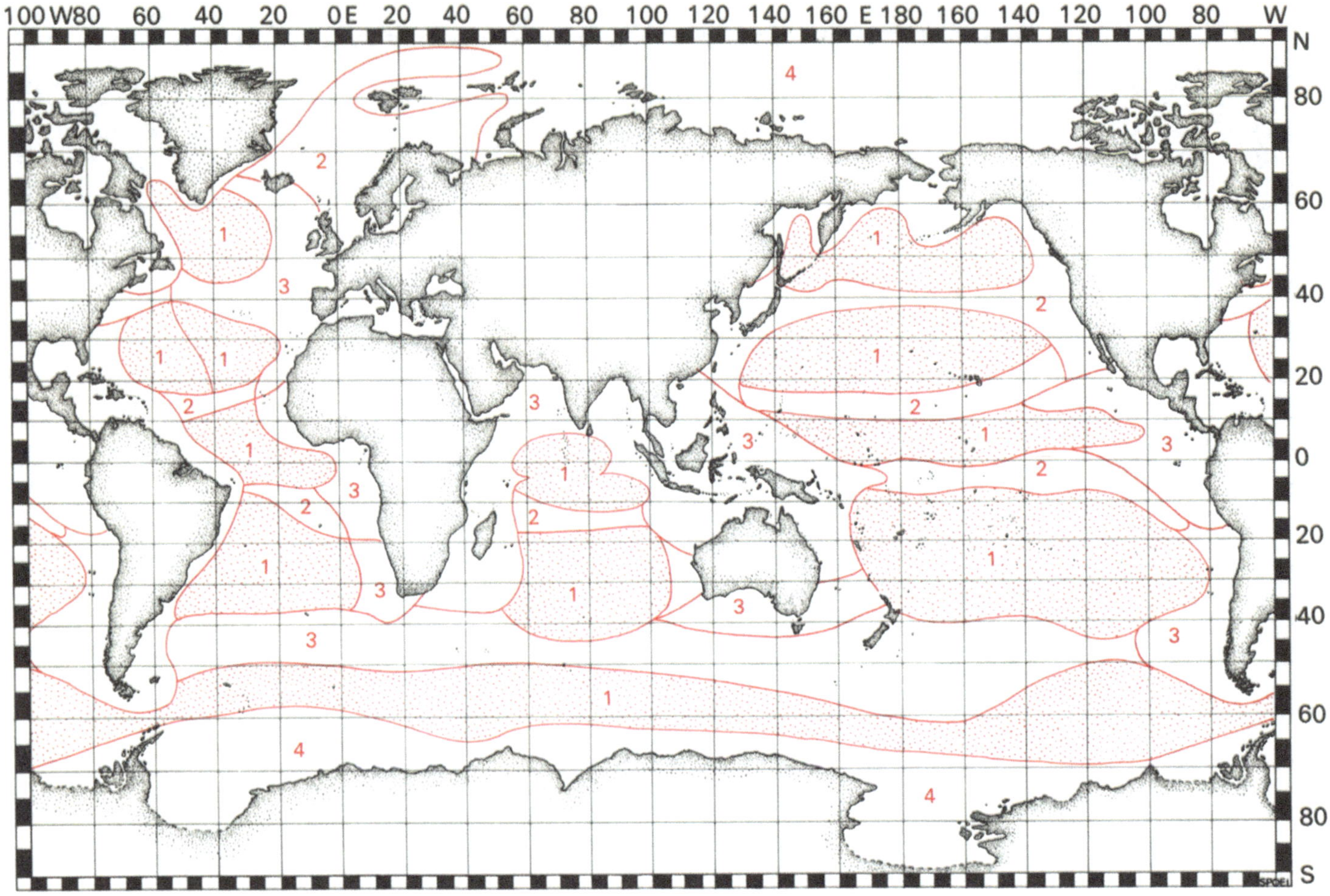

The faunal zones of the world ocean (fig. 4), with the more productive and the more impoverished areas, strongly differ from the latitudinal zones (fig. 2) and also from the subdivision in fig. 3. The more productive areas largely concern secondary production (zooplankton production), and they also provide an indication of the distribution of the distant-neritic taxa (fig. 7). The faunal zonation is in principle also latitudinal, but less detailed than that given in fig. 2.

The biogeographic zones given in fig. 5 have many points in common with those from fig. 4. The former, however, are somewhat more detailed and pay more attention to current influences modifying biogeographical patterns. In the present subdivision, the patterns of both phyto- and zooplankton are considered to establish biogeographical zones comparable with the terrestrial ones. The northern and southern limits of the Tropics are formed here by the 15 °C surface isotherm called 'Ortmann line'. The 'Dunbar line' is the 5 °C surface isotherm forming the northern border of the Temperate (= Boreal) zone. In the south, the 15 °C isotherm limits the Tropical zone in the vicinity of the Subtropical Convergence (fig. 17), while the Subantarctic and Cool-Temperate waters are separated from the Antarctic waters by the 2 °C surface isotherm,

FIG. 3 Pelagic communities of the world ocean. 1 = Cyclical water, 2 = Terminal water, 3 = Terminal-Cyclical water, 4 = Ice-Neritic water.

FIG. 4 Faunal zones of the world ocean with the more productive and the more impoverished areas. 1 = boundaries of Transitional zones, 2 = boundaries of Tropical zones, dotted = productive areas.

FIG. 5 Biogeographic zones. 1 = Arctic zone, 2 = Subarctic zone, 3 = Cool-Temperate zone, 4 = Warm-Temperate zone, 5 = Tropical zone, 6 = Temperate and Subtropical zone, 7 = Subantarctic zone, 8 = Antarctic zone, 9 = Dunbar line, 10 = Ortmann line, 11 = Antarctic Convergence.

which coincides with the Antarctic Convergence.

Most important in this map are the 5 °C and 15 °C surface isotherms in the north and the 15 °C and 2 °C surface isotherms in the south, all neatly coinciding with the major convergence zones.

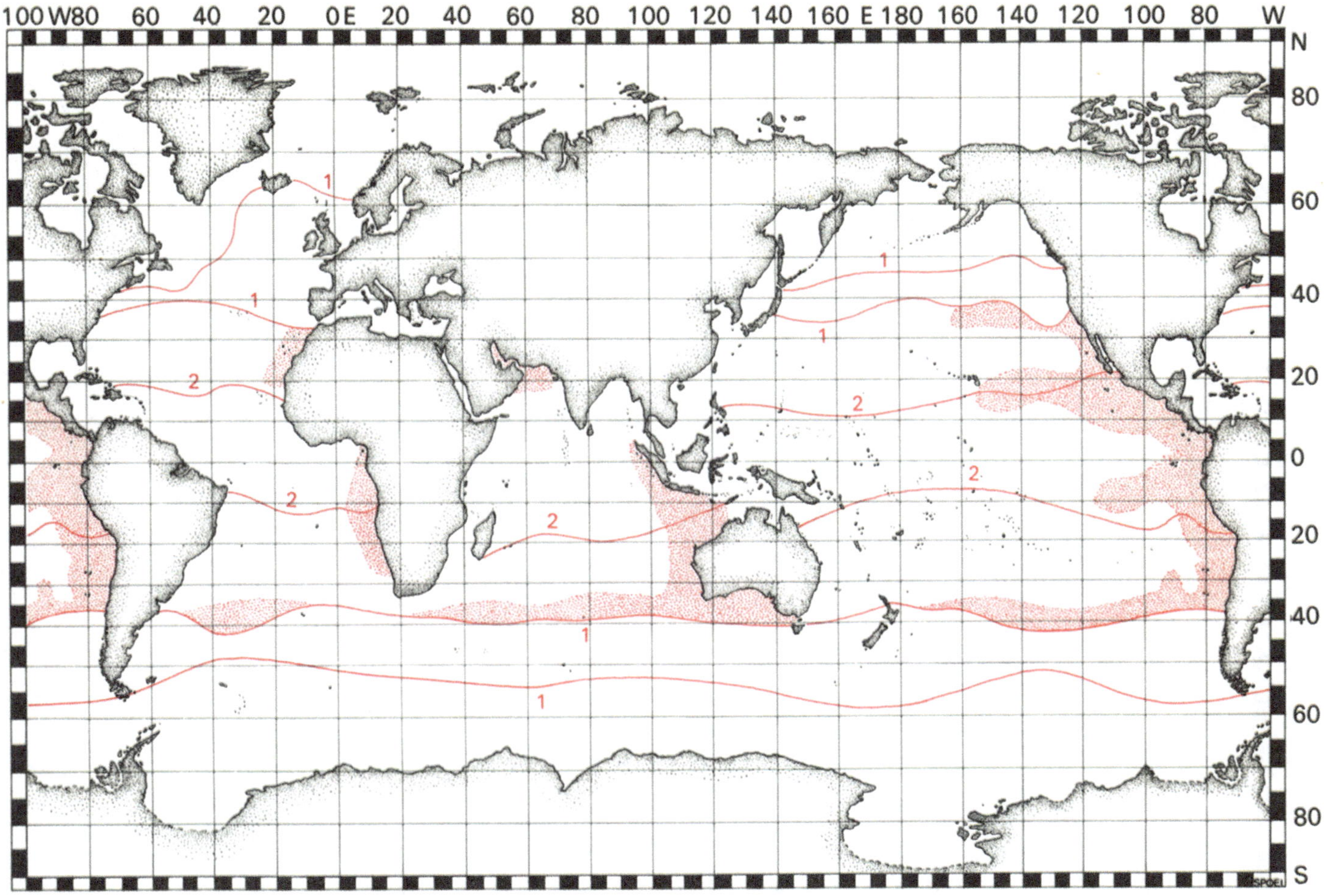
100 W80 60 40 20 0 E 20 40 60 80 100 120 140 160 E 180 160 140 120 100 80 W
N
80
60
40
20
0
20
40
60
80
S
1
2
SPOEL

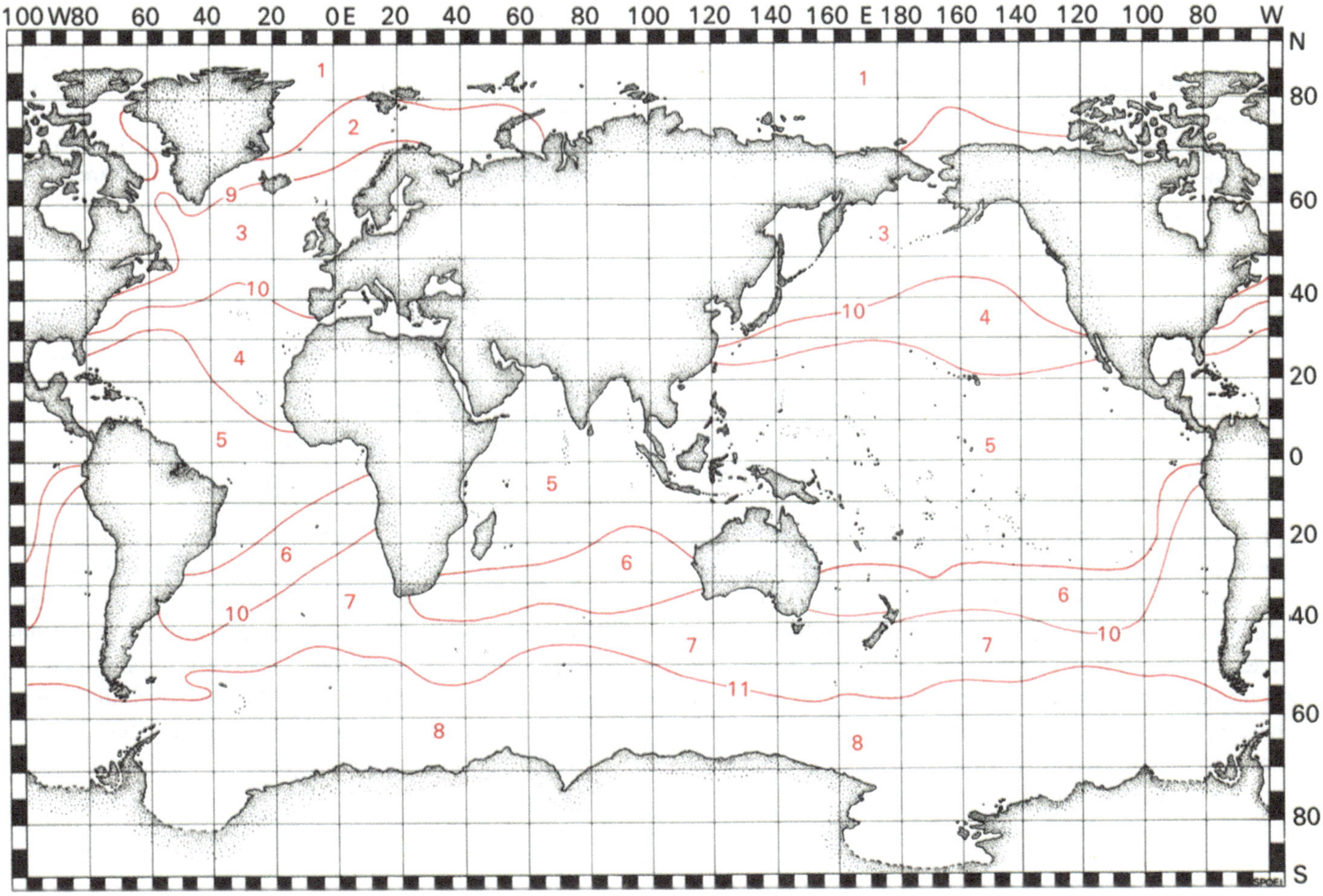
100 W80 60 40 20 0 E 20 40 60 80 100 120 140 160 E 180 160 140 120 100 80 W
N
80
60
40
20
0
20
40
60
80
S
1
2
3
4
5
6
7
8
9
10
11
SPOEL

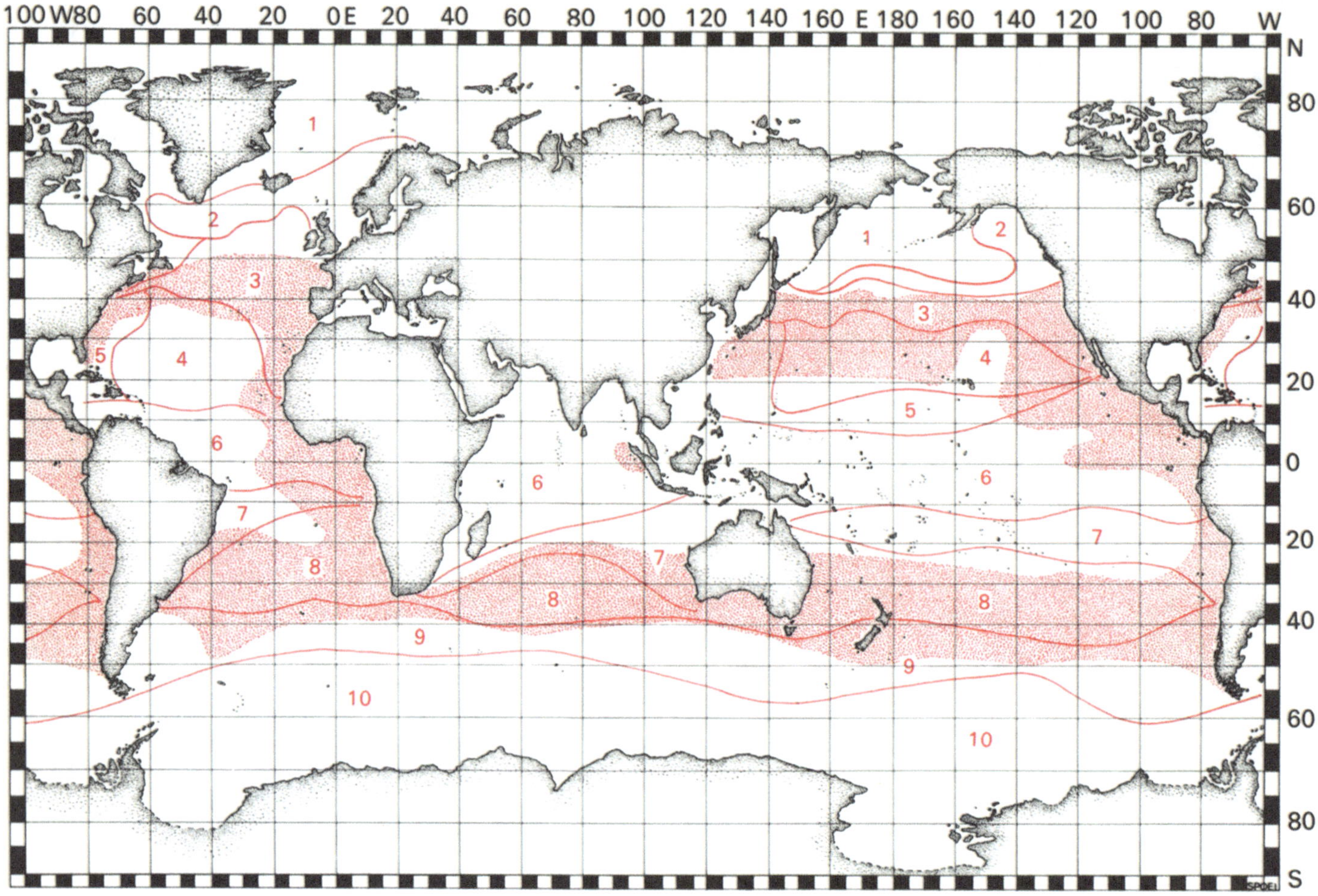

The zooplankton distribution patterns after Beklemishev et al. (1977) (fig. 6) are given to show, that for zoogeographic areas more detailed patterns can be given than for biogeographic ones (cf. figs 4 & 5). Strong mixing occurs, of course, in the Transitional areas (nos 3, 9). The marked difference of mixing intensity in the eastern and western boundary current areas warrants a subdivision of the ocean basins into eastern and western sections (cf. figs 7, 9 & 14).

Neritic species usually occur in shallow water along the coasts, although their distributions are not primarily determined by depth (the 200 m line, for example) but rather by the variability of the physico-chemical properties of the environment. When taking a somewhat broader view of the concept 'neritic', one can speak of 'distant neritic' species, which occur not only in the shallow waters along the coasts, but also in the adjacent oceanic areas. Their geographic distribution is outlined in fig. 7. The ice-covered areas, which are also neritic in nature, and the areas of the high latitudes in general, have been ignored (the transparent overlay gives these areas with a broken line).

FIG. 6 Zooplankton distribution patterns. 1 = Arctic area, 2 = Subarctic area, 3 = N. Temperate area, 4 = N. Subtropical area, 5 = N. Subtropical-Terminal area, 6 = Equatorial area, 7 = S. Subtropical-Terminal area, 8 = S. Subtropical-Central area, 9 = S. Temperate area, 10 = Antarctic area, dotted = areas of strongly mixing water masses.

FIG. 7 The distant-neritic areas.

FIG. 8 World distribution zones of planktonic Foraminifera. 1 = Arctic zone, 2 = Subarctic zone, 3 = N. Transitional zone, 4 = N. Subtropical zone, 5 = Tropical zone, 6 = S. Subtropical zone, 7 = S. Transitional zone, 8 = Subantarctic zone, 9 = Antarctic zone.

Several subdivisions of the world ocean are based on a single group of planktonic animals. To show the futility of such a subdivision, two examples are compared here. The Foraminifera (fig. 8) are chiefly epiplanktonic cold-tolerant species. The pattern of foraminiferal faunal zones is regular and resembles the biogeographical zones after Dunbar (fig. 5) though current influences, as represented in fig. 3, are also visible.

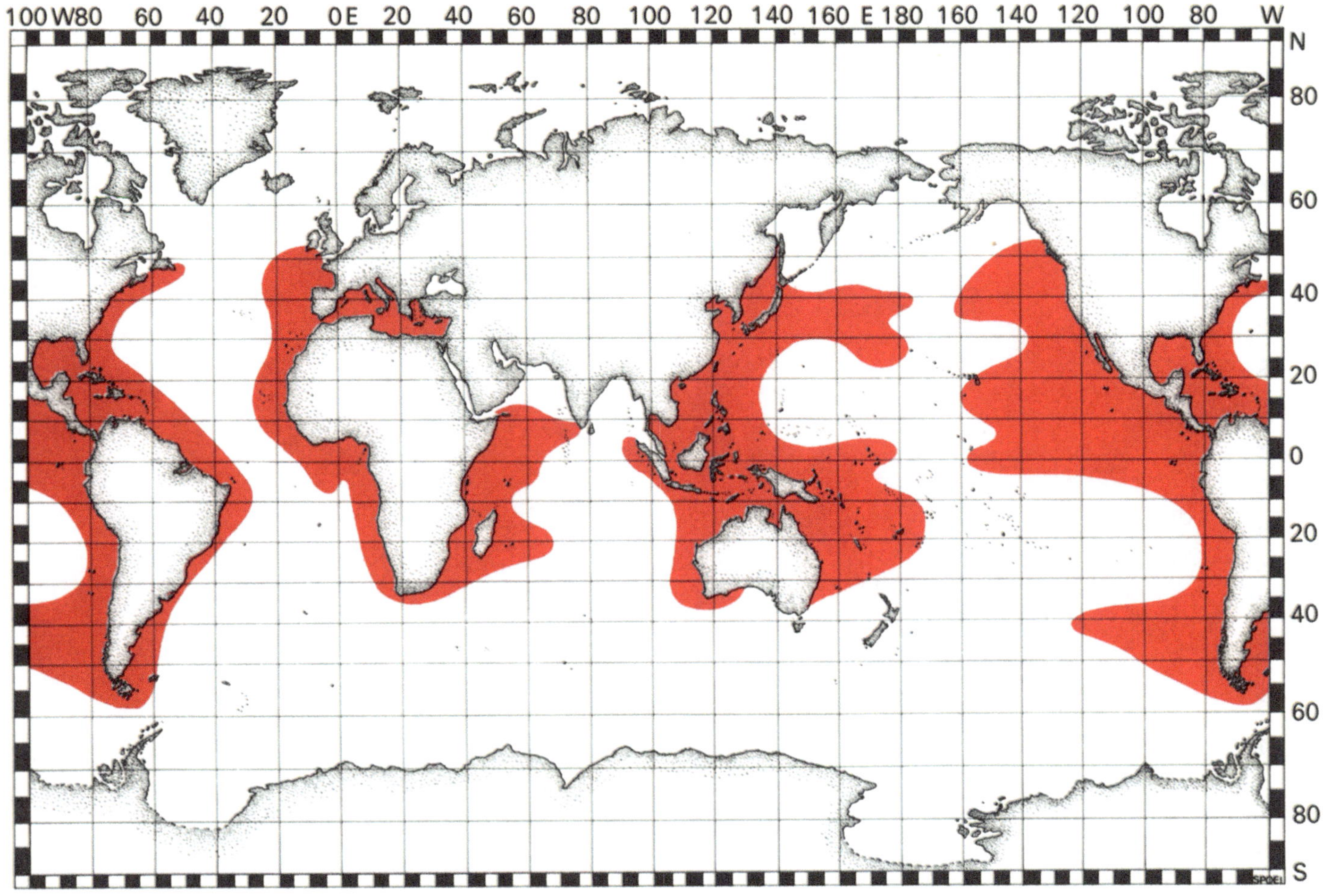
100 W 80
60
40
20
0 E
20
40
60
80
100
120
140
160
E 180
160
140
120
100
80
W
N
80
60
40
20
0
20
40
60
80
S

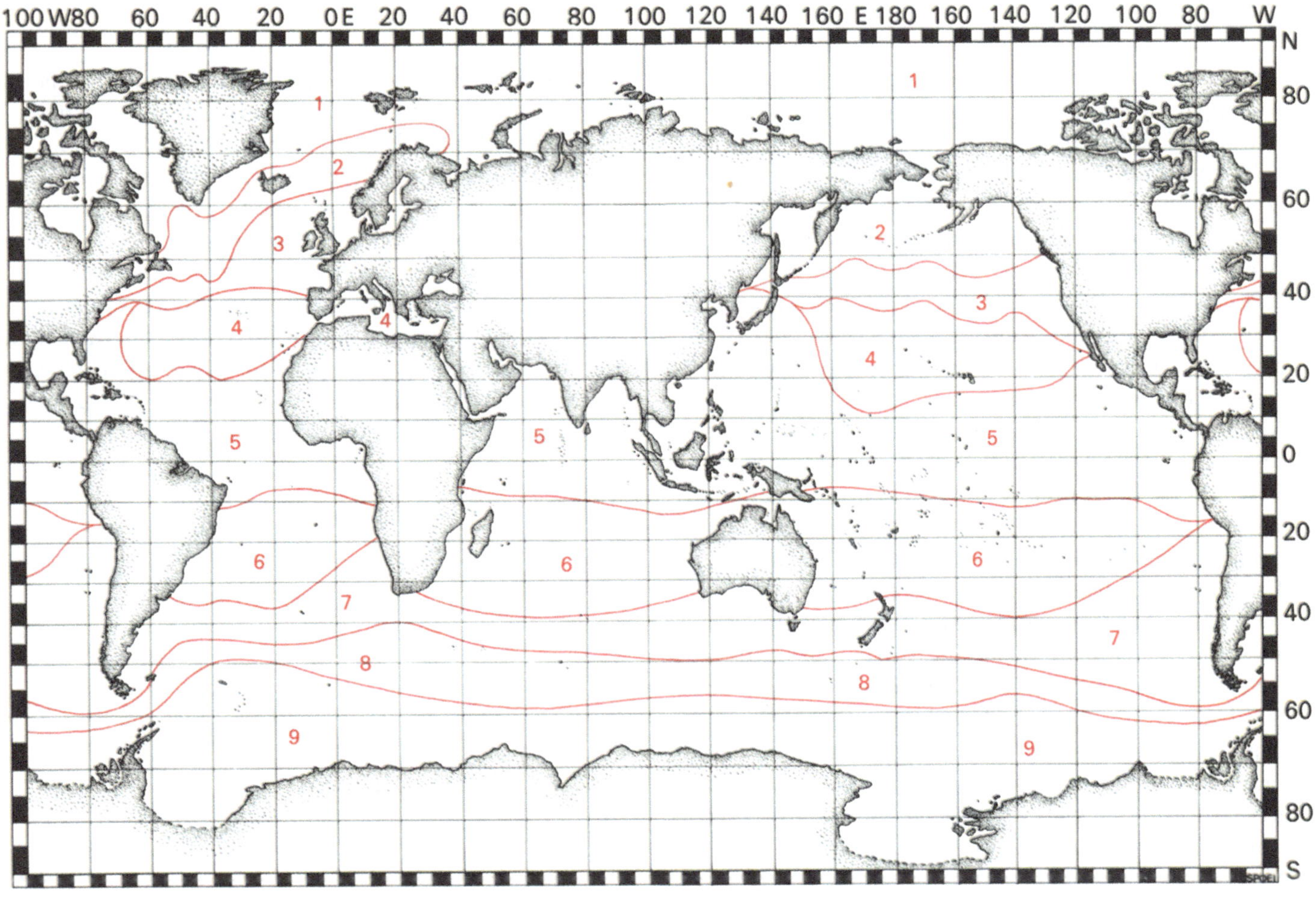
100 W 80
60
40
20
0 E
20
40
60
80
100
120
140
160
E 180
160
140
120
100
80
W
N
80
60
40
20
0
20
40
60
80
S
1
2
3
4
5
6
7
8
9

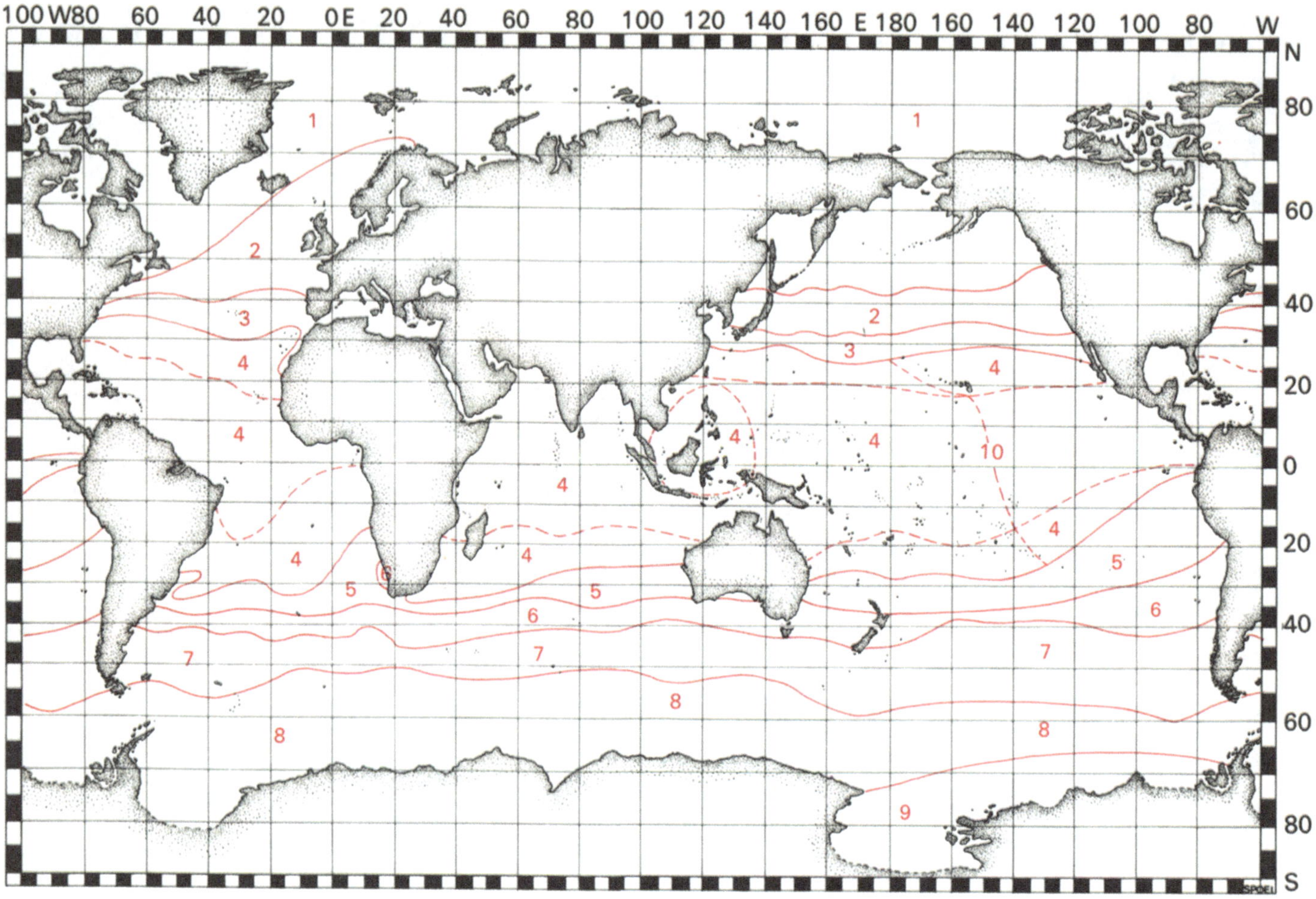

The present zones are all chiefly latitudinal and of roughly the same width. The biogeographic subdivision of the world ocean of fig. 9 is based on the epiplanktonic cold-intolerant group of the Thaliacea. Obvious is the compression of the latitudinal zones between the 20° and 45° latitudes, and the large extensions of the Antarctic and Arctic areas. The characteristics of this subdivision are due to the characters of the group, which is thermophylous and shows only little latitudinal speciation. It is probably the geological history which is responsible for the fact that in this zonation the Mediterranean Sea is considered to be 'Temperate' instead of 'Subtropical' as it was in the subdivision based on the Foraminifera. The division of the Tropical Pacific Ocean into an eastern and western section is evident in this group, while it is not found in the Foraminifera. Differences between eastern and western parts of ocean basins are also mentioned in relation to fig. 6 and in relation to the neritic distributions of fig. 7. The differences between the East and West Pacific have also been demonstrated by Sverdrup et al. (1942) and Van der Spoel & Pierrot-Bults (1979). The physical and chemical conditions are strongly different on both sides of this ocean, though several planktonic groups do not reflect these longitudinal discontinuities.

When comparing the phytoplankton geography of the world ocean (fig. 10) with the animal dispersal (cf. figs 3, 4 & 6) we can perceive some remarkable differences. Most striking is the strong tapering of the northern hemisphere belts in the west, the narrowness of the Transitional belts and the particular subdivision in the Tropical areas. In zooplankton, the supply of nutrient salts has only an indirect effect which, moreover, fades out because of the influence of currents and of the length of the foodchain. In phytoplankton, these influences are direct, so that in fertile areas and regions of strong mixing other patterns are found than for zooplankton.

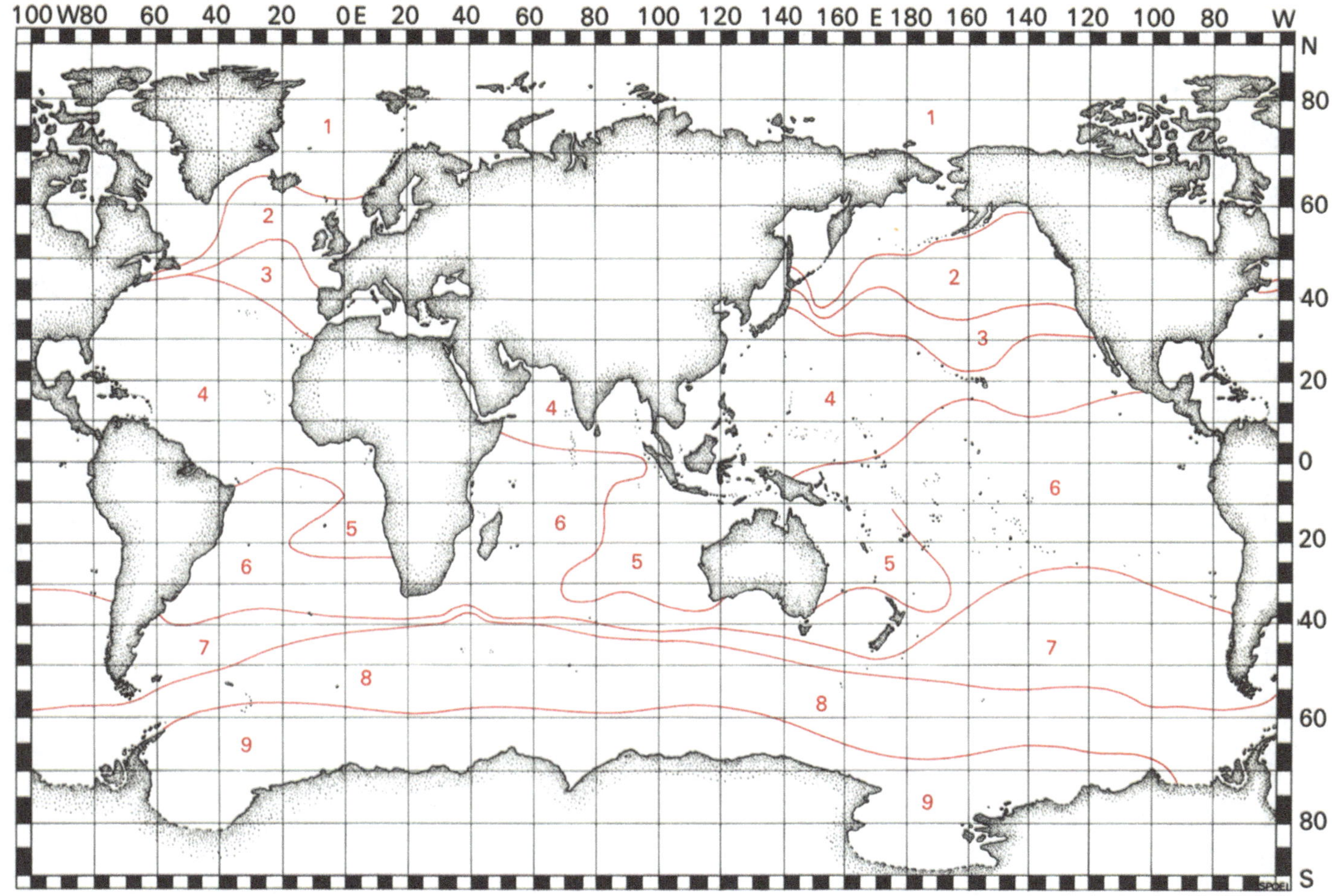

FIG. 9 Biogeographic subdivision of the world ocean based on Thaliacea and isotherms. 1 = Arctic zone, 2 = N. Temperate zone, 3 = N. Transitional zone, 4 = Tropical zone, 5 = S. Transitional zone, 6 = S. Temperate zone, 7 = Subantarctic zone, 8 = Antarctic zone, 9 = Pacific High Antarctic zone, 10 = separation between east and west Pacific Thaliacea faunas.

FIG. 10 Phytoplankton geography of the world ocean. 1 = Arctic area, 2 = Arctic-Boreal area, 3 = N. Transitional area, 4 = Tropical area, 5 = mixed Tropical area, 6 = Equatorial area, 7 = S. Transitional area, 8 = Subantarctic area, 9 = Antarctic area.

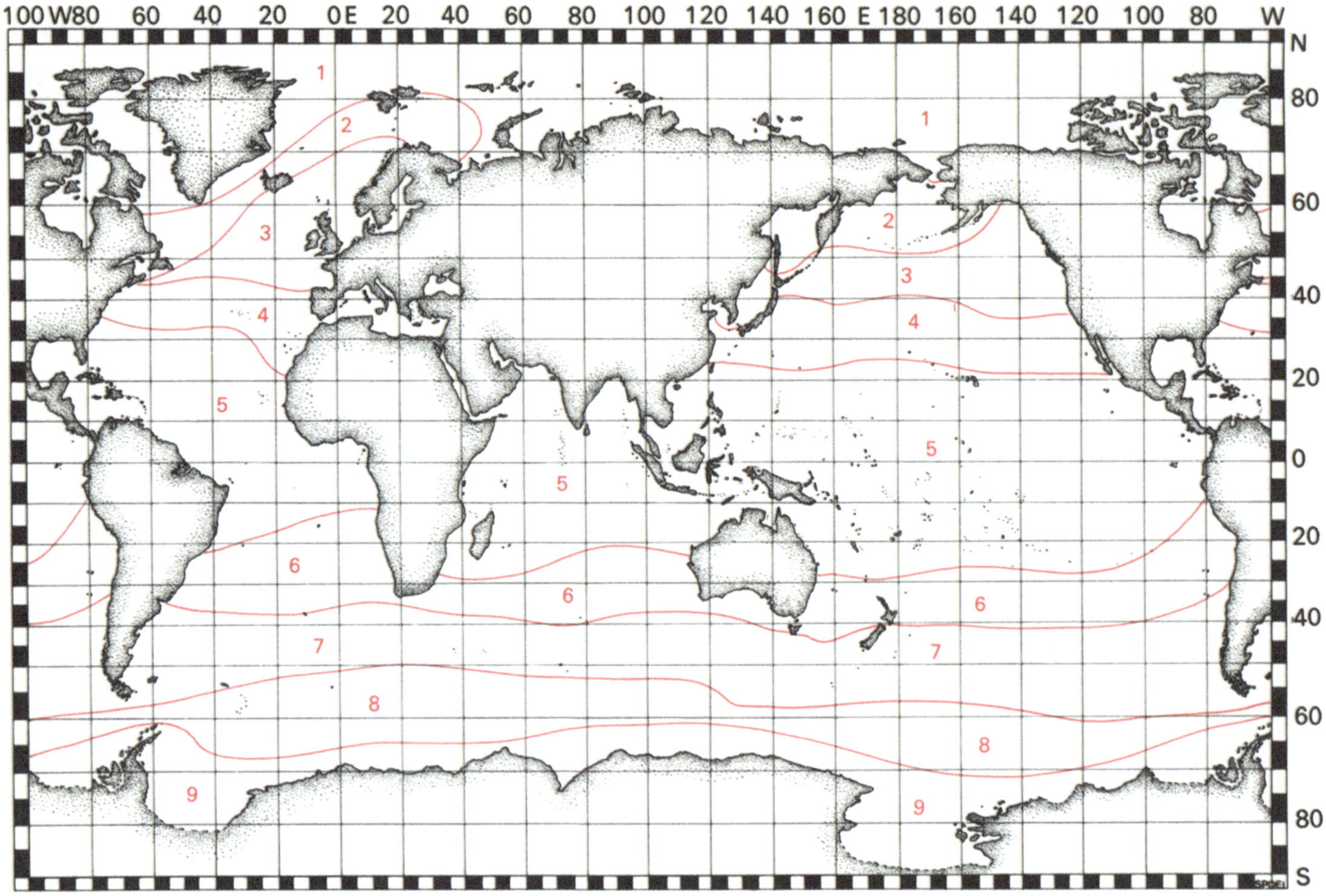

The subdivision into zones based on ichthyoplankton distributions (fig. 11) resembles most the one given by Dunbar (fig. 5) and that of the Foraminifera (fig. 6). The fact that meroplankton includes species which in the adult stage are capable of maintaining a geographically stable distribution, may explain this pattern very well. The adults return to particular, usually neritic, spawning areas. These places are found in specific latitudinal belts, largely influenced by climate, and consequently ichthyoplankton also tends to show latitudinal distributions.

From the very beginning of marine zoogeography, there has been a variety of interpretations of marine areas. The borders of 'warm' and 'polar' waters as distinguished in the beginning of this century are shown in fig. 12. Shaded (nos 1 & 2) are the areas in which the borders for polar and for warm water have thus far been situated by various authors. The width of the areas, reflecting the different interpretations of the concepts cold and warm water, strikingly corresponds with the seasonal fluctuations in sea surface temperature (cf. fig. 19). It is these seasonal changes in conditions which also make a physical and biological division of the oceans so difficult.

FIG. 11 Ichthyoplankton distributional zones in the world oceans. 1 = Arctic zone, 2 = Subarctic zone, 3 = N. Transitional zone, 4 = N. Subtropical zone, 5 = Tropical zone, 6 = S. Subtropical zone, 7 = S. Transitional zone, 8 = Subantarctic zone, 9 = Antarctic zone.

FIG. 12 The borders of 'warm' and 'polar' water as distinguished in the beginning of this century. 1 = border area of warm water, 2 = border area of polar-water, 3 = an extra subdivision of warm water by Dahl, 4 = borders after Meisenheimer, 5 = borders after Dahl, 6 = borders after Hesse.

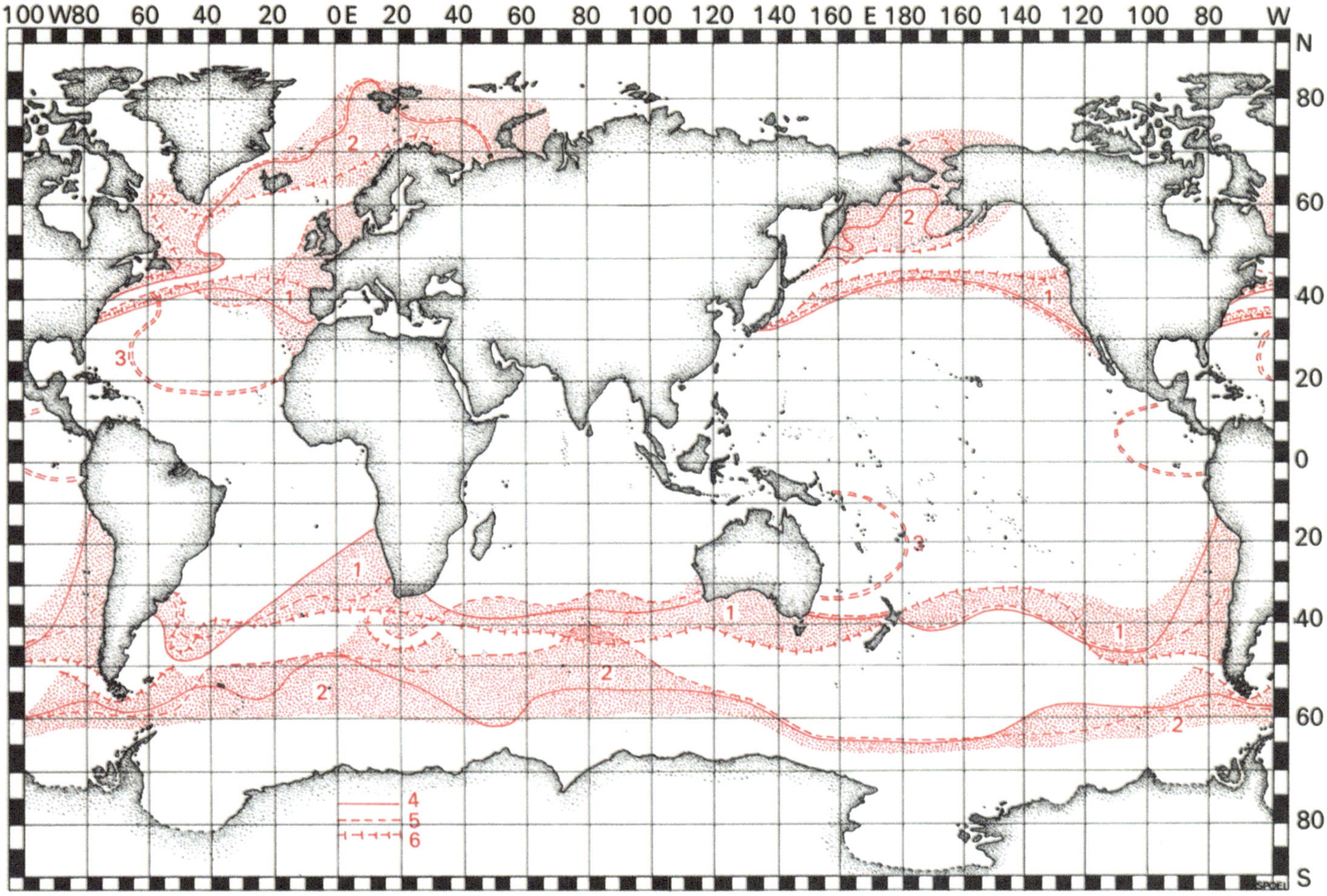
100 W 80 60 40 20 0 E 20 40 60 80 100 120 140 160 E 180 160 140 120 100 80 W
N 80 60 40 20 0 20 40 60 80 S
1
2
3
4
5
6
SPOE

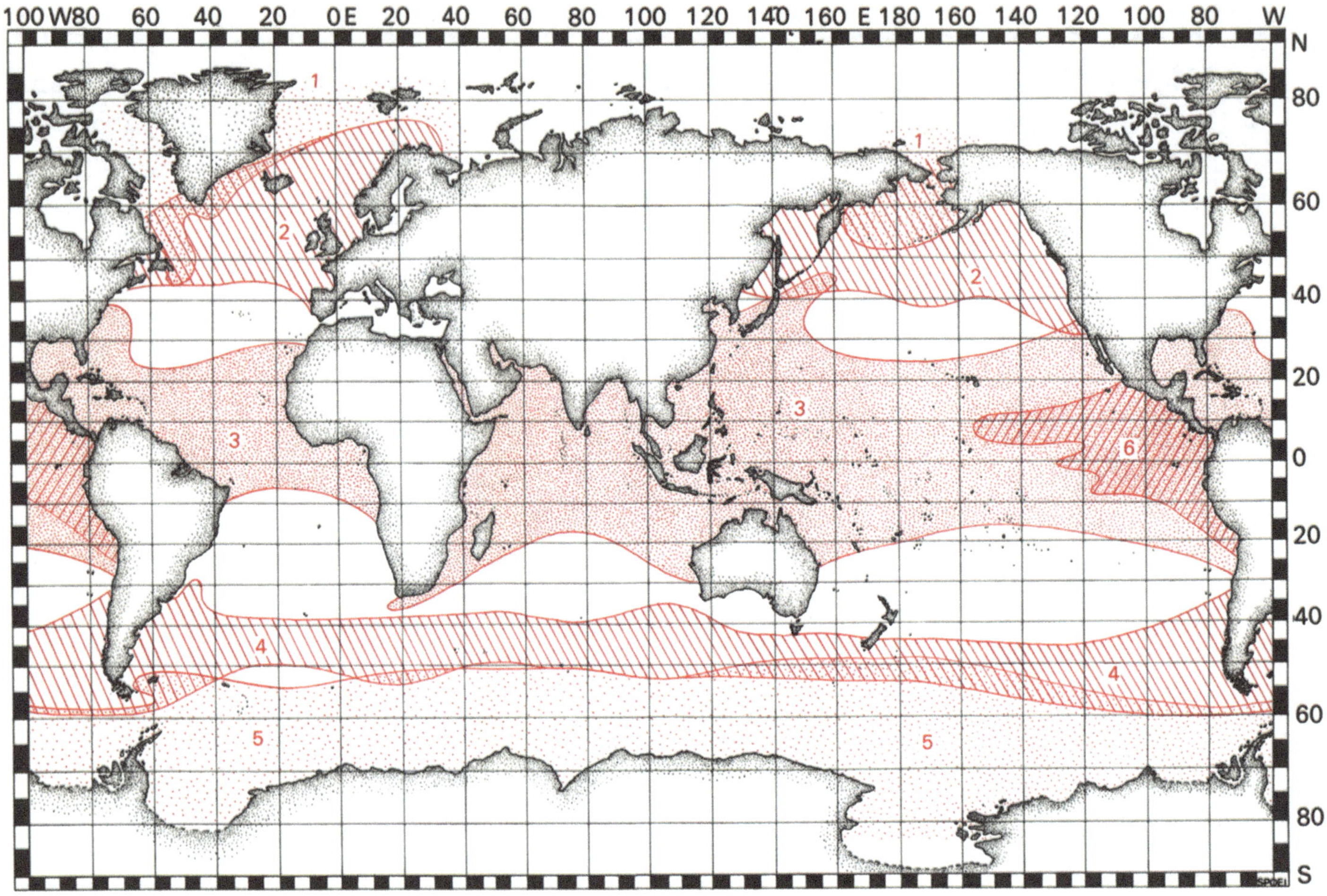

In the biogeographical provinces Goll proposed for plankton (figs 13 & 14), overlapping of provinces is accepted as a normal phenomenon. This is far from practical when biotopes have to be distinguished. When the provinces are considered to indicate the distributions of faunal elements belonging to a particular faunal centre, it may be a convenient subdivision, but then the faunal centres should be indicated too (cf. fig. 16). Most striking in figs 13 & 14 is the extent of penetration of the Transitional (no 7) and Equatorial (no 3) provinces into the Subtropical Central (no 8) areas.

Voronina (1978) discussed this phenomenon and explained it as penetration by expatriated specimens (cf. fig. 15 and p. 5 & 159). In fig. 15 the Equatorial (A) and Central (B) water masses are given diagrammatically with, as shaded areas, the penetration by the invasion of expatriated organisms from the surrounding water masses. If it is true that no autochtonous populations are concerned in this overlapping, the overlapping itself is not essential. However, many distributions show shapes similar to the areas given in figs 13 and 14.

FIG. 13 Biogeographic provinces of plankton in the world ocean after Goll, 1976. 1 = Arctic province, 2 = Subarctic province, 3 = Equatorial province, 4 = Subantarctic province, 5 = Antarctic province; (see fig. 14) 6 = E. Pacific province, 7 = Transitional province, 8 = Central province.

FIG. 14 Biogeographic provinces of plankton in the world ocean after Goll, 1976 (for explanations see fig. 13).

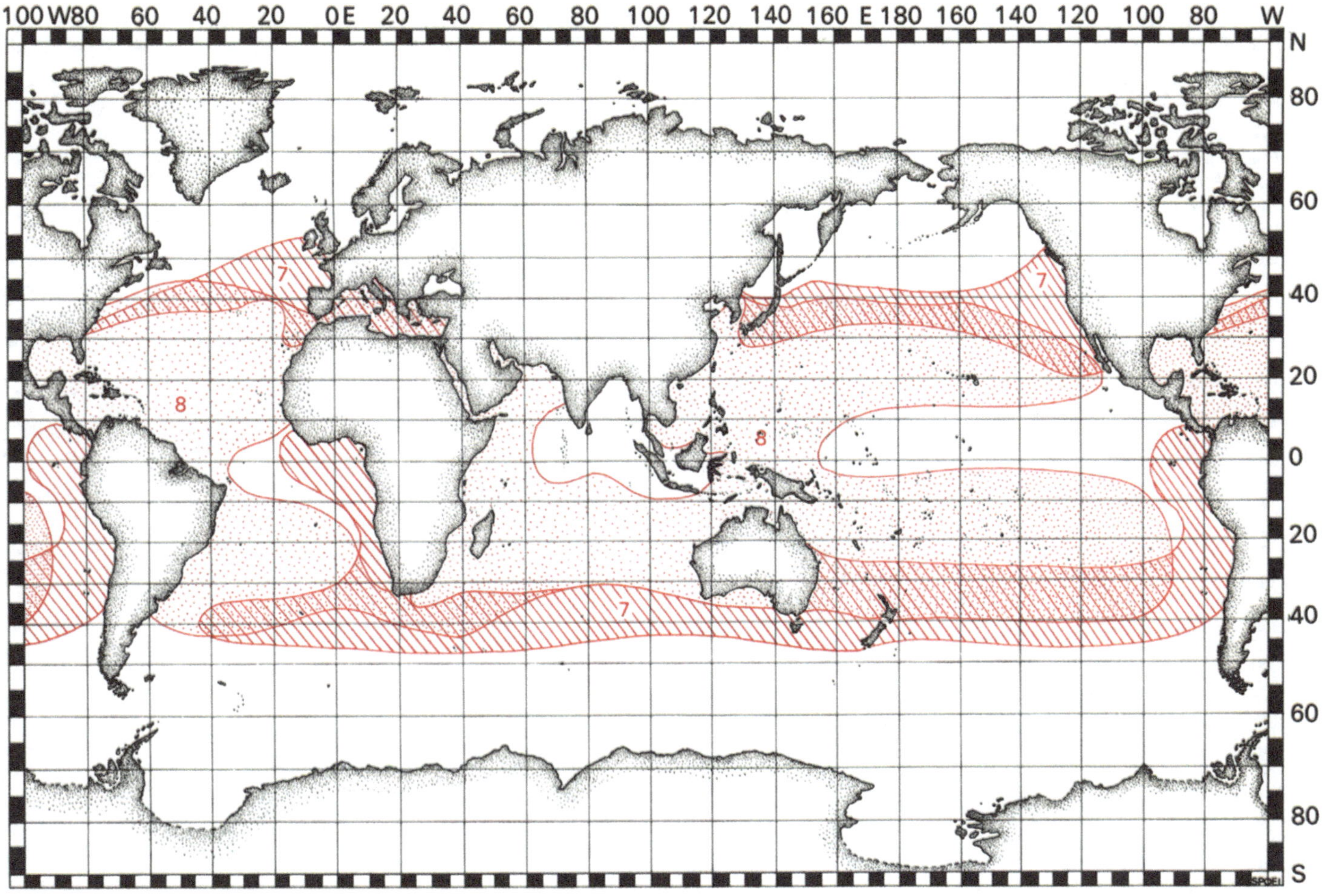

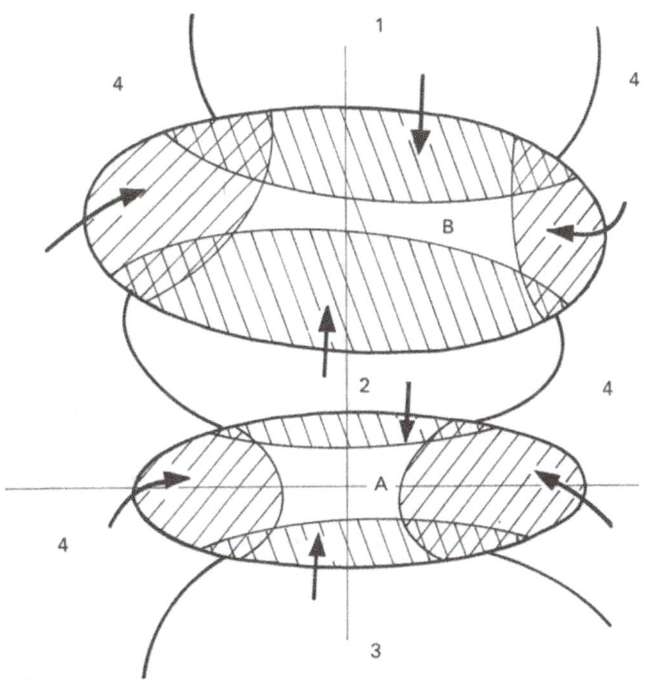

FIG. 15 Diagram of Equatorial (A) and Central (B) water masses with the penetration (hatched) of faunal elements from the surrounding Terminal water masses (1, 2, 3) and Boundary Currents (4).

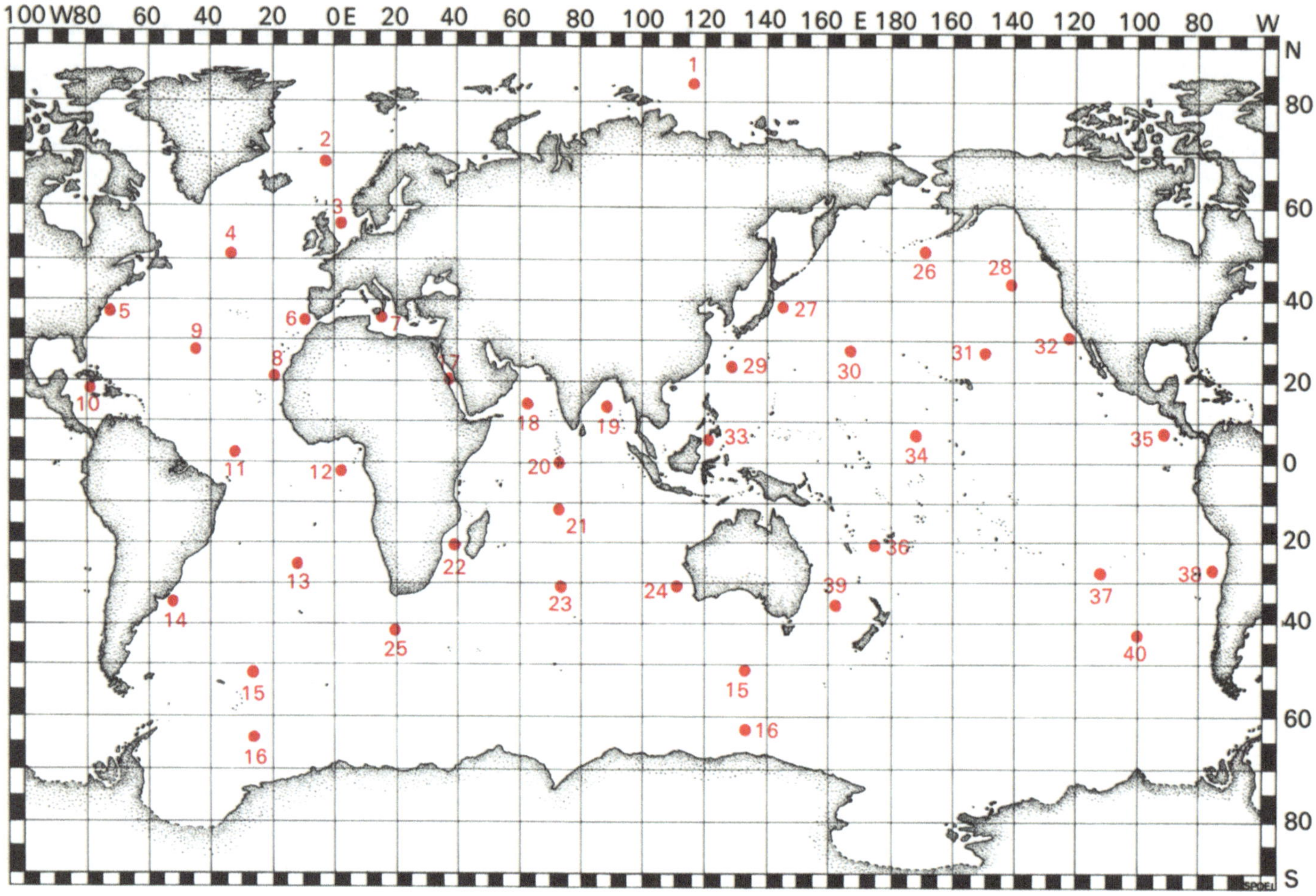

The faunal centres in the world ocean (fig. 16) (the transparent overlay gives the centres of this map with open circles) are given to enable one to compare ranges which have a different geographic shape, but are grouped around a fixed centre or area. For terrestrial biota, the faunal centres, from which after regressions taxa spread over larger distances, are geographically fixed areas, each with its distinct geological and climatological history. For planktonic animals, a faunal centre has to be situated in a water mass, and is as a consequence not geographically fixed. In the geological past it may have shifted so that its history is uncertain. Moreover, real regression of planktonic animals will have been slight as most planktonic taxa escape bad conditions by shifting their entire ranges.

Still, there are groups of planktonic taxa such as *Peraclis*, *Phronima*, some Cephalopoda, *Desmopterus* and *Diacria*, grouped around particular 'areas of speciation', and some of these taxa show more and some less expansion. The ranges of conditions tolerated by these taxa partly or largely overlap, so that the ranges are to a certain extent also sympatric. In such areas of sympatry the faunal centres are found.

A centre is then an area with conditions more or less characteristic of the faunal elements of that centre, and the taxa share the same ecological preferences or the same ancestral form. So, not all the centres represented here are centres of speciation or dispersal, a number of them are merely ecologically defined areas.

As a conclusion, one can state that there are many subdivisions of the ocean based on different biological parameters. A subdivision based on a single group may be of value for phylogenetic studies of that group but it usually does not contribute to general zoogeography. Depending on the problem studied, one can use ecological, physical or more climatological and hydrological subdivisions. For general zoogeographic studies, however, it seems most practical to use the synthetical subdivision presented in fig. 1, and its transparent overlay, which also fits in with the conditions discussed in the following paragraph.

Hydrological conditions

The area in which a particular biocoenosis lives, is characterized by physical, chemical and biological conditions. In the ocean, these areas are far from stable as they shift with the currents and they are only vaguely, if at all, limited, shading off into one another. The mentioned shift with the currents is not unlimited, of course, as the conditions smoothly change with the

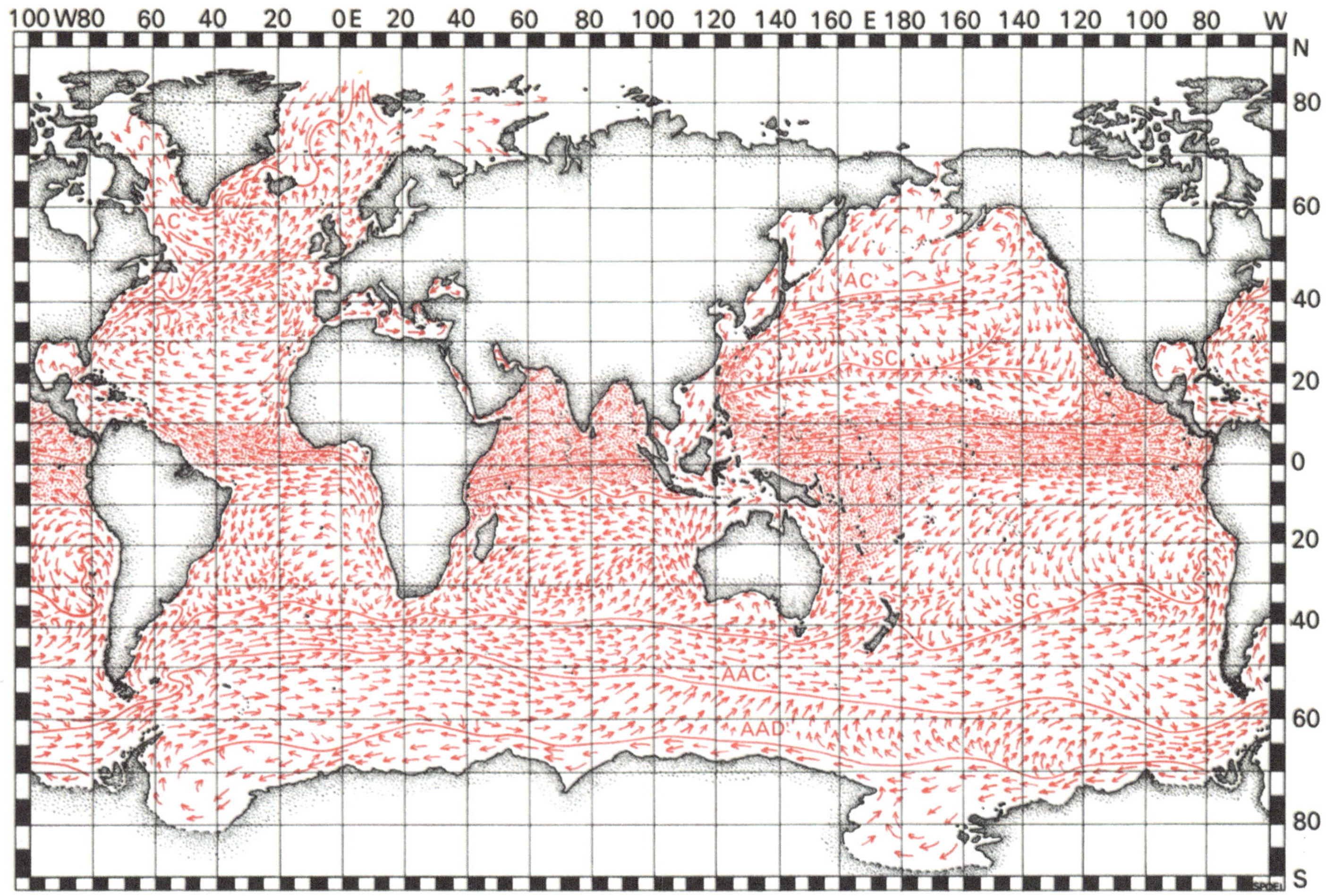

progress of the currents. These gradual transitions form the reason why there is no general agreement on the subdivision of the sea into zones, biotopes or whatsoever. As the organisms in the sea move, besides horizontally, also evidently vertically and as the energy, or biomass, flows along vertical foodchains, the biocoenosis is strongly dependent on vertical processes. As a consequence, biotopes change also when there are changes in the deeper layers, and when there are influences on the processes in the vertical direction.

Horizontal dispersal, and the biotopes as horizontal phenomena, will be discussed first, the vertical processes are discussed at the end of this atlas. For this purpose, the rough areal subdivision and terminology given in fig. 1 are applied in combination with the faunal centres as given in fig. 16 and the distant-neritic range as given in fig. 7. For the comparison of ranges, the major physico-chemical conditions should be taken into account.

Surface currents (fig. 17) are not constant in areas where strong seasonal climate changes are common. In the Equatorial area, especially in the Indian Ocean, current direction depends on monsoon effects, and so does the velocity of the boundary currents. Climate changes also markedly influence the currents in the higher latitudes.

FIG. 16 Faunal Centres of the world ocean. 1 = Arctic C., 2 = Subarctic Atlantic C., 3 = North Sea C., 4 = N. Transitional Atlantic C., 5 = N.W. Atlantic C., 6 = Atlanto-Mediterranean C., 7 = Mediterranean C., 8 = N.W. African C., 9 = N. Subtropical Central Atlantic C., 10 = Caribbean C., 11 = Equatorial Atlantic C., 12 = E. Equatorial Atlantic C., 13 = S. Subtropical Central Atlantic C., 14 = Argentina C., 15 = Subantarctic C., 16 = Antarctic C., 17 = Red Sea C., 18 = Arabian C., 19 = Bay of Bengal C., 20 = Equatorial Indian C., 21 = S. Tropical Terminal Indian C., 22 = Madagascar C., 23 = S. Subtropical Central Indian C., 24 = W. Australian C., 25 = Agulhas C., 26 = Subarctic Pacific C., 27 = Japan C., 28 = N.E. Transitional Pacific C., 29 = Kuroshio C., 30 = N.W. Subtropical Central Pacific C., 31 = N.E. Subtropical Central Pacific C., 32 = California Current C., 33 = Indo-Malayan C., 34 = Equatorial Pacific C., 35 = E. Equatorial Pacific C., 36 = S.W. Subtropical Central Pacific C., 37 = S.E. Subtropical Central Pacific C., 38 = Peru C., 39 = Tasman Sea C., 40 = S.E. Transitional Pacific C.

FIG. 17 Surface currents with the areas of strong seasonal current changes. AAC = Antarctic Convergence, AAD = Antarctic Divergence, AC = Arctic Convergence, SC = Subtropical Convergence, dotted = areas with strong seasonal current changes.

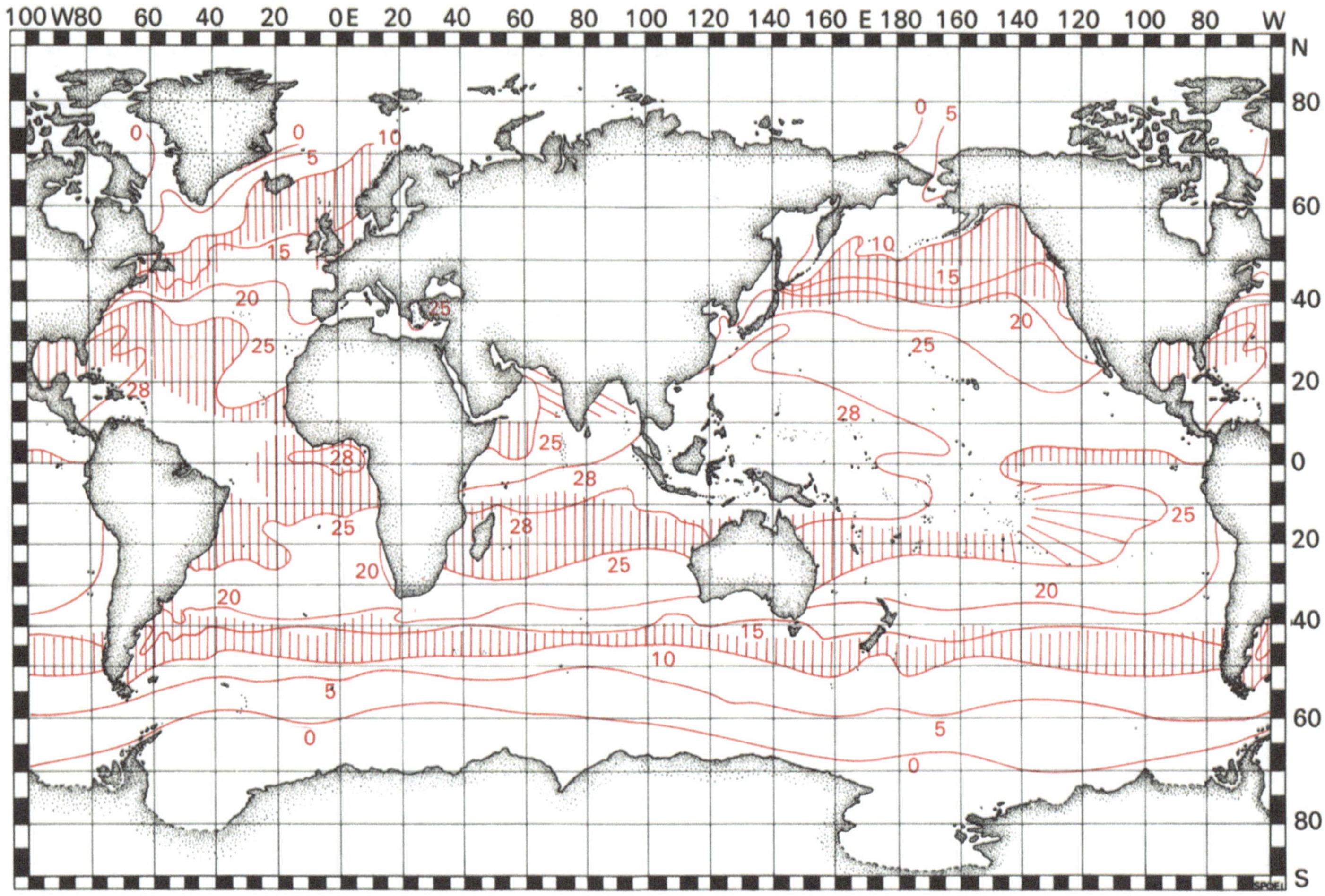

Sea-surface temperatures, too, change seasonally. In fig. 18 the conditions are given in the northern hemisphere for the northern summer and in the southern hemisphere for the southern summer. The warm water area in the Atlantic is clearly smaller than in the other oceans and in all the oceans the warm-water area is broader in the west than in the east. As a rule, the latitudinal temperature gradients in the eastern parts are less steep than those in the western parts of the oceans.

The temperature changes can be expressed by shifts of isotherms, but more impressive are the seasonal sea-surface temperature fluctuations at one locality (fig. 19). As the stability of the environment is largely dependent on temperature, it is clear from this map that the highest and lowest latitudes are the most stable ones with temperature fluctuations over 4-0 °C, while near 40° north and south the greatest instabilities are found.

FIG. 18 Surface isotherms for the summer period, with the shift of the 10 °C and 25 °C isotherms in the winter period hatched.

FIG. 19 Seasonal sea surface temperature fluctuations (in degrees Celcius).

FIG. 20 Sea-surface temperatures in August 18000 BP. hatched = ice-covered and dry areas.

In the geological past, temperature fluctuations due to the Pleistocene Ice Ages presented another picture, as the fluctuations and instabilities were then greatest at the high latitudes (fig. 20). In this map also the closed situation of the Bering Strait, and of most passages in the Indo-Malayan Archipelago, as well as the nearly closed Drake Passage are shown. The warm-water areas are small, and the temperature gradients are all much steeper than at present.

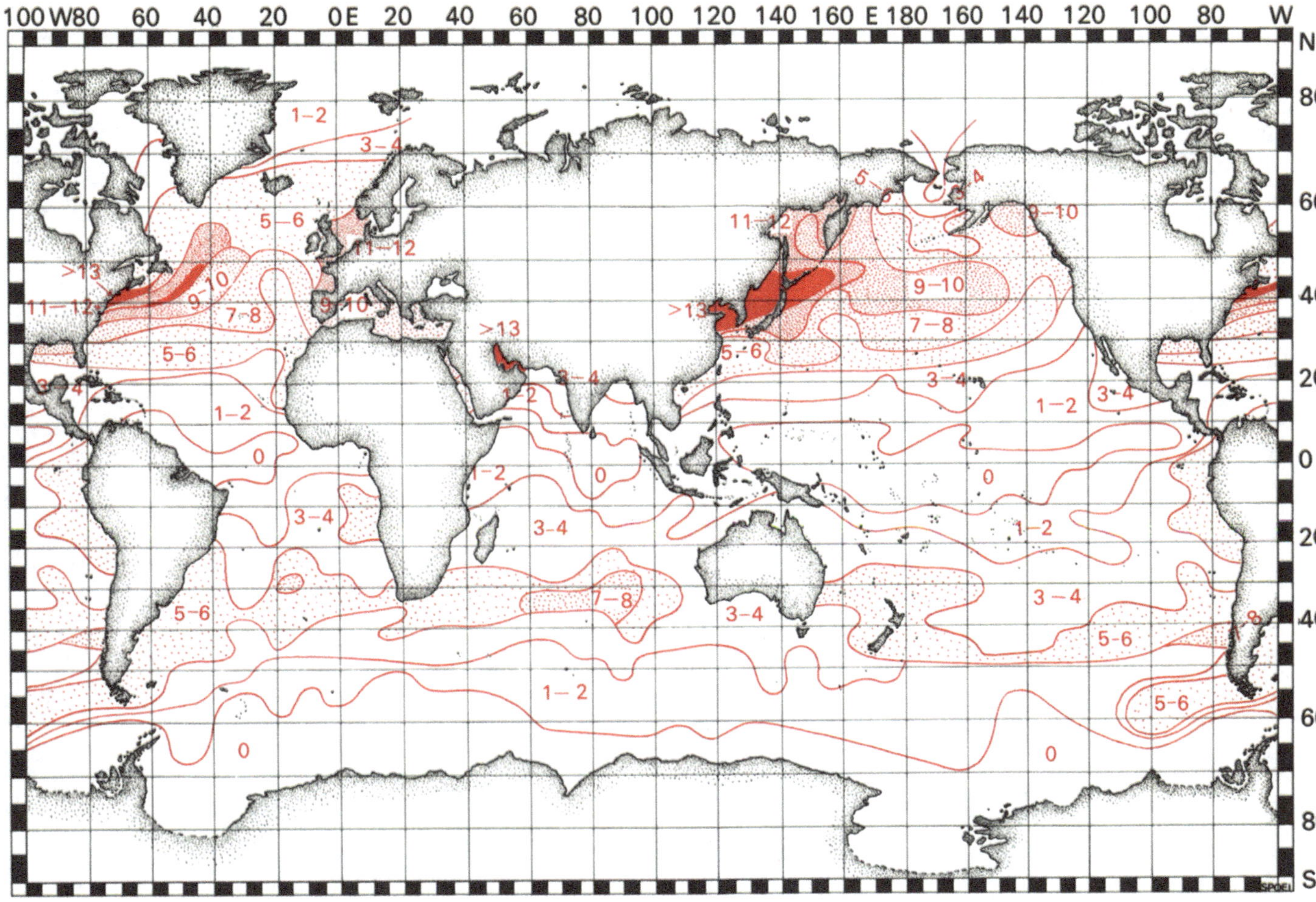
100 W 80 60 40 20 0 E 20 40 60 80 100 120 140 160 E 180 160 140 120 100 80 W
N 80 60 40 20 0 20 40 60 80 S
1–2
3–4
5–6
11–12
>13
9–10
7–8
5-6
1–2
0
3–4
5-6
>13
1–2
0
3–4
7–8
3–4
1–2
0
11–12
>13
5-6
9–10
7–8
3–4
1–2
3-4
0
1–2
3 – 4
5-6
5-6
0

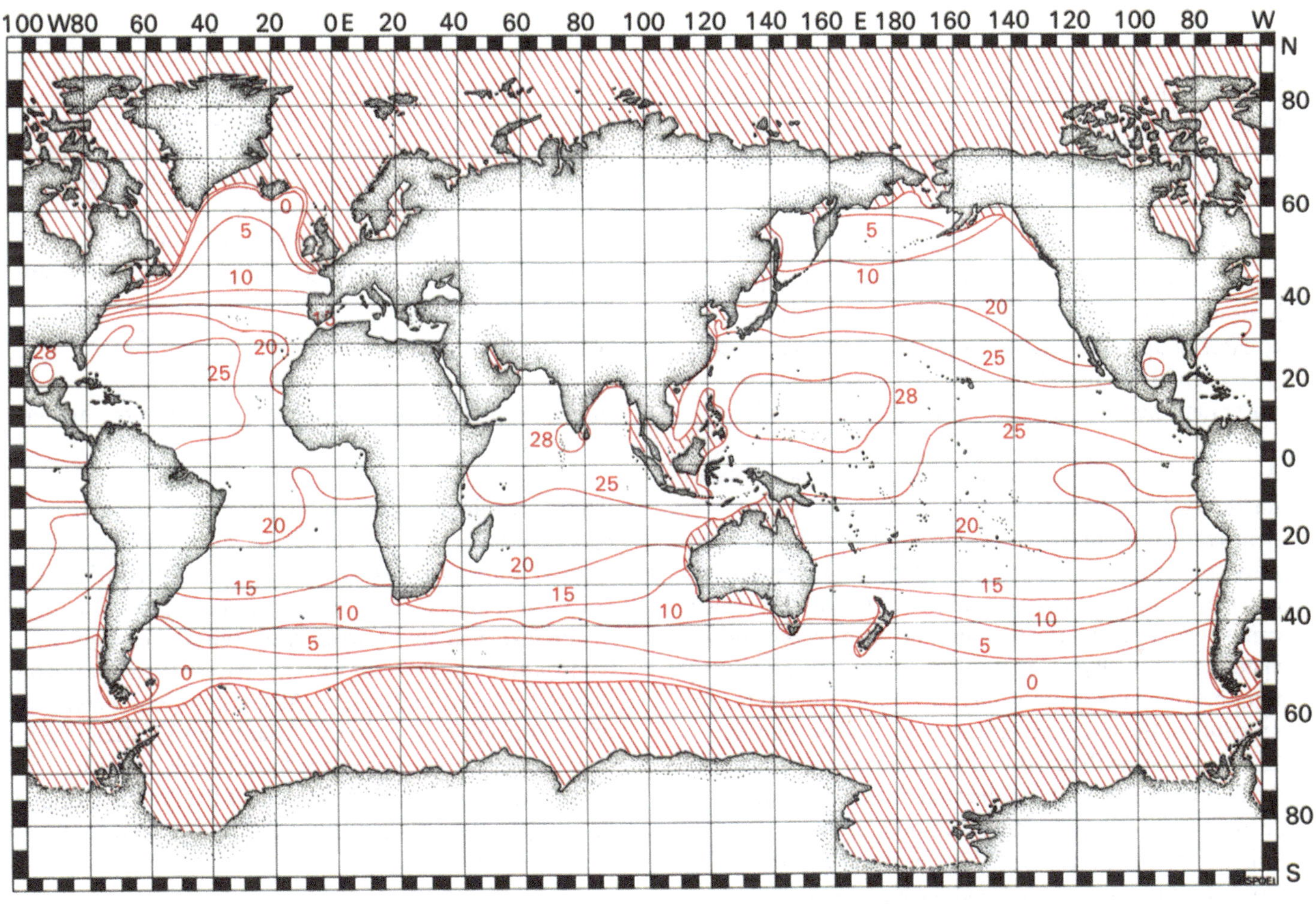
100 W 80 60 40 20 0 E 20 40 60 80 100 120 140 160 E 180 160 140 120 100 80 W
N 80 60 40 20 0 20 40 60 80 S
0
5
10
20
25
28
20
15
10
5
0
28
25
20
15
10
5
10
20
25
28
25
20
15
10
5
0

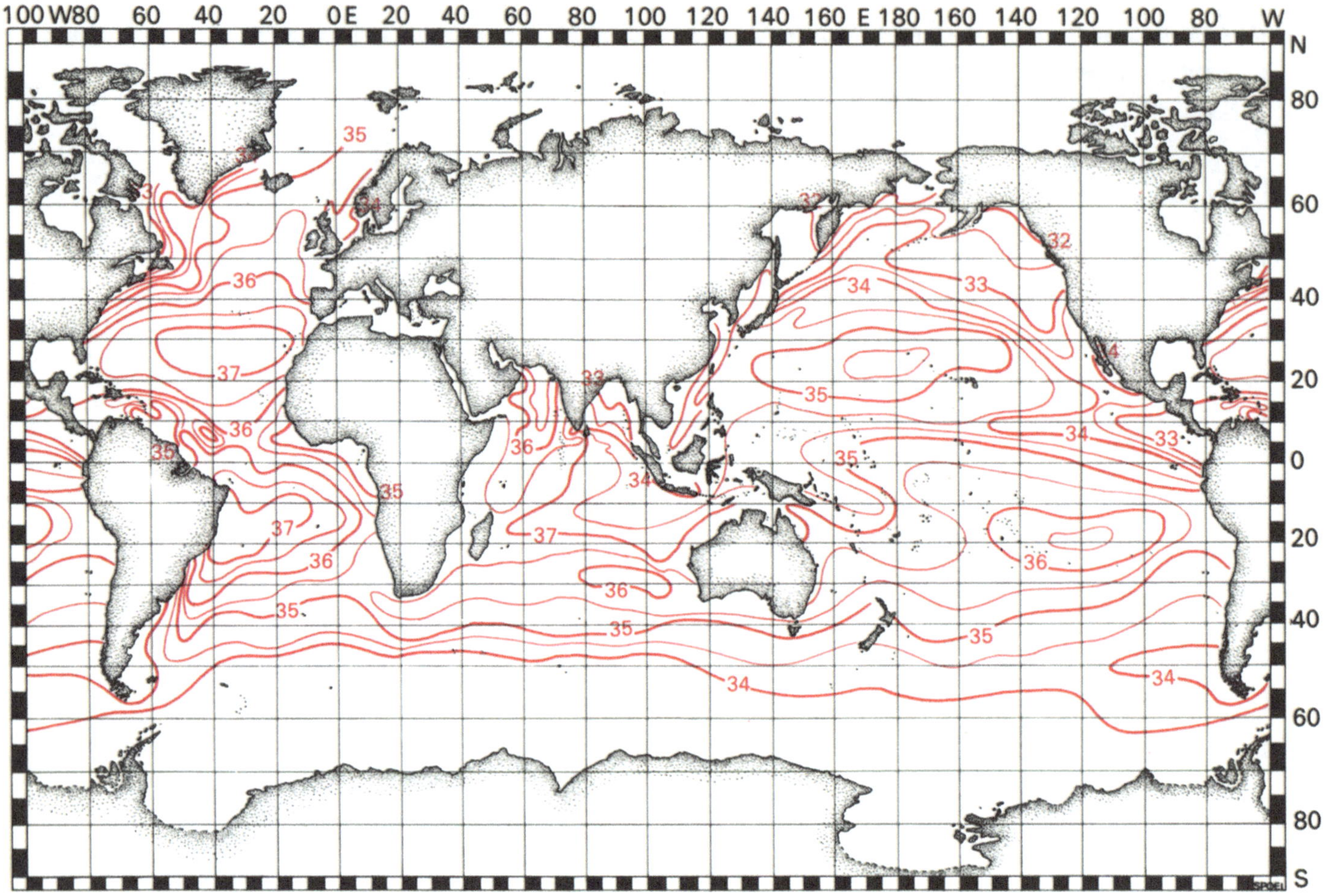

FIG. 21 Sea surface isohalines with 0.5‰ S intervals.
FIG. 22 Diagram of the fertility of the world ocean.
dotted = areas of high fertility, blanc = areas of low fertility.
FIG. 23 Inorganic phosphate-phosphorus and oxygen concentrations at 100 m depth in the world ocean.
dotted = areas with less than 2.0 ml/l O_2 dissolved,
black = areas with less than 0.5 μg-at/l inorganic phosphate-phosphorus.

The pattern of isohalines (fig. 21) parallels the latitudes much less than that of the isotherms. The major pattern is that of the highly saline Central waters surrounded by less saline water masses.

The diagram of the fertility of the sea (fig. 22) reflects the nutrients available for primary production, which is hydrographically largely dependent on coastal and equatorial upwelling in the Tropics and on river outflow (northern hemisphere) and upwards flow of bottom water (southern hemisphere) in the higher latitudes.

A good correlation of this pattern is possible with that of nutrients (fig. 23) but not with those of salinity, temperature or currents. A comparison with fig. 4 shows great differences between the fertility (phytoplankton) and the productive areas (zooplankton). This difference can be explained by the fact that the primary production is measured as phytoplankton productivity, which may temporarily be very high and low at other moments, for example in the northern latitudes during the winter. Zooplankton, with usually a longer life-span, tends not to fluctuate so strongly in abundance, which is only high in areas where a constant high amount of food (phytoplankton) is present. This fits in well with the greater correlation of phyto- and zooplankton biomass in the Tropics as opposed to the situation in the higher latitudes.

The inorganic phosphate-phosphorus and oxygen concentrations at 100 m depth (fig. 23) show that the Central waters are oligotrophic and the Equatorial, coastal and high latitude waters are eutrophic. To distinguish sharply between these types of ecological environments is difficult but to separate them by the 0.5 μg-at/l line seems adequate. All oxygen-poor waters are of the eutrophic type. In the Atlantic and Indian Oceans these oxygen-poor waters are found at places with a decreasing neritic fauna, in the East Pacific the area of low oxygen values is characterized by a special fauna with many endemic forms.

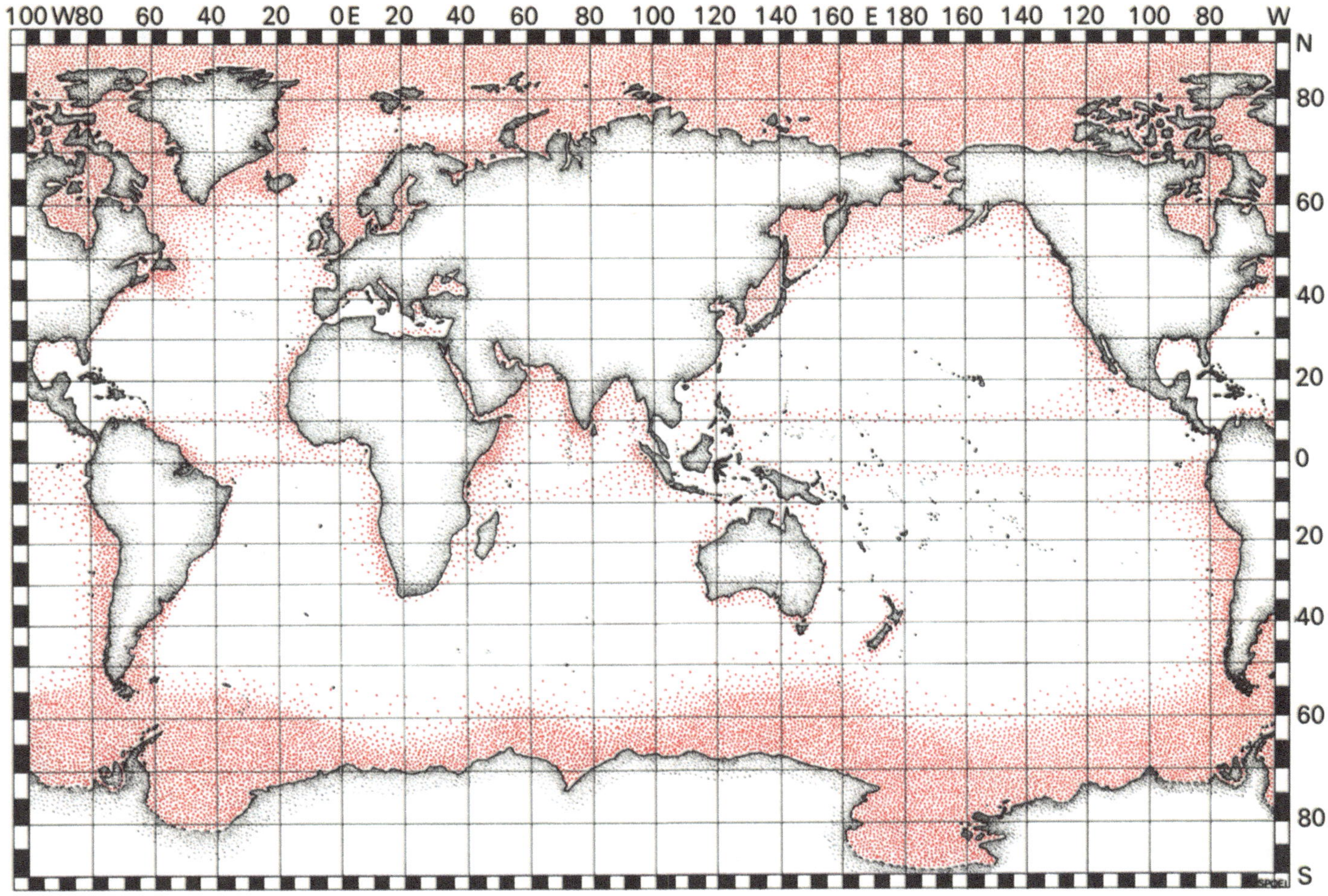
100 W80 60 40 20 0E 20 40 60 80 100 120 140 160 E 180 160 140 120 100 80 W
N
80
60
40
20
0
20
40
60
80
S

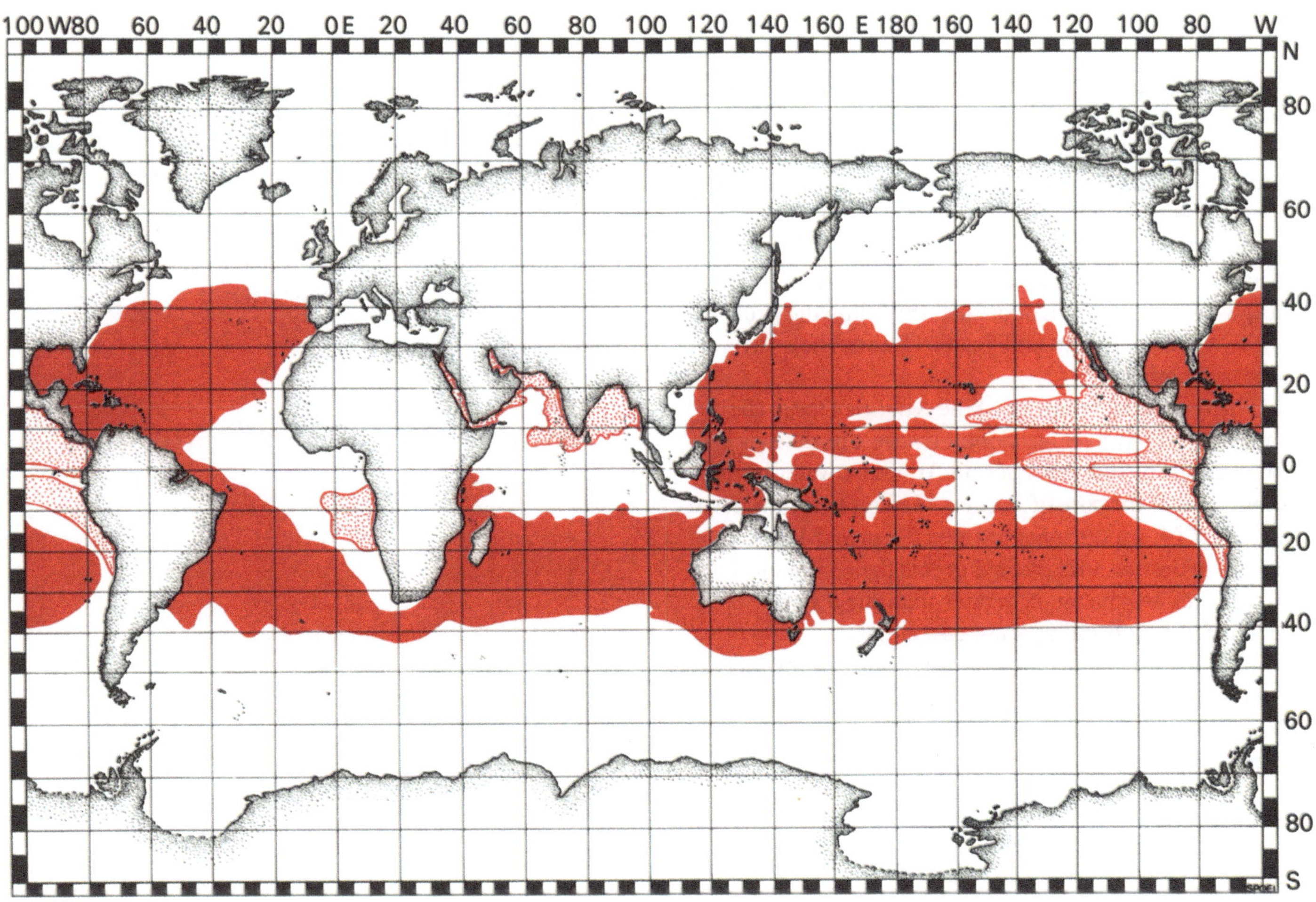
100 W80 60 40 20 0E 20 40 60 80 100 120 140 160 E 180 160 140 120 100 80 W
N
80
60
40
20
0
20
40
60
80
S

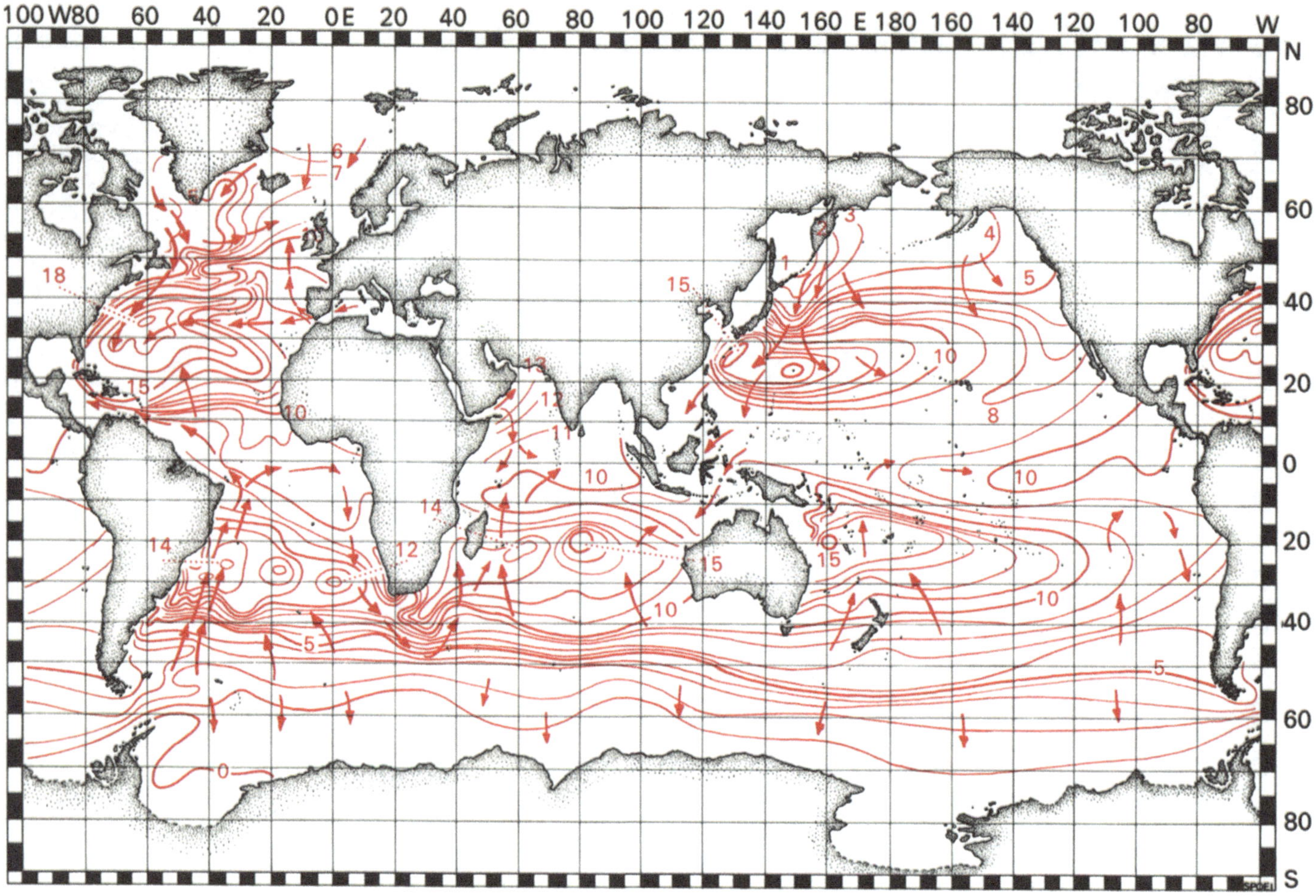

Figure 24 gives the temperatures at about 400 m depth and the intermediate flow between 400 and 1000 m depth. The currents at these depths are only vaguely known. In this map the origin of Intermediate waters are supposed to be found near 50° S and near 60°–70° N in the Atlantic, near 50° N in the Pacific and near the Red Sea and Mediterranean Sea. To give an impression of the vertical zonation of oceans, the following profiles of the Atlantic, Indian and Pacific Oceans show the average conditions.

The temperature profile of the Atlantic Ocean (fig. 25) shows the yearly average along a transect in the axis of the ocean. The influence of cold Antarctic water in the south, and of the core of warmer Mediterranean water at about 40° N, are the most obvious features. Minor irregularities are not reflected in this figure and local differences may be much greater than suggested.

The figures 26 and 27 show two almost simultaneously made temperature profiles of the North Atlantic at only 25° longitude distance of each other. They highlight local differences and the influences of currents and eddies on a profile.

FIG. 24 Temperatures at about 400 m depth and intermediate flow between 400 and 1000 m. arrows = current direction, lines = isotherms with 1° C interval.

FIG. 25 Temperature profile of the Atlantic Ocean, isotherms with 1° C interval.

FIG. 26 Temperature profile along 30° W in the N. Atlantic, from April 1980. A = core of N. Atlantic Current, B = southern border of Subarctic water.

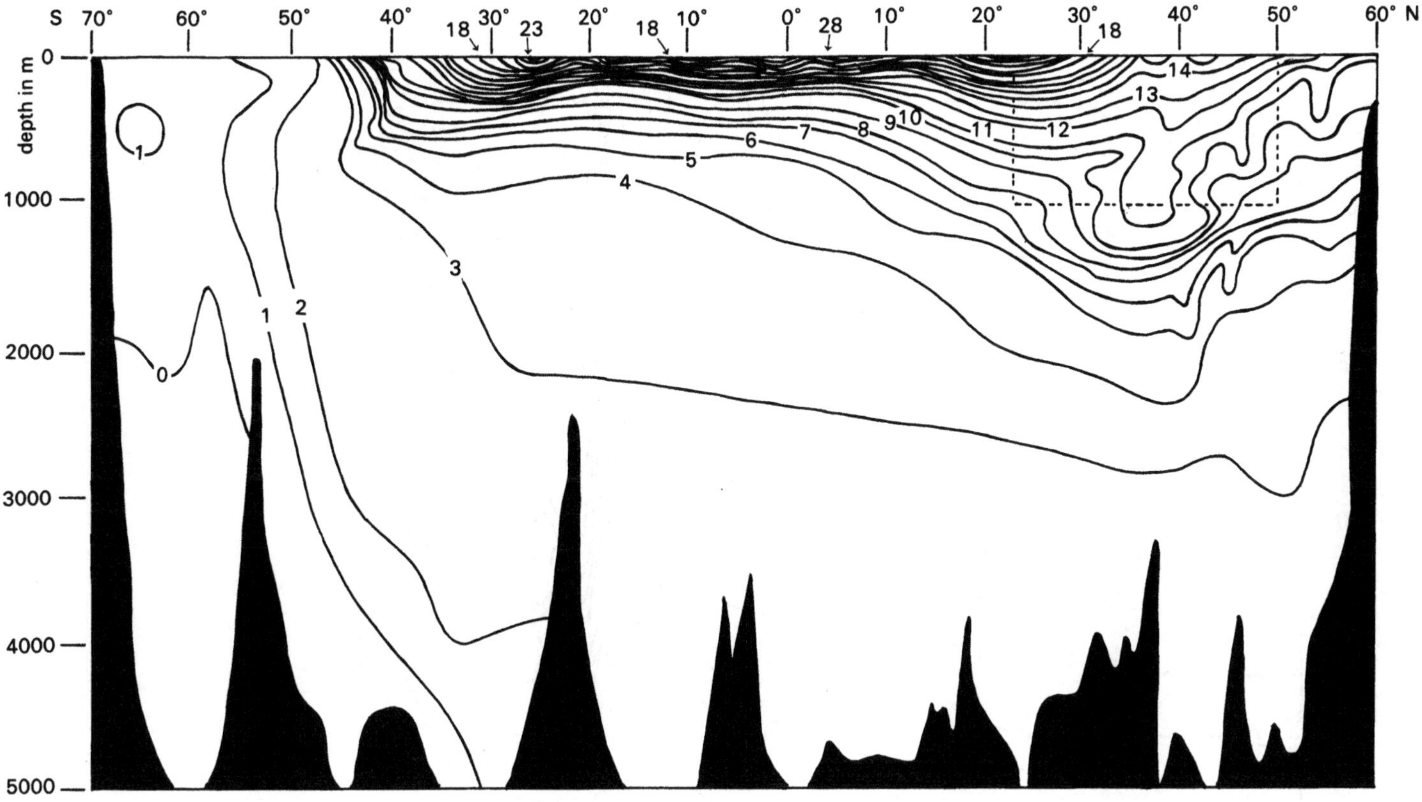

S 70° 60° 50° 40° 30° 20° 10° 0° 10° 20° 30° 40° 50° 60° N
depth in m
0
1000
2000
3000
4000
5000
18 23 18 28 18
14 13 12 11 10 9 8 7 6 5 4 3 2 1 0

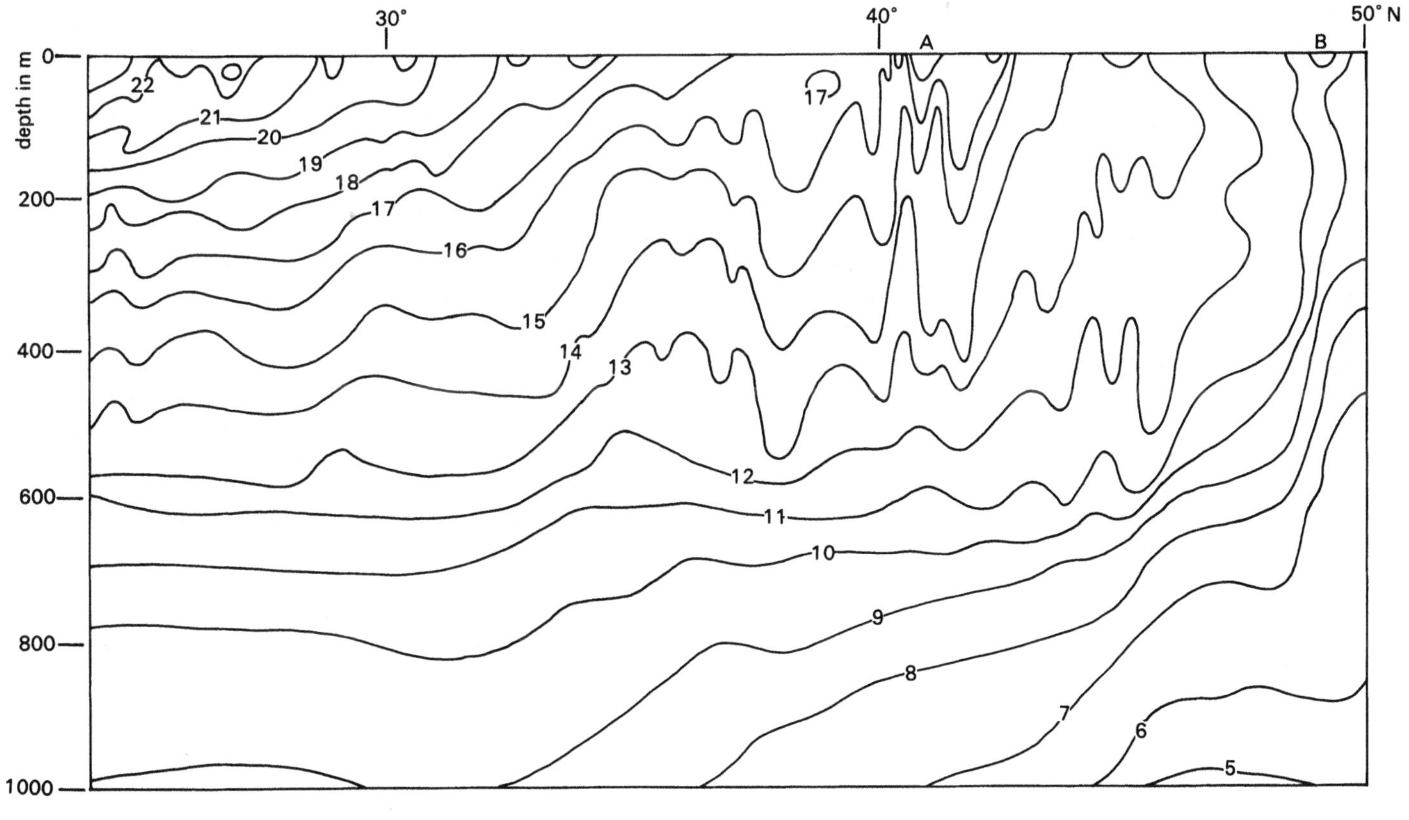

30° 40° 50° N
A B
depth in m
0
200
400
600
800
1000
22 21 20 19 18 17 16 15 14 13 12 11 10 9 8 7 6 5

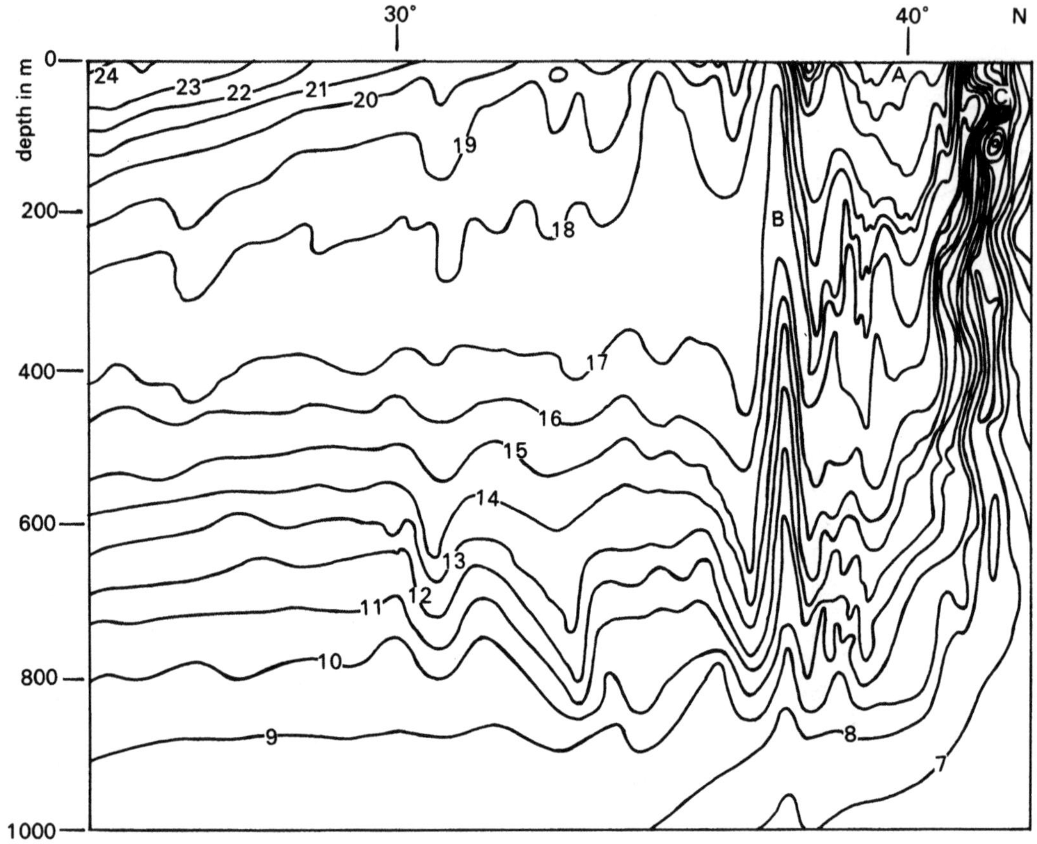

In the salinity profile of the Atlantic (fig. 28), the influence of the Mediterranean outflow is very clear near 40°N at 1200 m depth. Antarctic Intermediate and Bottom water as well as Arctic deep water determine most of the deeper salinity conditions.

The division into water masses in the vertical profile of fig. 29 is also based on physicochemical conditions. The division into water masses is not given for the surface waters in the maps, and are added here only to show the influences of waters of different latitudes on the deep waters of the oceans.

FIG. 27 Temperature profile along 55°W in the N. Atlantic, from April 1980. A = core of N. Atlantic Current, B = cold core ring, C = core of Labrador Current.
FIG. 28 Salinity profile of the Atlantic Ocean, isohalines with 0.1‰ S interval.
FIG. 29 Profile of the Atlantic Ocean giving the division into water masses. 1 = Antarctic Slope water, 2 = Circum Antarctic Deep water, 3 = Antarctic Bottom water, 4 = Atlantic Deep water, 5 = Atlantic & Indian Deep water, 6 = Antarctic Intermediate water, 7 = Antarctic Surface water, 8 = Intermediate Mixed Antarctic water, 9 = Subantarctic Intermediate water, 10 = Subantarctic Surface water, 11 = Subtropical water, 12 = Tropical water, 13 = Mediterranean water, 14 = N. Atlantic Intermediate Mixed water, 15 = Arctic water, 16 = N. Atlantic Surface water, 17 = N. Pacific Deep water, 18 = Tropical Intermediate water, 19 = N. Pacific Intermediate water, 20 = N. Pacific Surface water, 21 = Central Indian water, 22 = N. Indian water, 23 = Red Sea & Persian Gulf water, 24 = N. Indian Deep water, AAC = Antarctic Convergence, AC = Arctic Convergence, AD = Antarctic Divergence, SC = Subtropical Convergence, T = Thermal front.

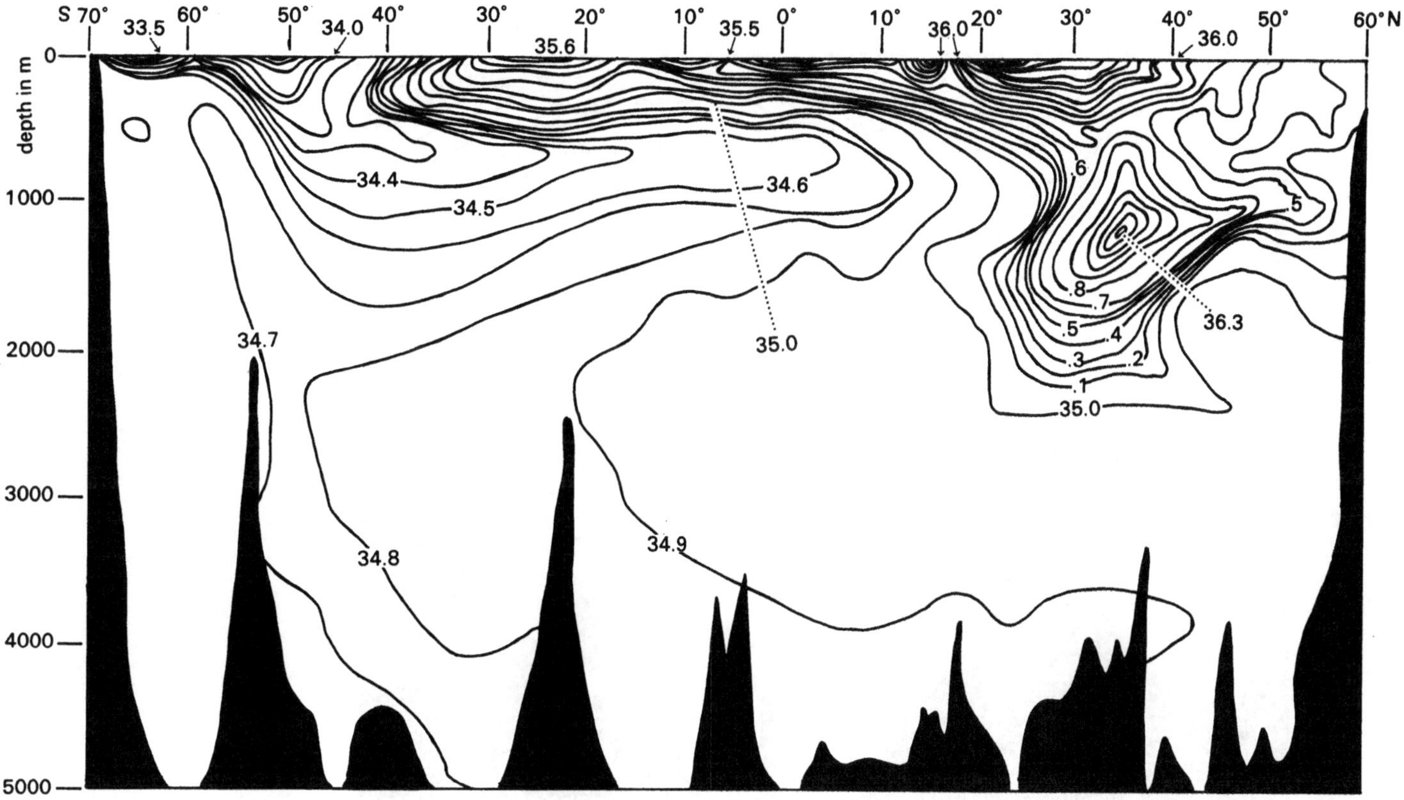
S 70°
60°
50°
40°
30°
20°
10°
0°
10°
20°
30°
40°
50°
60°N
33.5
34.0
35.6
35.5
36.0
36.0
depth in m
0
1000
2000
3000
4000
5000
34.4
34.5
34.6
34.7
34.8
34.9
35.0
36.3
.6
.5
.8
.7
.5
.4
.3
.2
.1
35.0

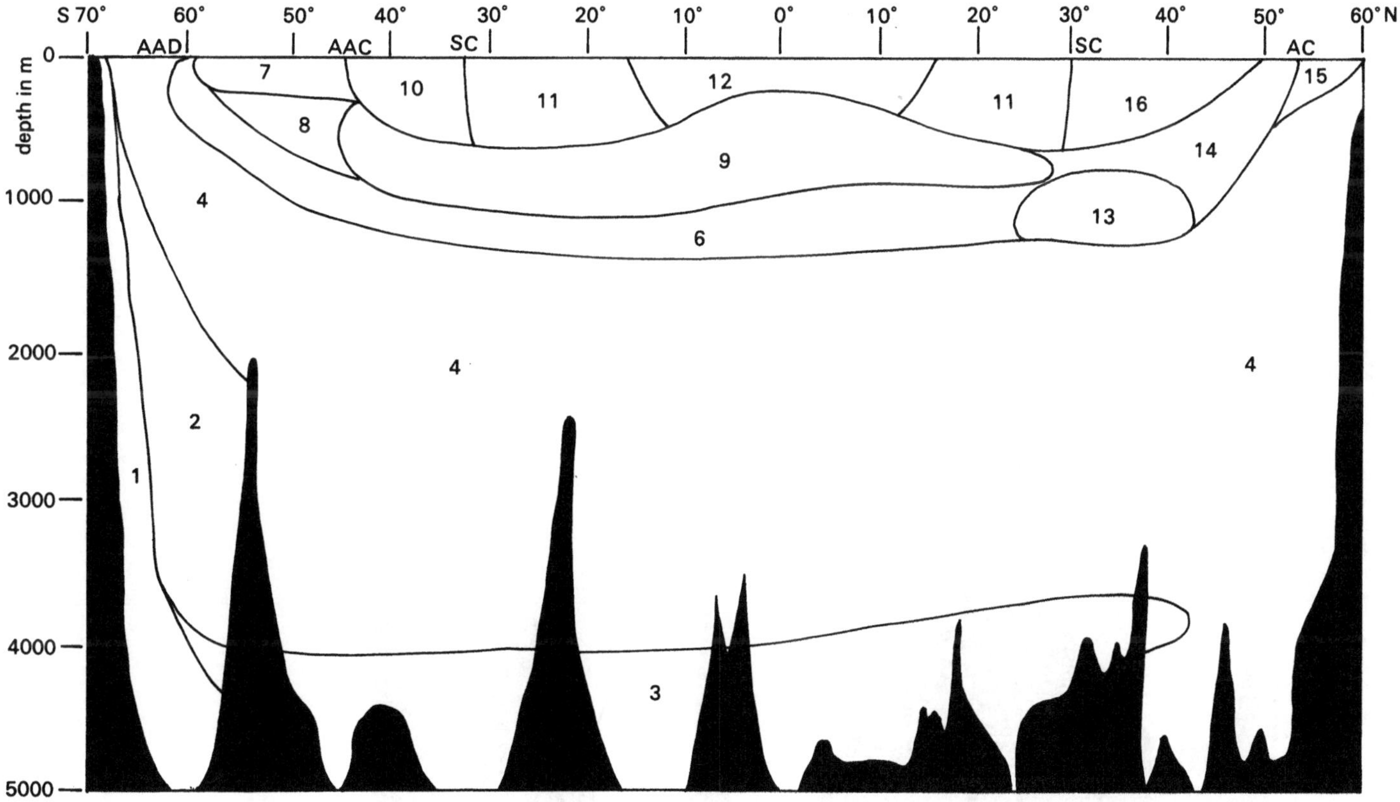
S 70°
60°
50°
40°
30°
20°
10°
0°
10°
20°
30°
40°
50°
60°N
AAD
AAC
SC
SC
AC
depth in m
0
1000
2000
3000
4000
5000
7
10
11
12
11
16
15
8
9
14
4
13
6
4
4
2
1
3

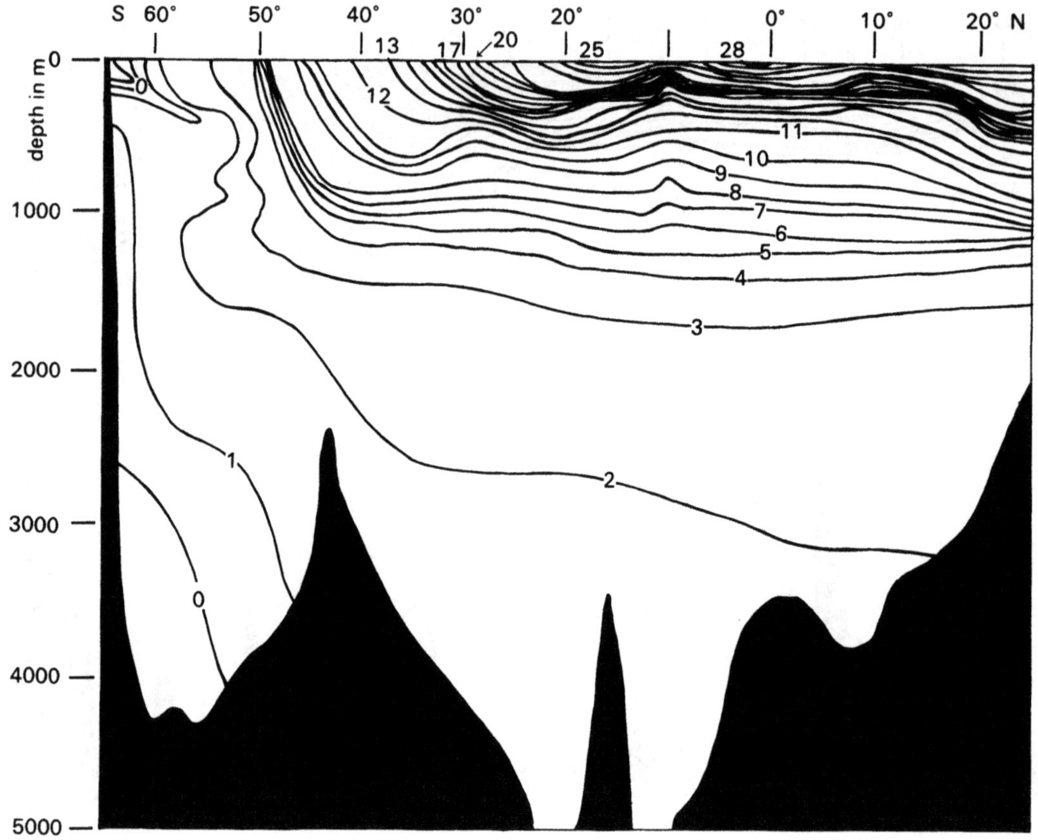
S 60°
50°
40°
30°
20°
0°
10°
20° N
13
17
20
25
28
depth in m
0
1000
2000
3000
4000
5000
0
12
11
10
9
8
7
6
5
4
3
2
1
0

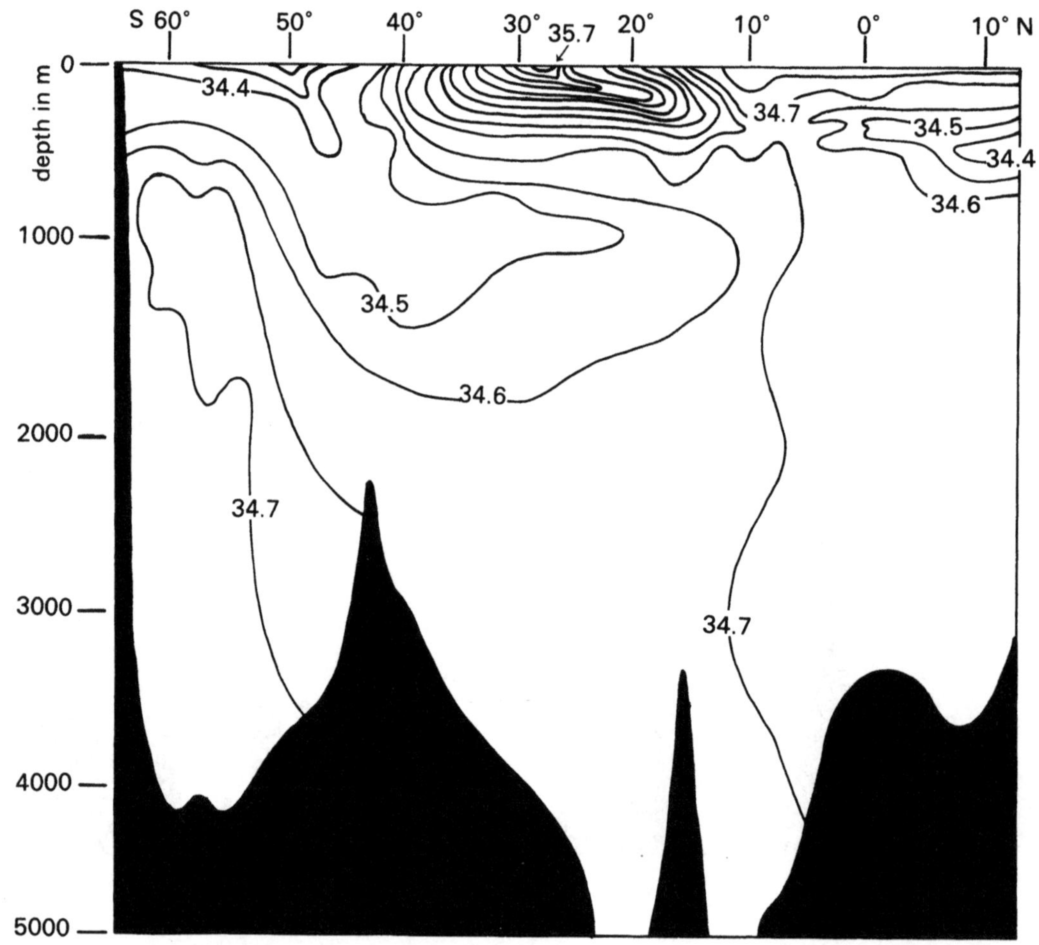
S 60°
50°
40°
30°
35.7
20°
10°
0°
10° N
depth in m
0
1000
2000
3000
4000
5000
34.4
34.7
34.5
34.4
34.6
34.5
34.6
34.7
34.7

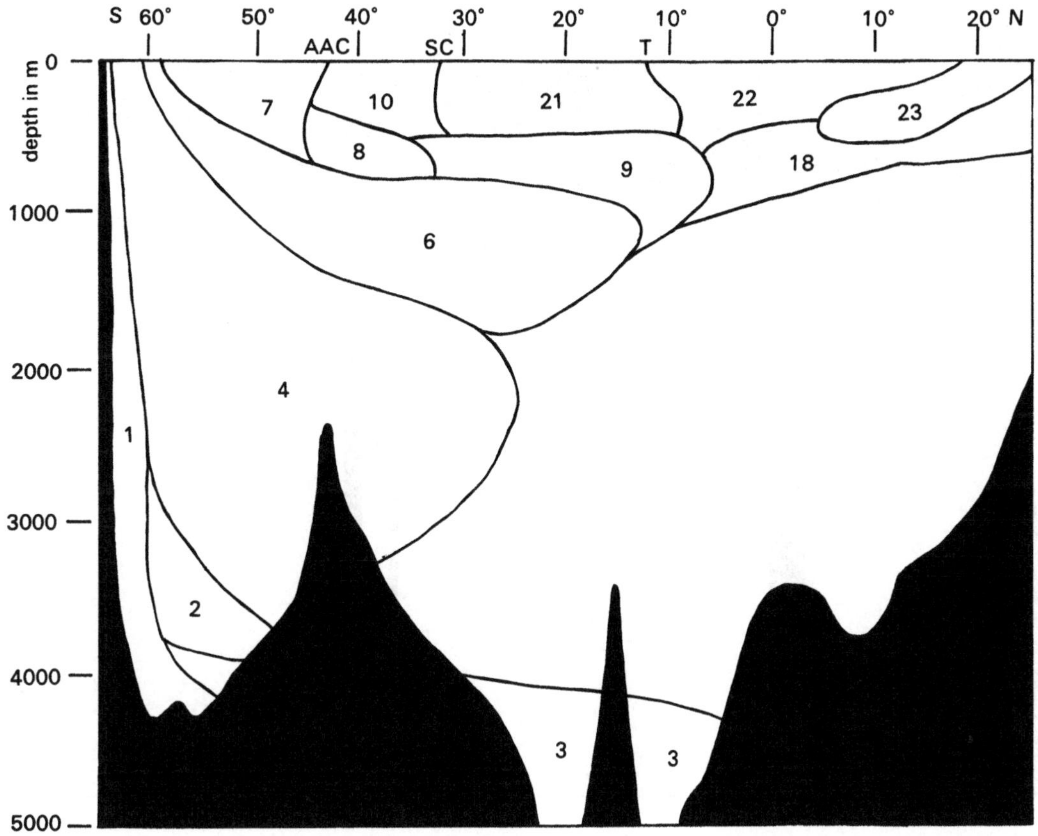

The profiles for the three oceans are roughly identical, only the Indian Ocean lacks cold influences in the north (fig. 30). A deepening of the isotherms in the north is due to outflow from the Red Sea and the Persian Gulf.

These outflows are more prominently reflected by the salinity profile (fig. 31). The water masses of the Indian Ocean (fig. 32) are characterized by strong influences of the Atlantic Ocean, and of the Red Sea in combination with the Persian Gulf.

FIG. 30 Temperature profile of the Indian Ocean, isotherms with 1°C interval.
FIG. 31 Salinity profile of the Indian Ocean, isohalines with 0.1‰ S interval.
FIG. 32 Profile of the Indian Ocean giving the division into water masses (for explanation see fig. 29).

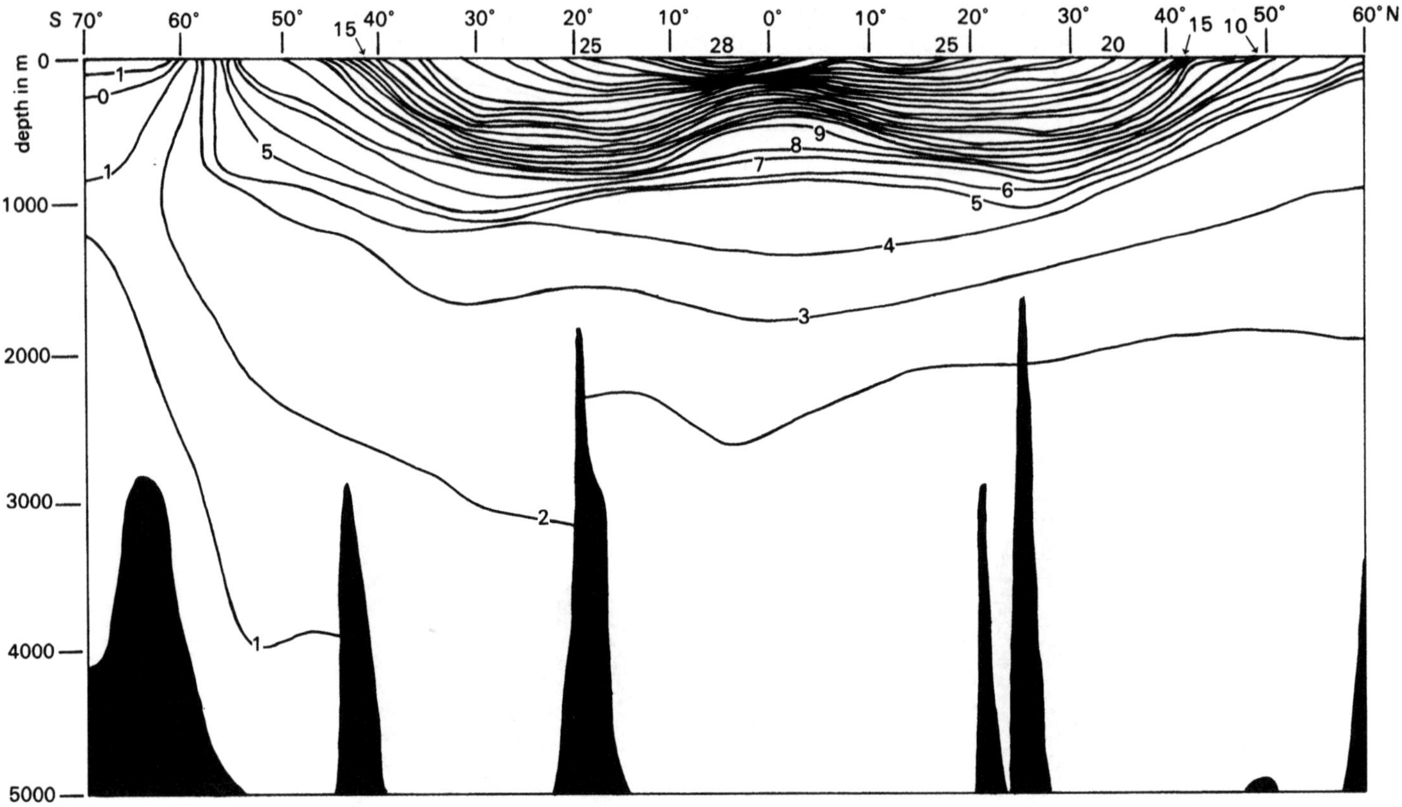

S 70°
60°
50°
40°
30°
20°
10°
0°
10°
20°
30°
40°
50°
60° N
depth in m
0
1000
2000
3000
4000
5000
15
25
28
25
20
15
10
1
0
1
5
9
8
7
6
5
4
3
2
1

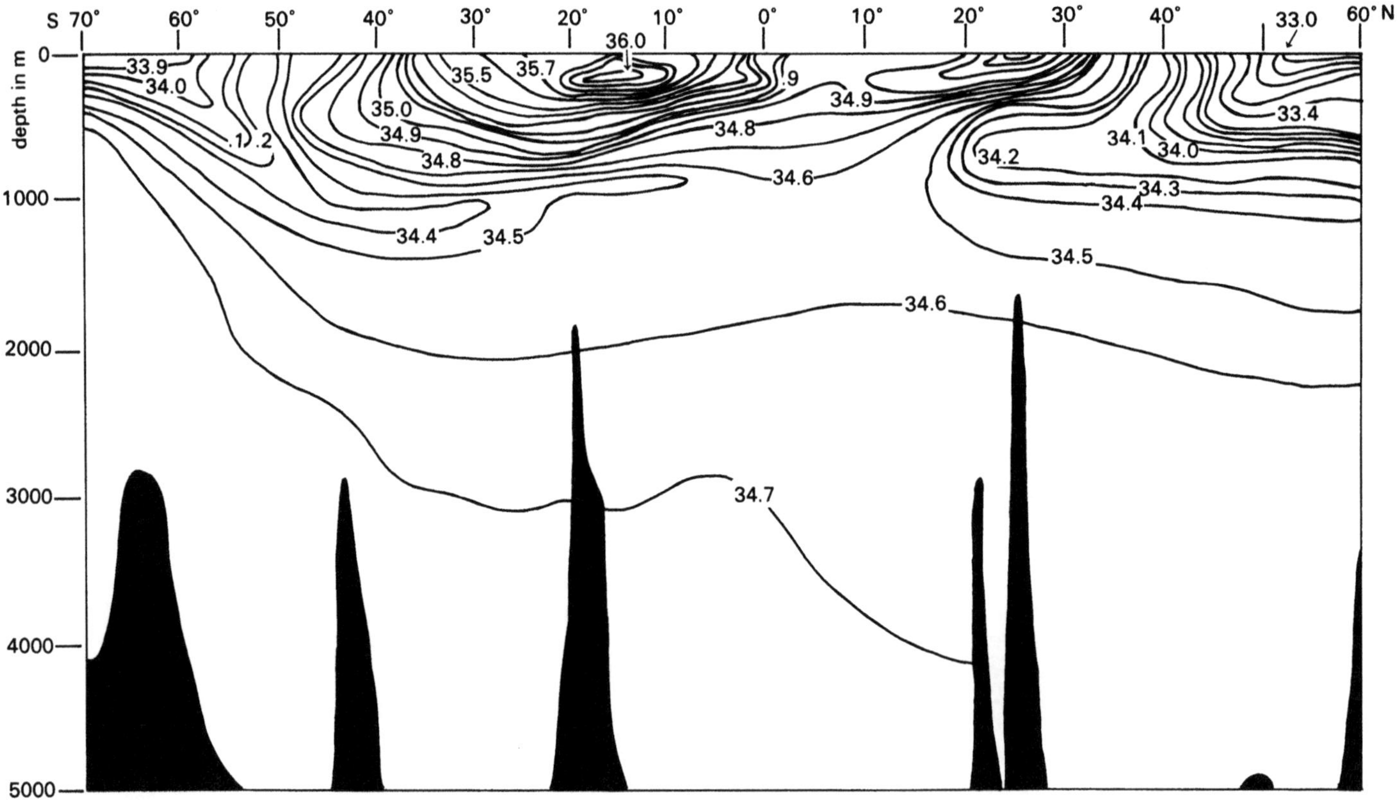

S 70°
60°
50°
40°
30°
20°
10°
0°
10°
20°
30°
40°
33.0
60° N
depth in m
0
1000
2000
3000
4000
5000
36.0
33.9
34.0
35.5
35.7
.9
34.9
33.4
35.0
34.9
34.8
.1
.2
34.8
34.6
34.1
34.0
34.2
34.3
34.4
34.4
34.5
34.5
34.6
34.7

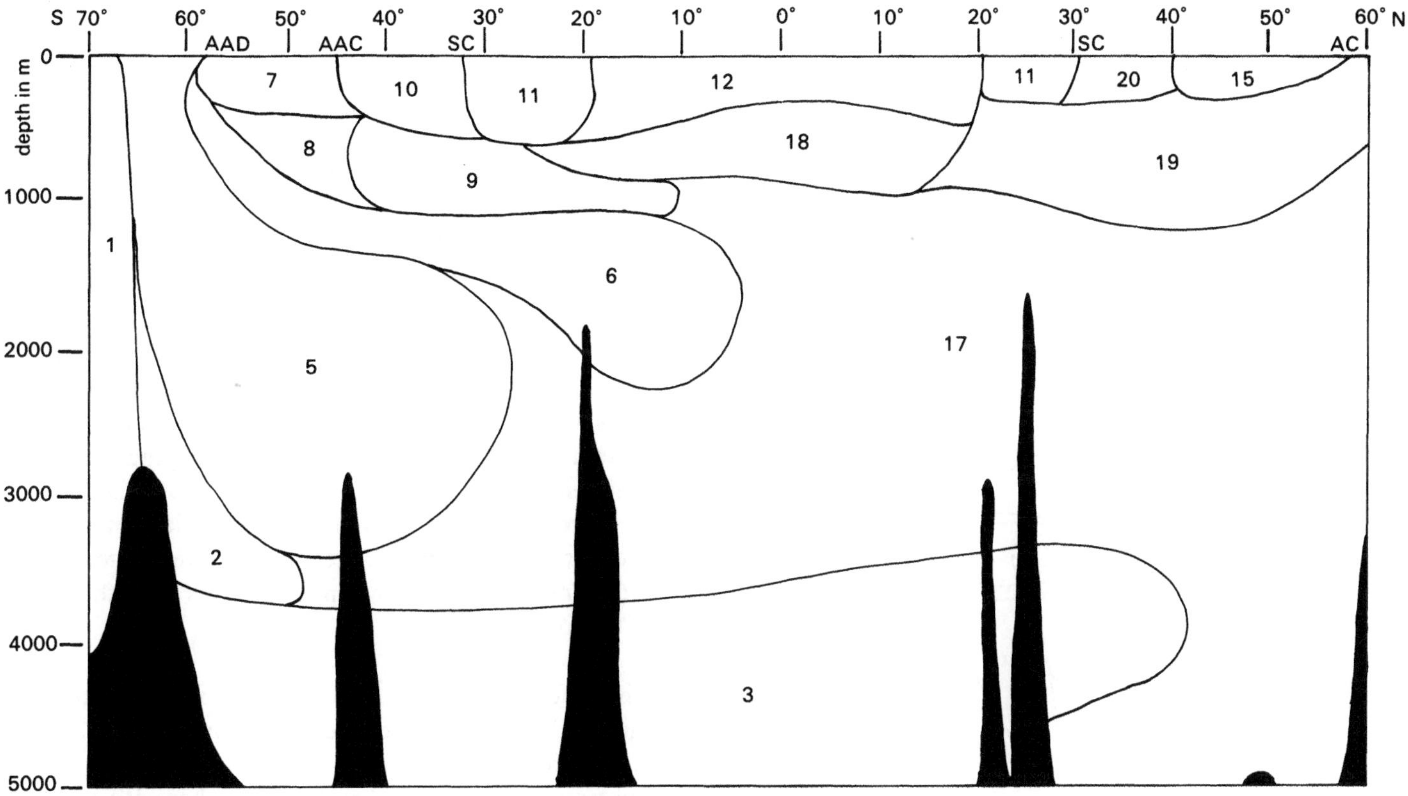

The temperature profile of the Pacific Ocean (fig. 33) is very regular and symmetrical compared with those of the other oceans. This is caused by the restricted Arctic influences in the deep North Pacific, and by the Atlantic and Indian Ocean waters, which restrict the Antarctic influences in the South Pacific.

The salinity profile (fig. 34) shows the same regular division.

The division of water masses of the Pacific Ocean (fig. 35) based on physico-chemical conditions, shows the Indo-Atlantic influence in the south and the simple structure of the North Pacific.

In short, it is clear that the conditions strongly fluctuate with the seasons and have fluctuated in the geological past. Temperature differences are largely latitudinal, salinity differences and conditions at greater depth rather reflect the current patterns.

FIG. 33 Temperature profile of the Pacific Ocean, isotherms with 1° C interval.

FIG. 34 Salinity profile of the Pacific Ocean, isohalines with 0.1‰ S interval.

FIG. 35 Profile of the Pacific Ocean giving the division into water masses (for explanation see fig. 29).

Ranges and their interpretations

All the localities where a species occurs do not necessarily belong to its reproductive area. As plankton, per definition, depends in its movements on water circulation, currents may easily transport specimens also outside this area. This expatriation may take place in horizontal and in vertical directions. In the non-sterile expatriation area specimens reproduce, but they cannot support a population without influx of specimens from

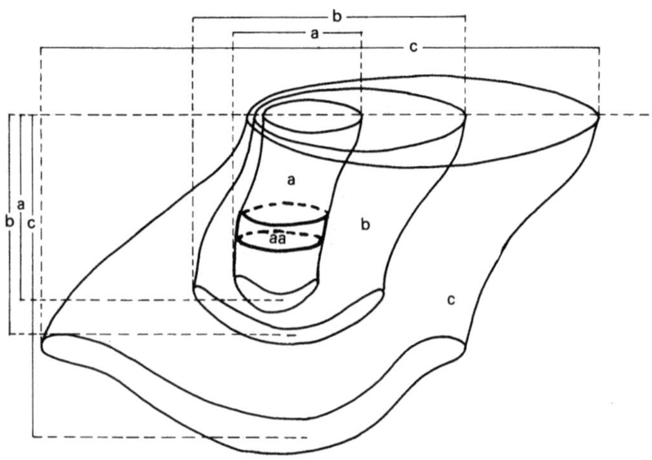

the reproductive area. The total distribution of a species thus consists of the reproductive range, the non-sterile expatriation range and the sterile expatriation range (fig. 36). In chapter 3, the geographically different distributions will be given, here attention is focused on the more essential differences in the occupation of the ranges. For taxonomic studies and zoogeography only the reproductive range (cf. p. 38), and to a lesser degree the non-sterile expatriation range, is important. With the commonly used sampling techniques, it is, however, very difficult to discriminate between the different sections of a range. Many species show vertical migration which gives their ranges a vertical component. Many taxa are known only to reproduce in a particular depth layer of their ranges (cf. p. 133), so that the actual reproductive range (fig. 36, aa) is frequently a horizontally and vertically restricted sheet of water.

Restricted expatriation is a common phenomenon, but in some cases the transport bridges extremely long distances, distinctly influencing the overall distributional patterns. In fig. 37 the expatriation of cold- and warm-water diatoms is given as an example of long-distance expatriation.

To this point, only living plankton has been considered, but many zoogeographic studies incorporate distributional data from sediment material. Ranges in sediments always differ from living plankton ranges, as:

FIG. 36 Diagram of the three-dimensional range of a plankton taxon. a = reproductive range, aa = actual area where reproduction is found, b = non-sterile expatriation area, c = sterile expatriation area.
FIG. 37 Expatriation in diatoms. dotted = warm-water populations, black = cold-water populations.
FIG. 38 Distribution of *Globorotalia hirsuta*. hatched = surface sediment distribution, black = living plankton distribution.

a) sedimentation is of a variable intensity in different areas due to differences in bottom structure and differences in mortality rates, b) during dying and sedimentation on the sea-floor transport over large distances may occur, and c) as decomposition may prevent fossilisation in some areas. In some cases the sediment range is smaller, in some larger than the range of the living species.

The differences between the larger sediment range and the smaller living plankton distribution of *Globorotalia hirsuta* (fig. 38) are most important in the N. Atlantic and Subtropical-Central Indian Oceans. These differences may be due to transport of living specimens or of specimens in the period between dying and sedimentation. The sediment range also resembles the living plankton patterns of species such as *Eucalanus hyalinus* (fig. 58), which points to the fact that the sediment area could indeed be the expatriation area of *Globorotalia hirsuta*.

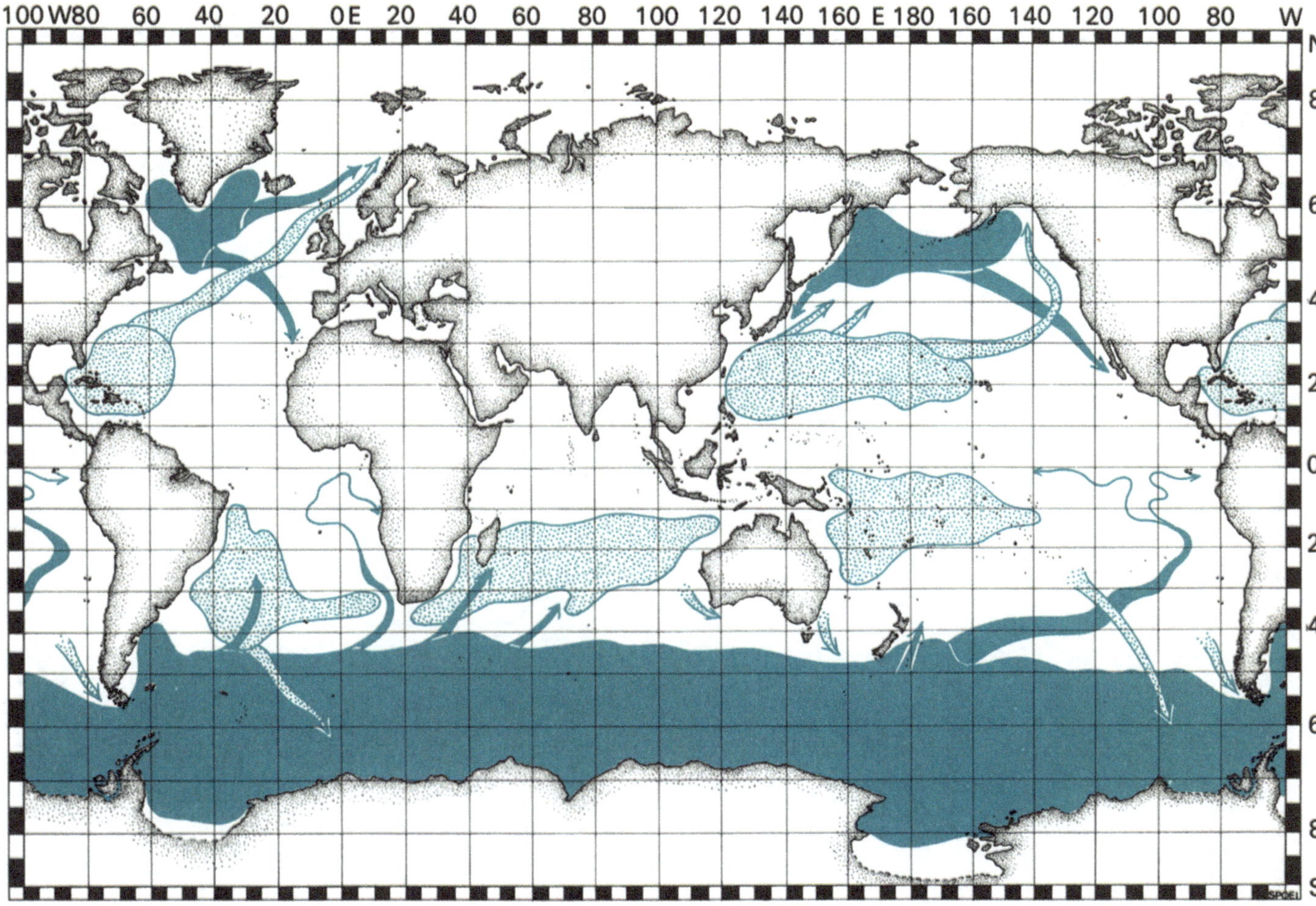

100 W 80 60 40 20 0 E 20 40 60 80 100 120 140 160 E 180 160 140 120 100 80 W
N 80 60 40 20 0 20 40 60 80 S

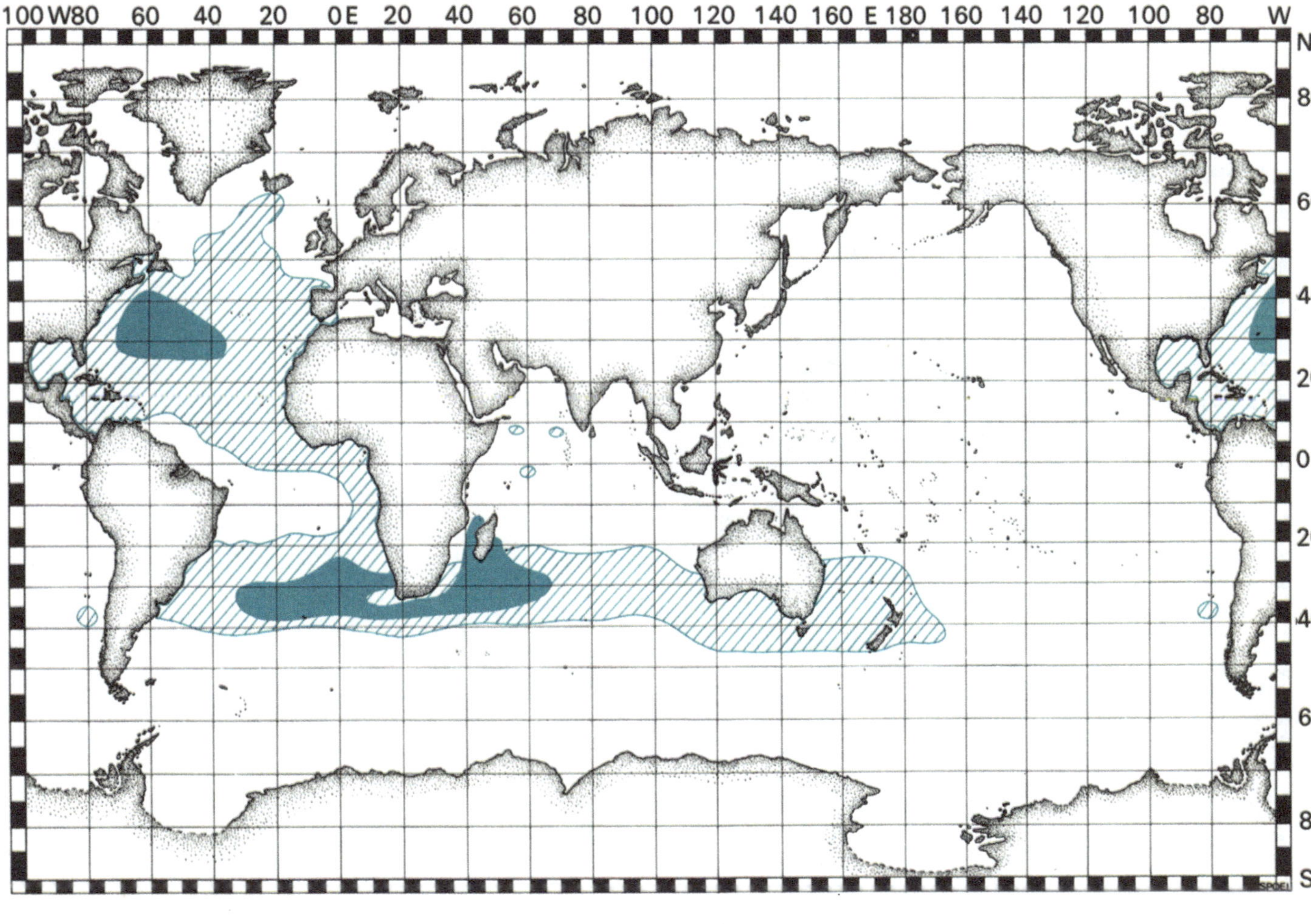

100 W 80 60 40 20 0 E 20 40 60 80 100 120 140 160 E 180 160 140 120 100 80 W
N 80 60 40 20 0 20 40 60 80 S

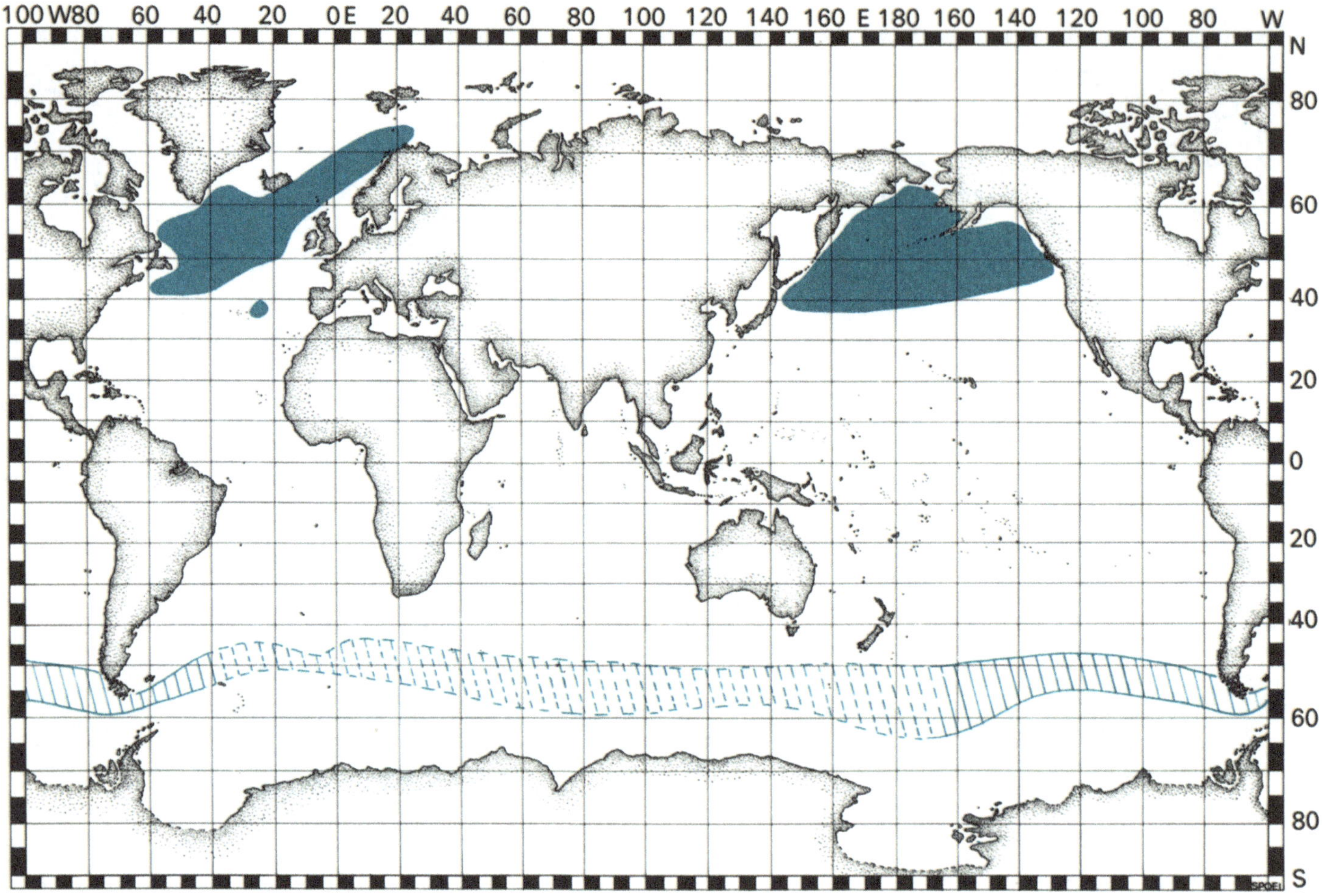
100 W80
60
40
20
0 E
20
40
60
80
100
120
140
160
E 180
160
140
120
100
80
W
N
80
60
40
20
0
20
40
60
80
S

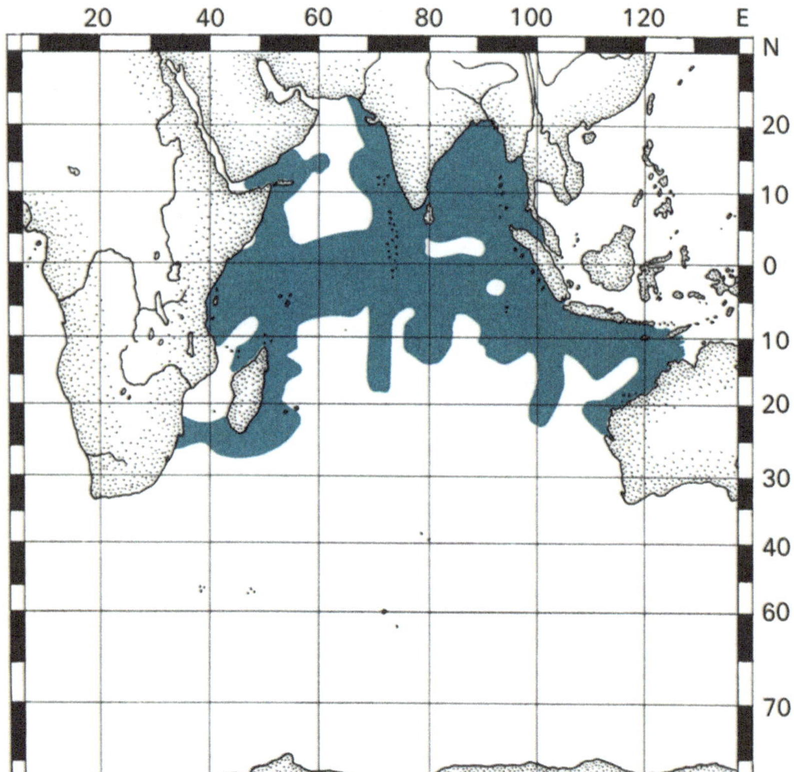
20
40
60
80
100
120
E
N
20
10
0
10
20
30
40
60
70
S

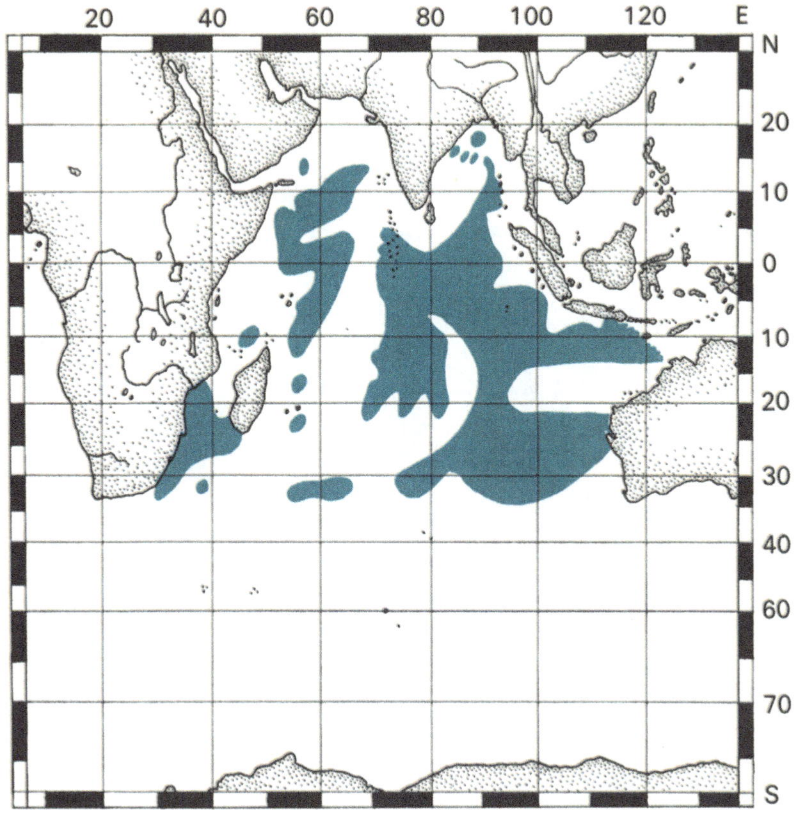

Coccolithus pelagicus shows a bisubpolar pattern in surface sediments (fig. 39). At present it is, however, restricted to the northern hemisphere in waters of 6° to 14°C. In the Southern Ocean, the belt of 6°-14°C is very narrow and easily disturbed by temperature fluctuations. According to McIntyre et al. (1970) the Post-Glacial period of warming up at about 8000 B.C. resulted in such a disturbance that the species disappeared from the southern hemisphere. If this hypothesis is correct, it also explains why Subpolar species were not able to cross the equator after about 8000 B.C. This map is especially given to show that sediment areas which are larger than living plankton areas may also be caused by extinction of the species in parts of their former ranges.

Determination of ranges can also be difficult due to the patchy distribution of species. Some species, such as *Candacia catula* (fig. 40) occur in continuous areas, others, such as *C. ethiopica* (fig. 41) live in interrupted ranges, whereas some, such as *C. longirana* are really patchily distributed (fig. 42).

All the ranges presented in this atlas as continuous, equally covered areas are usually no more than areas in which the taxon concerned occurs more or less frequently in scattered patches of variable abundance

FIG. 39 Distribution of *Coccolithus pelagicus*. black = occurrence in surface sediment and plankton, hatched + black = occurrence in surface sediment.

FIG. 40 *Candacia catula* in the Indian Ocean. black = areas of positive hauls in the period 1962-1965, white = areas of negative hauls in the same period.

FIG. 41 *Candacia ethiopica* in the Indian Ocean (for explanation see fig. 40).

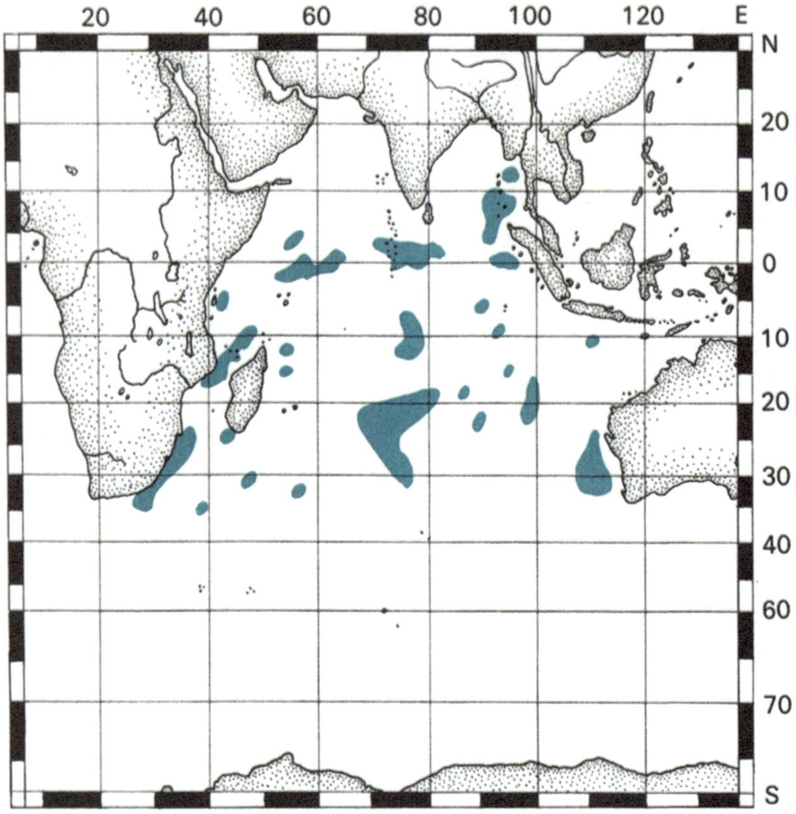
20
40
60
80
100
120
E
N
20
10
10
0
10
20
30
40
60
70
S

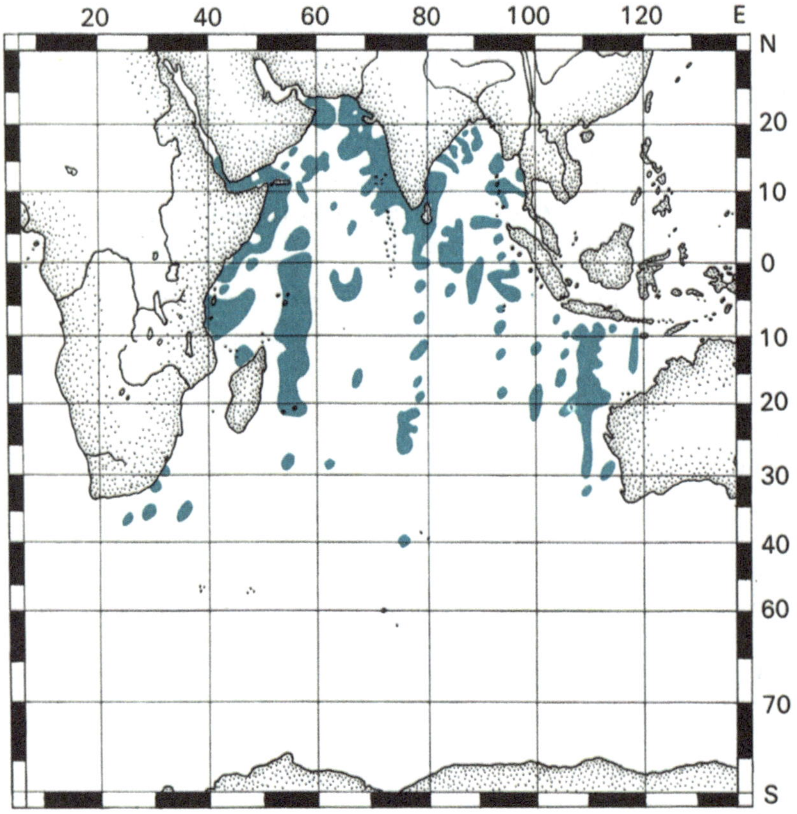
20
40
60
80
100
120
E
N
20
10
0
10
20
30
40
60
70
S

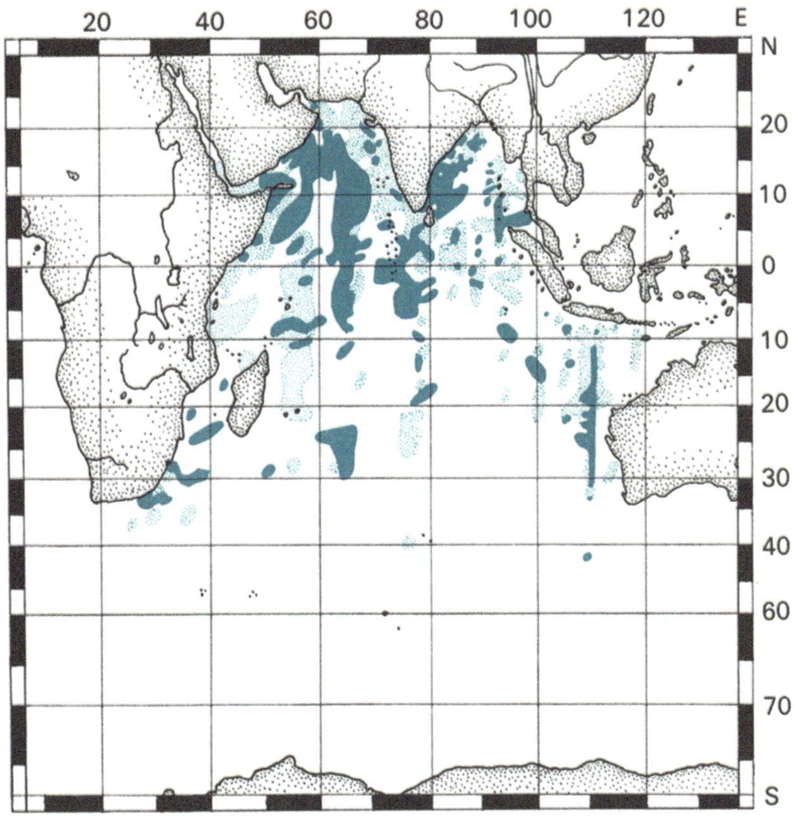

and size. Studies of the Indian Ocean distribution of Candaciidae show that species with more or less similar overall ranges may occur, even in material collected over longer periods, in a large continuous patch (fig. 40), in a few larger and smaller patches (fig. 41) or in numerous small, widely scattered, patches (fig. 42).

A more synoptic sampling than represented in the figs 40 to 42, such as that of *Desmopterus papilio* (figs 43 & 44), also shows a very patchy pattern. The distributions of the patches in the two figures differ strongly, due to climatological changes. The differences given as dotted in fig. 44 are most clear in upwelling and equatorial areas where currents reverse under the monsoon influence (cf. also fig. 17).

Not only the patchiness of a distribution, but also the extreme borders of occurrence are a result of seasonal influences.

FIG. 42 *Candacia longirana* in the Indian Ocean (for explanation see fig. 40).
FIG. 43 Distribution of *Desmopterus papilio* in the Indian Ocean during the period 16-X-1965 to 16-IV-1966 (cf. fig. 44).
FIG. 44 Distribution of *Desmopterus papilio* in the Indian Ocean. black = distribution in the period 16-IV to 16-X-1966, dotted = the pattern of fig. 43 for comparison.

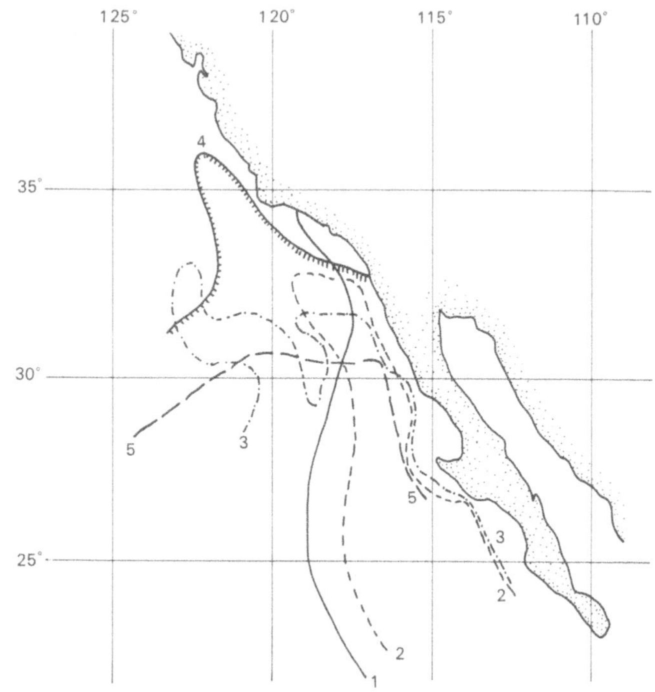

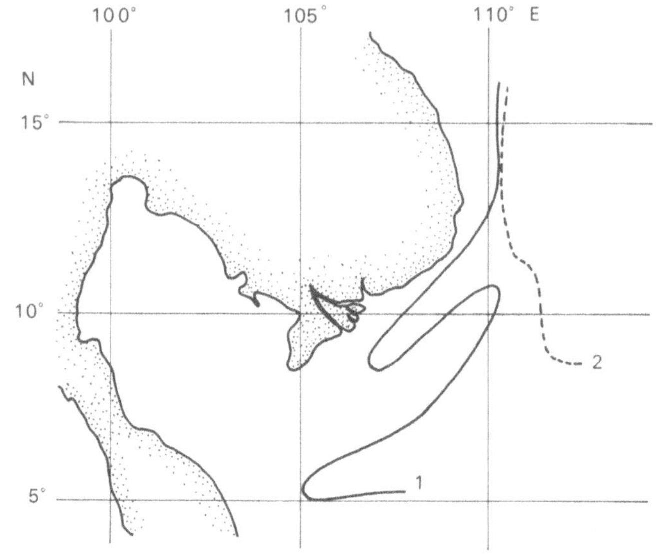

FIG. 45 Northern borders of *Cavolinia inflexa* off California. 1 = October border, 2 = August border, 3 = June border, 4 = April border, 5 = November border.

FIG. 46 Western borders of *Creseis virgula* in the S. China Sea. 1 = during N.E. Monsoon, 2 = during S.W. Monsoon.

FIG. 47 Distribution of *Limacina lesueuri*. black = areas with high abundance, hatched = areas with low abundance.

FIG. 48 Distribution of *Clio cuspidata*. black, = areas with high abundance, hatched = areas with low abundance.

Figure 45 gives an example of seasonal fluctuations of the northern distribution limits of *Cavolinia inflexa*. In this species, it is chiefly the California Current which induces the shifts. The strength of this current is variable due to the latitudinally varying climatic conditions, which results in varying positions of the distributional borders of many species living near 30°N in the E. Pacific. In the maps of this atlas no attention is paid to such fluctuations with a yearly (or smaller) periodicity.

The shift of the western borders of *Creseis virgula* in the S. China Sea (fig. 46) is entirely due to the biannual shift in the direction of the prevailing winds. This phenomenon is common in all areas with a monsoon regime and especially well visible in the North Indian Ocean, Indo-Malayan area and East Pacific Ocean (cf. figs 43 & 44). Besides the variation in the position of borders, also the abundance of a species may change periodically due to climatological and hydrological changes.

For the interpretation of a range, also the abundances of populations should be taken into account. Thus, for example, the distribution of *Limacina lesueuri* (fig. 47) is of the EUPHAUSIA BREVOIDES type (cf. fig. 89), when only the areas of high abundance are considered. The other parts of the range, with relatively low abundances, stretching from 45°N to 45°S, are then considered to be expatriation areas of great extent. Probably, reproducing adults being found here also, they are of the non-sterile type, so that *L. lesueuri* may alternatively be considered to show the STYLOCHEIRON CARINATOIDES type of distribution (fig. 63). It depends on the value one attaches to abundance data, and the biological consequences one connects with the abundance, whether one proposes the one or the other type of distribution as characteristic of a species.

The total range of *Clio cuspidata* (fig. 48), like that of the preceding species, is of the STYLOCHEIRON CARINATOIDES type (cf. fig. 63). The area with high abundance, however, shows another pattern than in *Limacina lesueuri*, being found in the boundary currents, the N. Transitional area and the S. Subtropical-Terminal areas. In the Atlantic Ocean, an abundant eastern and western population can be distinguished, which resembles the distribution of distant-neritic species. These areas with high abundance are somewhat similar to the CARDIAPODA PLACENTOIDES type of distribution (cf. fig. 128).

It is clear that one and the same overall distribution can mask differences in the abundance pattern, and thus essential differences are not elucidated. For most plankton species, sampling is too incidental to discriminate properly between such differences shown by reproductive range or area of high abundance, and expatriation areas or areas with low abundance. When enough data are available one should, however, discriminate especially between patterns with and those without strong expatriation tendencies.

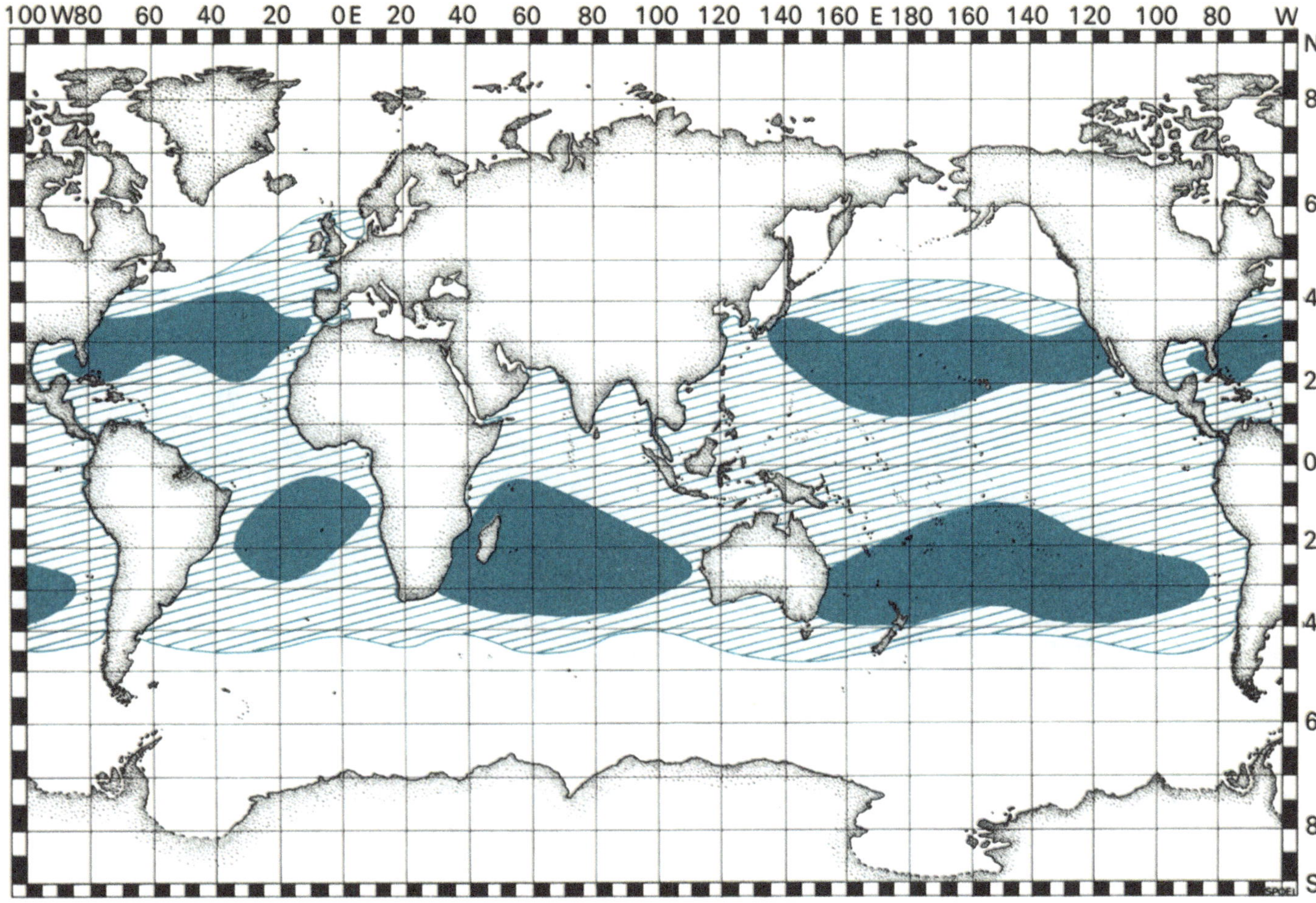
100 W80 60 40 20 0E 20 40 60 80 100 120 140 160 E 180 160 140 120 100 80 W
N 80 60 40 20 0 20 40 60 80 S

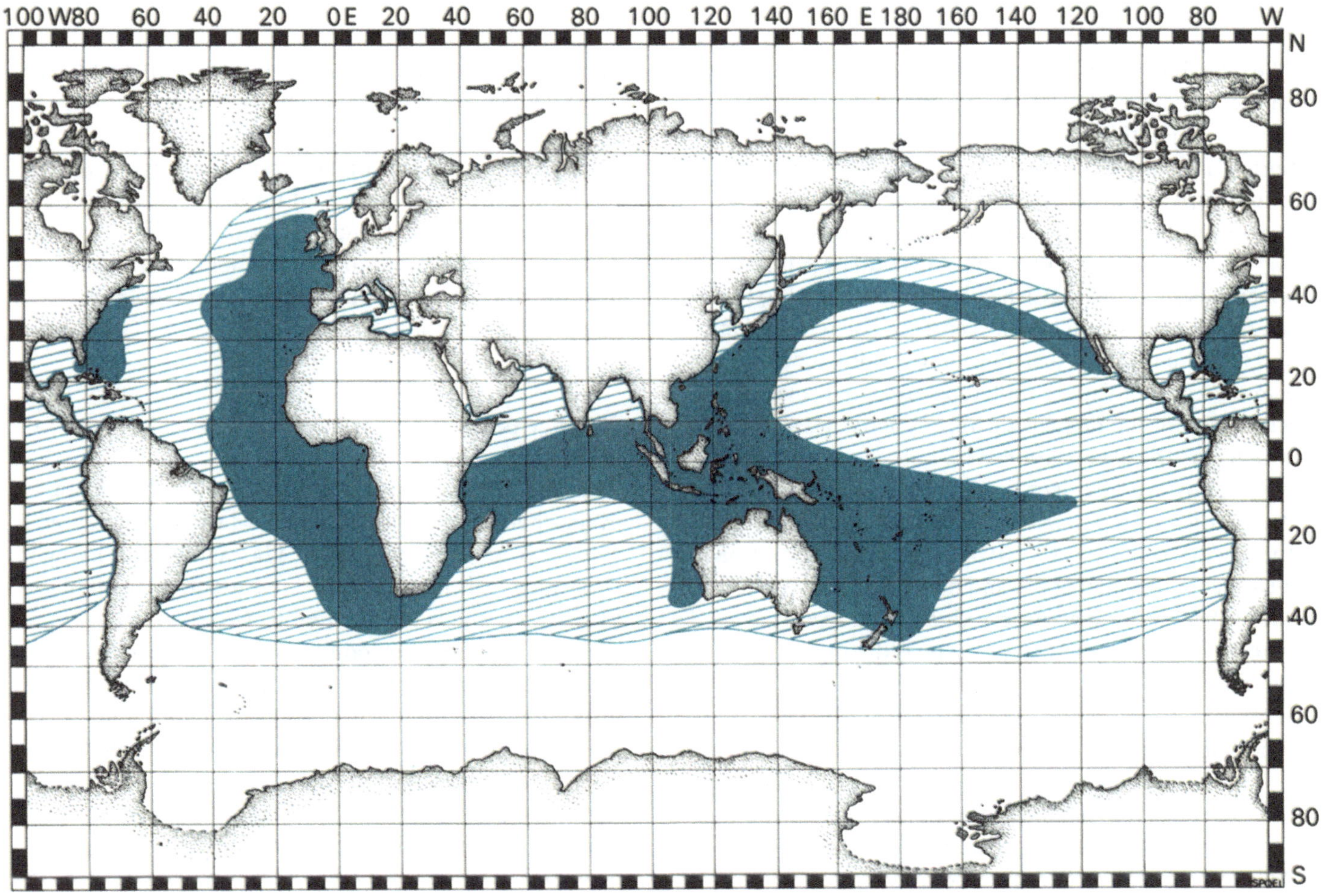
100 W80 60 40 20 0E 20 40 60 80 100 120 140 160 E 180 160 140 120 100 80 W
N 80 60 40 20 0 20 40 60 80 S

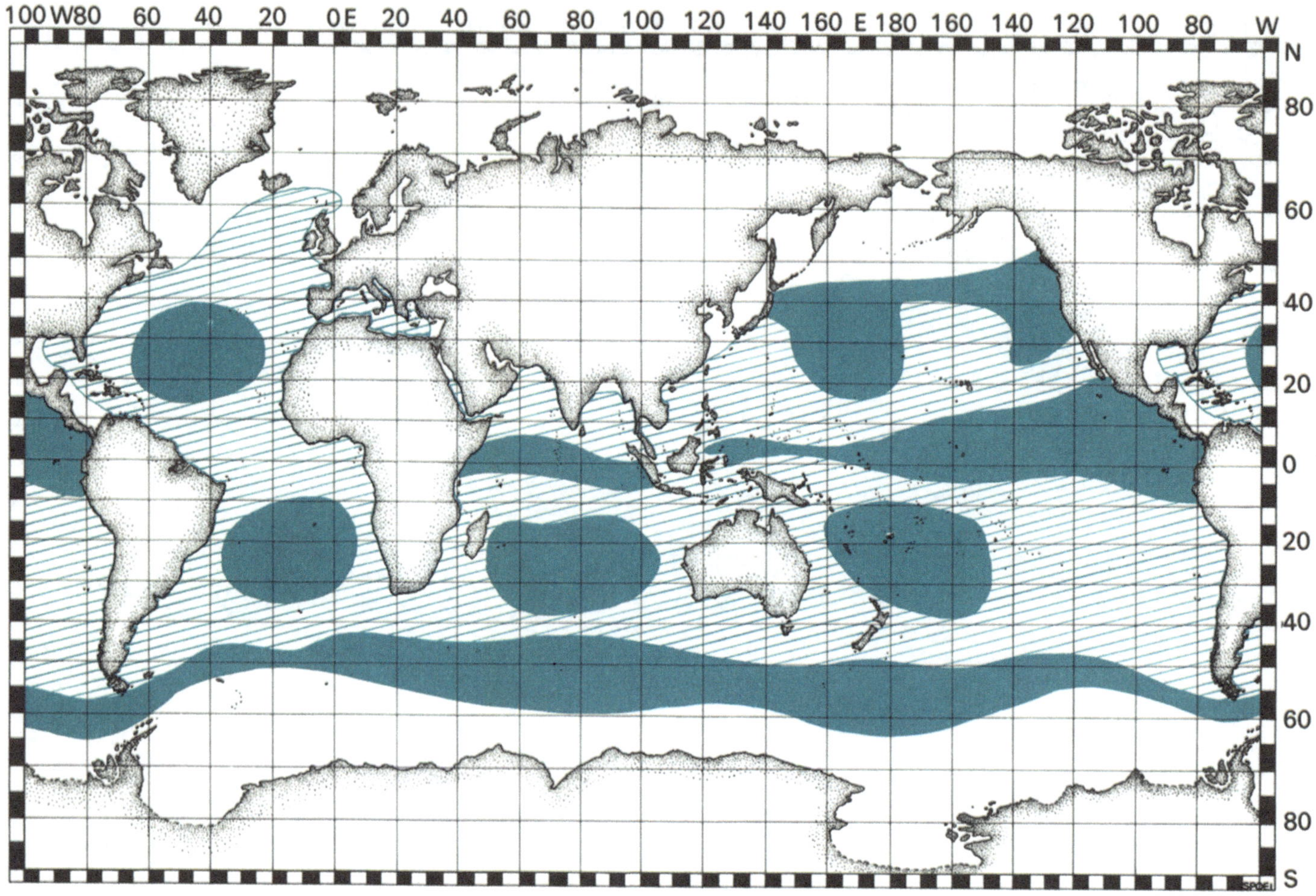

The overall range of *Phronima sedentaria* (fig. 49) is a normal circum-global 60°N-60°S distribution. The areas to which this species 'is more related' according to Shih (1969: 66) do not provide a pattern comparable with that of any other species. The reproductive range proper, probably formed by the most abundant populations, is found throughout cyclical or primary water masses as distinguished by Beklemishev (1971) (cf. fig. 3). If this conclusion is correct, the terminology and treatment of this distribution by Shih (1969) are also correct, but data are still too scarce, to prove exactly which part of the range is the reproductive area and which part the expatriation area.

It is, however, evident that the 60°N-60°S distribution of *Phronima sedentaria* differs strongly from the RHIZOSOLENIA ALATOIDES type (cf. fig. 70), which also occurs between 60°N and 60°S, but which avoids the cyclical water masses. For this reason a separate PHRONIMA SEDENTAROIDES type is distinguished here (cf. fig. 49).

The distribution of *Histioteuthis bonelli* and *Symplecoteuthis onalaniensis* (fig. 50) is given to show the identicity of some planktonic and nektonic distributions. Of the nektonic Cephalopoda, *H. bonelli* shows the GLOBOROTALIA HIRSUTOIDES type (fig. 38) and *S. onalaniensis* shows the CAVOLINIA GLOBULOSOIDES type (fig. 98), both planktonic types of distribution.

It has been made clear that the range, especially the reproductive range, important for taxonomic studies and zoogeographical comparisons, is difficult to distinguish. Interpretation of sediment distribution and of abundance is useful but sufficient data on these phenomena are not yet available.

The plankton species

That the distribution of a taxon is not a stable range, everywhere similar in character and evenly populated is evident from the above maps. The taxon itself, living in these ranges, is neither stable, nor identical in all places. Usually a taxon is represented by taxonomically different specimens or populations at different places, due to geographic and ecophenotypic variation. Besides geographic variation (different forms in different areas) and ecophenotypic variation (different forms in different environmental conditions) one distinguishes: seasonal variation (different forms in different seasons) chronological variation (different forms in different geological periods) and ethological variation (different

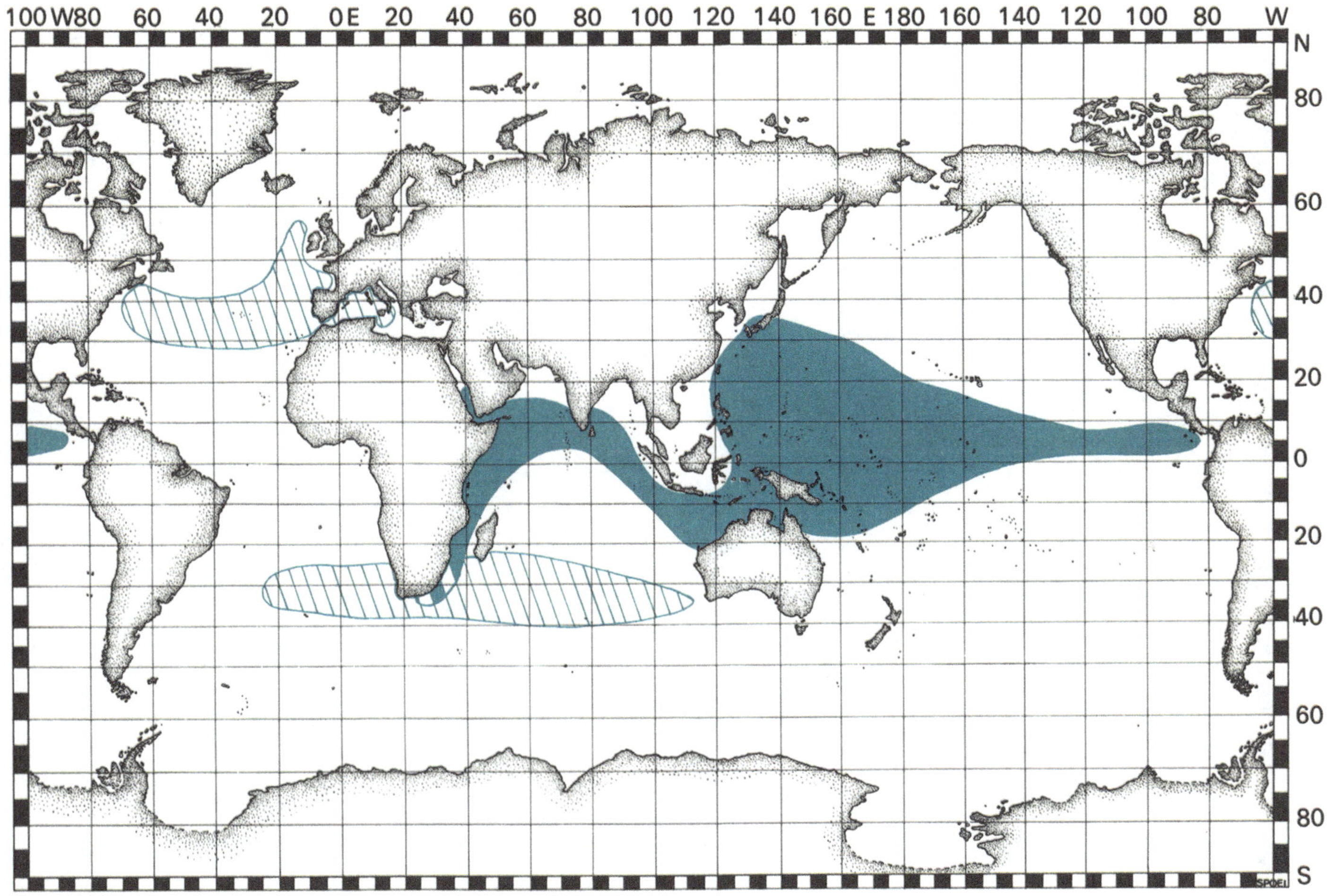

forms with different behaviour); the latter three types will be omitted for the time being. The ecological and geographical variations shade off into one another, the smaller the area studied or the wider the ecological conditions are spread; a situation rather common in the oceans. Both are therefore treated together as 'variation'.

Variation may be discontinuous (subspecies, semispecies) or continuous (clinal variation). As a rule, discontinuous variation only developes when gene flow between populations is strongly or absolutely blocked, so when geographic isolation is present. When such isolation lasts long enough, speciation may succeed variation through a process of allopatric speciation, giving rise to new species. In the ocean, however, isolation by barriers, or distances, is rare, and so discontinuous variation is not frequently found. Only taxa restricted to the Tropical belt (25°N-25°S) encounter the isolating influences of continents. Other isolations may be caused by distance barriers, e.g. between bipolar taxa and between taxa restricted to the east and west sides of an ocean. The populations separated by isolation usually first develop into subspecies.

When isolation is absent or weak, dissimilarities between populations may also develop; in this case it is the difference in selective pressure in various parts of the range which induces such differences to develop. This type of continuous variation is the most common type in plankton. Though in most modern theories on speciation, sympatric speciation (in completely overlapping populations or within a mixed group of animals) is not accepted, there is a possibility of speciation without isolation. The parasympatric speciation (in partly overlapping or adjacent, but not geographically separated, populations) is the process of a continuously diverging development of populations, so that they finally lose genetic contact. Depending on the age of the group and the speciation rate of the taxon the variation may result in different populations, different infraspecific taxa or in different species (Endler, 1977).

FIG. 49 Distribution of *Phronima sedentaria*. black = areas to which the species is more related according to Shih, hatched = areas with low abundance.
FIG. 50 Distribution of *Histioteuthis bonelli* (hatched) and *Symplectoteuthis onalaniensis* (black).

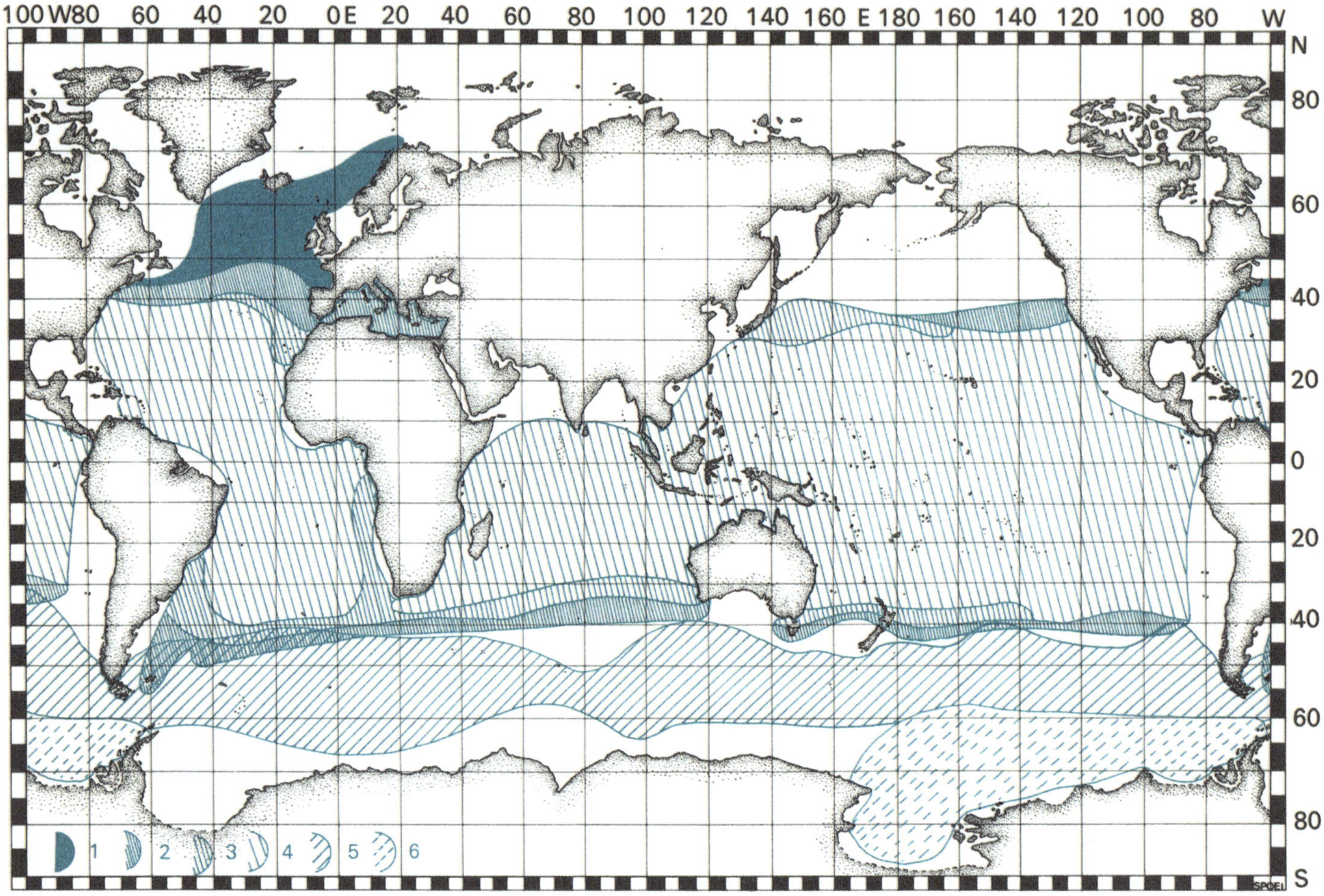

In the species *Salpa fusiformis* (fig. 51), north-south variation is found in the numbers of fibres in the body muscles of the solitary and aggregate forms. Most typical is the greater penetration of, and variation in, the N. Atlantic Ocean, compared with the N. Pacific, and the higher degree of speciation in the high latitudes of the southern hemisphere compared with the northern one. The variation in *S. fusiformis* is correlated with water temperature or density. Latitudinally changing climatological and hydrological conditions are usually the major factors which determine the variation of the north-south clinal type shown in fig. 51.

Besides N-S variation also E-W variation is found in some species. In *Eucalanus subtenuis* (fig. 52) the within-ocean variation is smaller than the inter-ocean variation. Its variation is, like the preceding one, clinal, here influenced by different selective pressures in different basins, and restricted gene flow in between. This species lives in Tropical and Subtropical waters between 30°N and 30°S, which distribution is related with a strong reduction of inter-ocean contacts (cf. figs 75 & 76). The latitudinal differences are, in this range, too small to provide the stimulus for north-south variation.

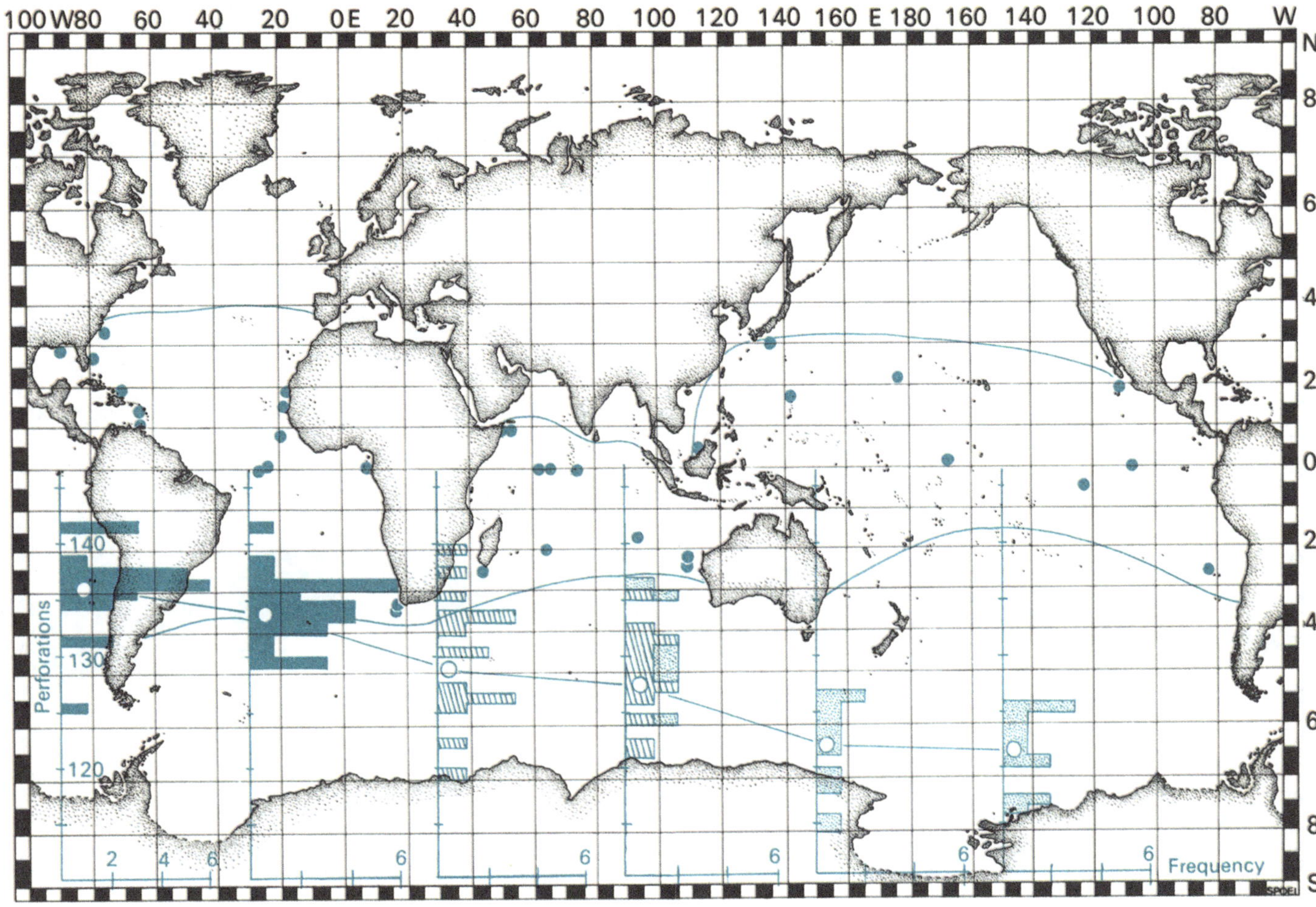

FIG. 51 Distribution of the *Salpa fusiformis* group.
1 = *S. fusiformis* with 60-70 fibres/muscle, 2 = *S. fusiformis* with 50-60 fibres/muscle, 3 = *S. fusiformis* with 40-50 fibres/muscle, 4 = *S. fusiformis* with 30-40 fibres/muscle, 5 = *S. thompsoni*, 6 = *S. gerlachei*.

FIG. 52 Distribution of *Eucalanus subtenuis*, represented by its northern and southern borders (lines) and by the localities used in the variation studies (black points). Variation is represented by histograms giving frequencies (horizontal axis) of the total number of perforations of integument per specimen (vertical axis) in adult females collected within 60° segments of longitude separated by the 30°, 90° and 150° long., the means of each segment are indicated by open circles. black, = Atlantic specimens, hatched = Indian specimens, dotted = Pacific specimens.

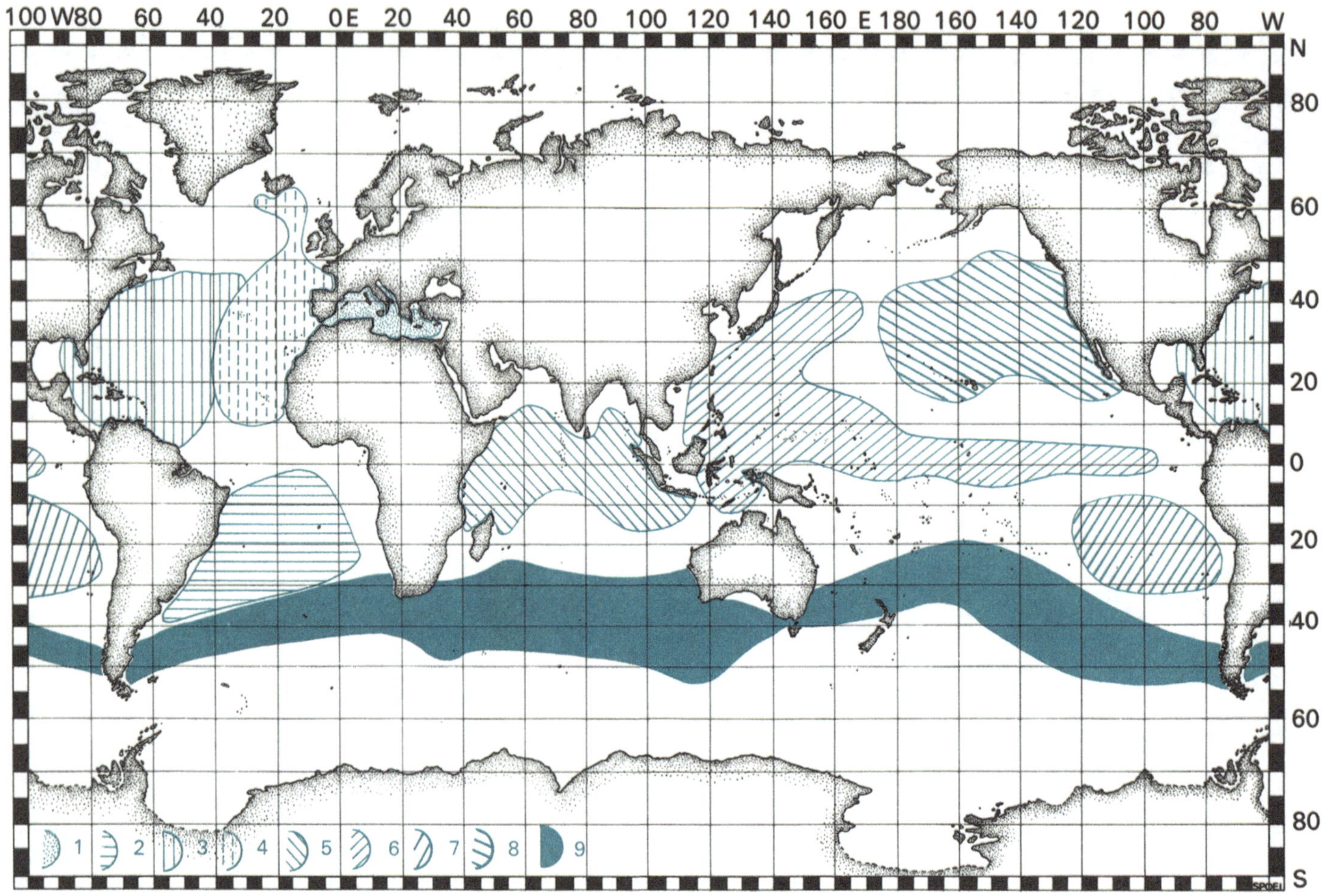

The variation in the polytypic species *Cavolinia tridentata* (fig. 53) does not show a clinal gradient, so that for this pattern the separate 'CAVOLINOIDES' type is proposed. Being bound to water masses, these forms show more provincialism than clines do. Consequently, the legenda can also be read as: 1 = Mediterranean, 2 = S. Central, 3 = N.W. Central, 4 = N. Central, 5 = Equatorial, 6 = Equatorial, 7 = S.E. Central, 8 = N.E. Central, 9 = S. Transitional.

A certain degree of isolation between the oceans, as found also in the preceding species, is here accompanied by isolations between water masses. The boundaries between the forms partly coincide with the lines along which stronger variation is found in clines of polytypic species, and also with borders of some nonvariable species. Many of the endemic forms of *Cavolinia tridentata* are chosen as examples of restricted ranges (p. 98–100).

As speciation proceeds, the infraspecific groups become species-groups. These species-groups are composed of closely related, usually monophyletic species, with partly overlapping distributions. The patterns of the taxa in a species-group also show some regularities, which are dealt with below.

In the species-group *Eucalanus attenuatus* (fig. 54) the distribution of *E. sewelli* may be more extensive than indicated in this map, and in origin more of the type of the *Clio pyramidata* forma *lanceolata* distribution (fig. 57). The evidently disjunct distribution, however, provides this range with a special character so that the EUCALANUS SEWELLOIDES type is introduced for it (p. 95).

This species-group developed a circum-global warm-water species, *E. langae*, an Indo-Pacific warm-water species, *E. attenuatus*, and a northern and a southern Transitional water taxon, *E. parki* and *E. sewelli* resp.

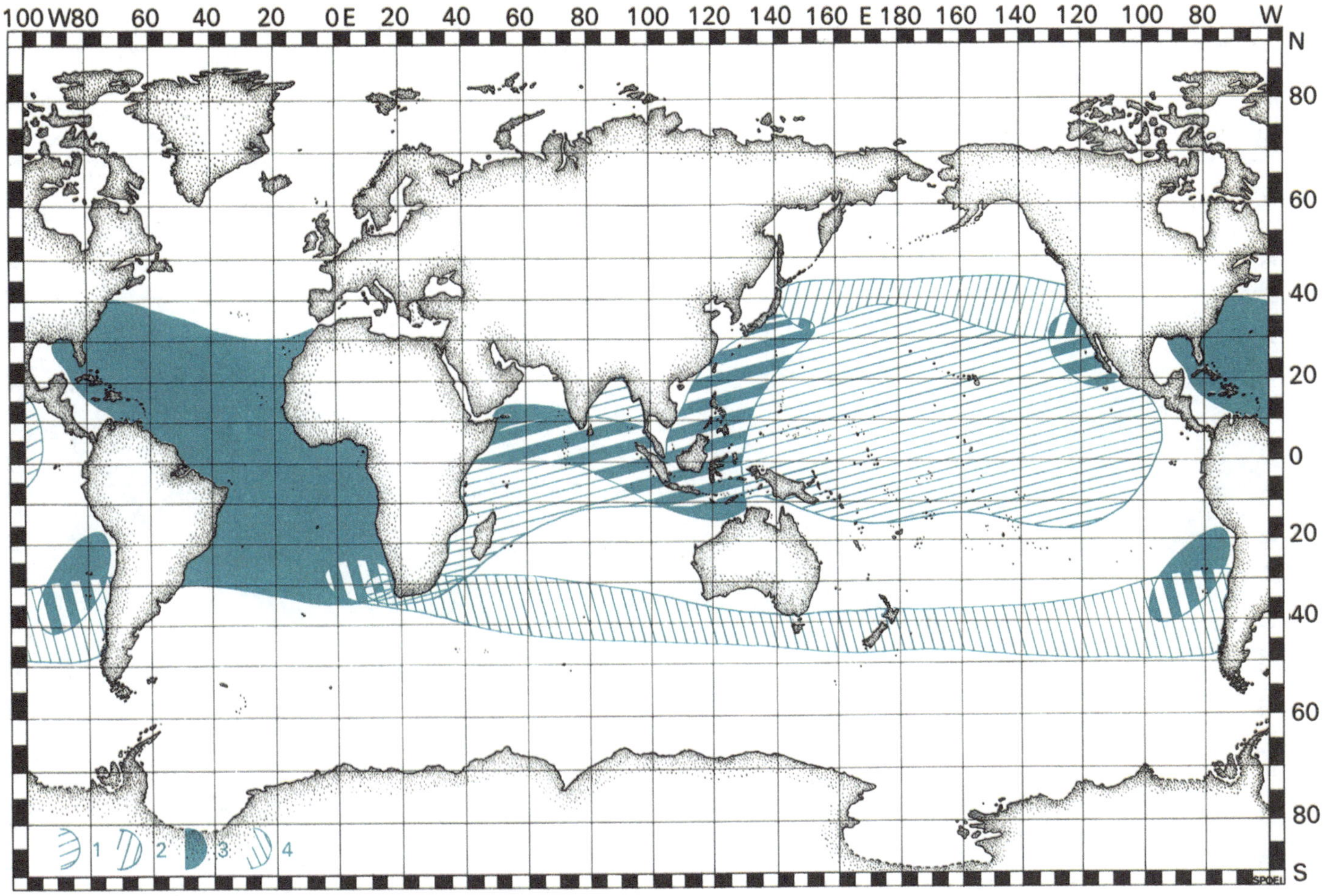

FIG. 53 Distribution of the *Cavolinia tridentata* group. 1 = *C.t.* forma *tridentata*, 2 = *C.t.* forma *atlantica*, 3 = *C.t.* forma *bermudensis*, 4 = *C.t.* forma *dakarensis*, 5 = *C.t.* forma *danae*, 6 = *C.t.* forma *teschi*, 7 = *C.t.* forma *affinis*, 8 = *C.t.* forma *occidentalis*, 9 = *C.t.* forma *australis*.

FIG. 54 Distribution of the species-group *Eucalanus attenuatus*. 1 = *E. attenuatus*, 2 = *E. parki*, 3 = *E. sewelli*, 4 = *E. langae*.

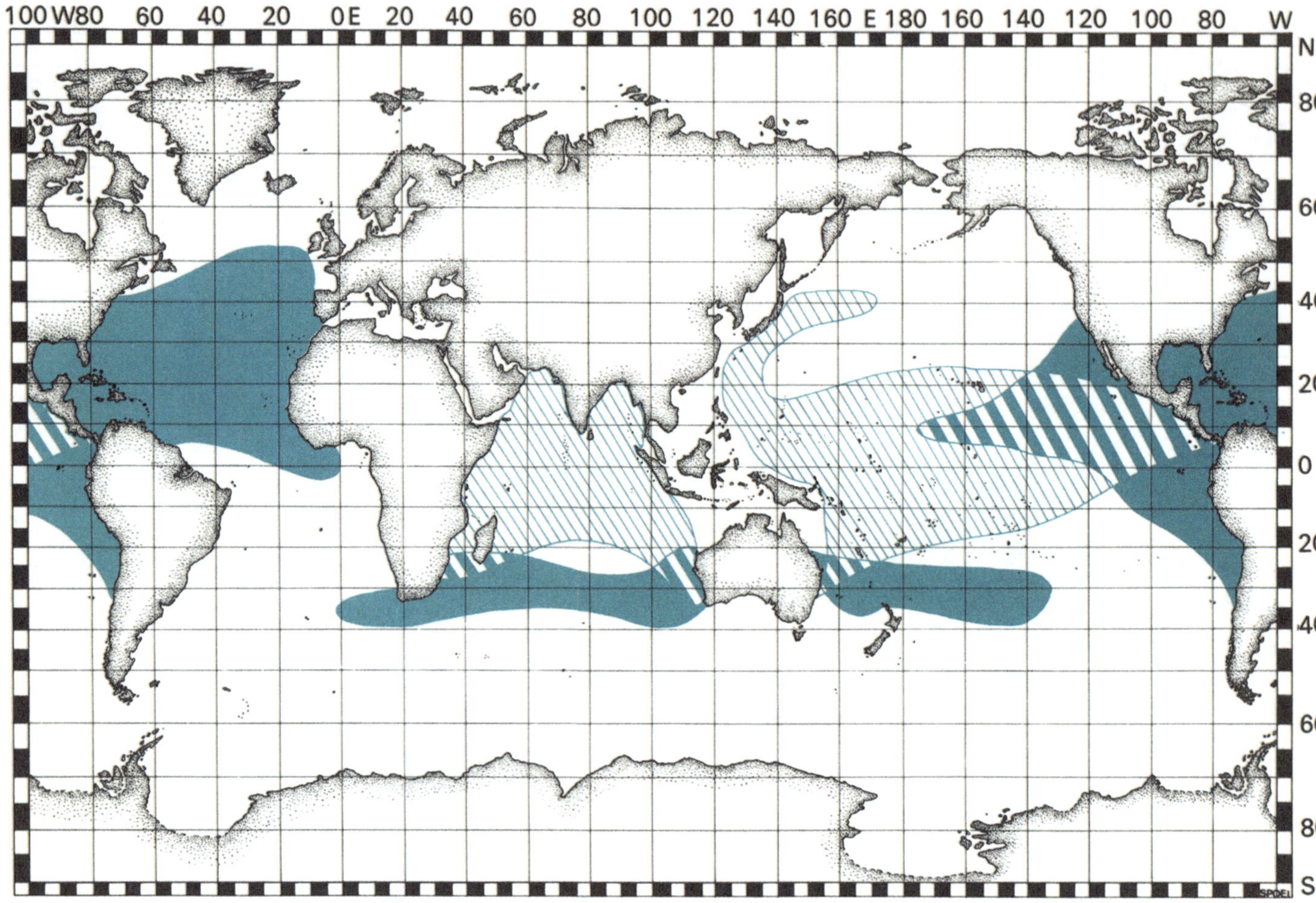

In the species-group *Clausocalanus arcuicornis* (figs 55 & 56) sympatry is frequent, but the almost sympatric *C. paupulus* and *C. arcuicornis* (cf. the STYLOCHEIRON SUHMOIDES type in fig. 90) differ in being a warm-water species (40°N-40°S) and a Central water species, resp. *C. farrani* is an Indo-Pacific restricted warm-water species (30°N-30°S). The species *C. jobei* and *C. minor* show nearly the same distributions as *C. arcuicornis* and *C. farrani*, and are considered to be less related to the other three members of the group.

In the polytypic species *Clio pyramidata* (fig. 57), characters of the species-group are present, as shown by the development of *Clio convexa*, and characters of the SALPA FUSIFORMOIDES type of distribution (cf. fig. 51), while also resemblances with the preceding species-group are evident. *C. p.* forma *lanceolata* shows the warm-water distribution (40°N-40°S), *C. convexa* shows the restricted Indo-Pacific pattern (30°N-30°S). The other three taxa are the result of latitudinal variation, as also shown in fig. 51. The formas *pyramidata* and *lanceolata* together form exactly the SALPA FUSIFORMOIDES type of distribution.

FIG. 55 Distribution of the species-group *Clausocalanus arcuicornis*. black = *C. jobei*, hatched = *C. minor* (see also fig. 56).

FIG. 56 Distribution of the species-group *Clausocalanus arcuicornis*. 1 = *C. paupulus*, 2 = *C. arcuicornis*, 3 = *C. farrani* (see also fig. 55).

FIG. 57 Distribution of the polytypic species *Clio pyramidata*. 1 = *C. p.* forma *pyramidata*, 2 = *C. convexa*, 3 = *C. p.* forma *lanceolata*, 4 = *C. p.* forma *antarctica*, 5 = *C. p.* forma *sulcata*.

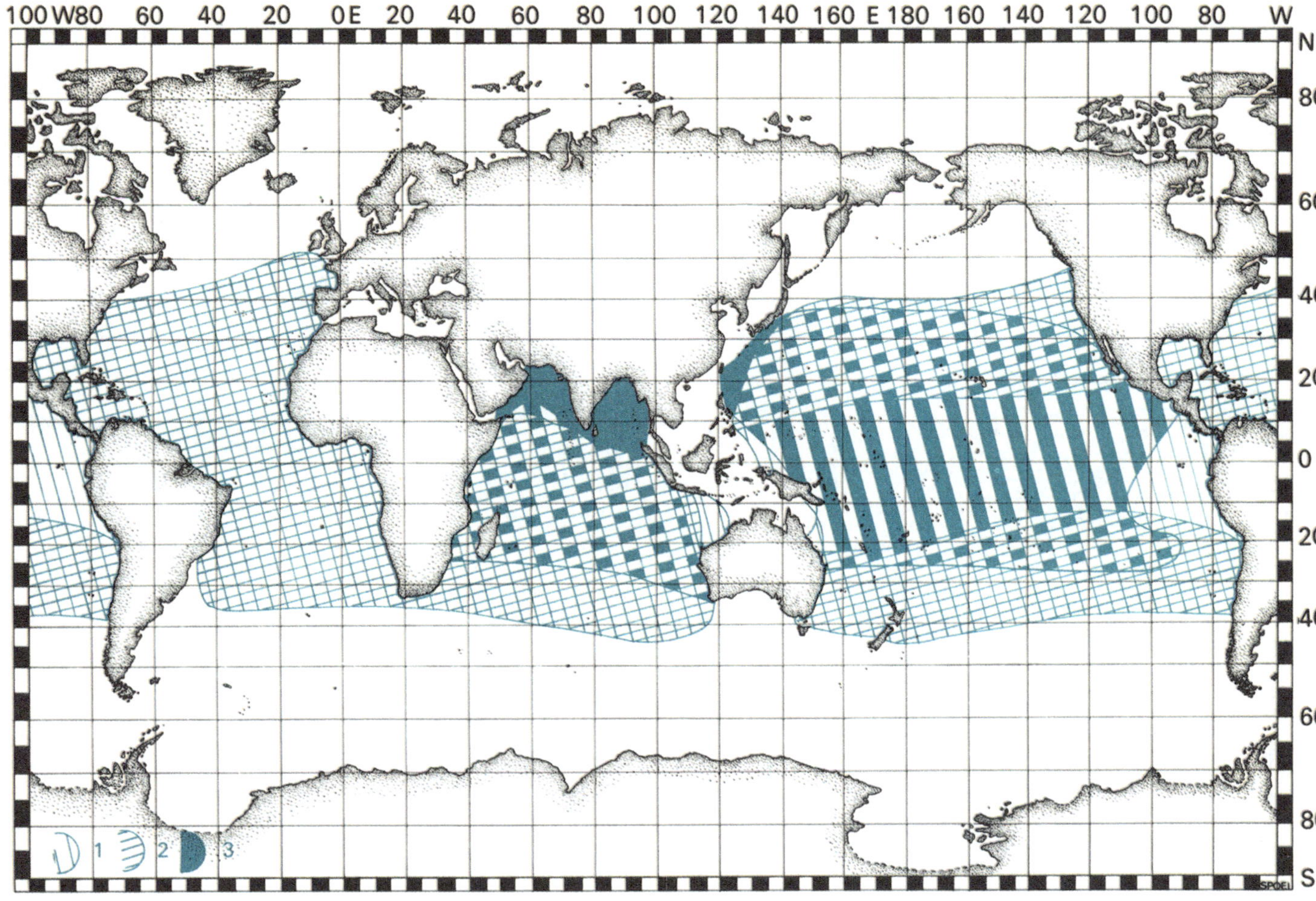
100 W 80 60 40 20 0 E 20 40 60 80 100 120 140 160 E 180 160 140 120 100 80 W
N 80 60 40 20 0 20 40 60 80 S
1 2 3
SPOEL

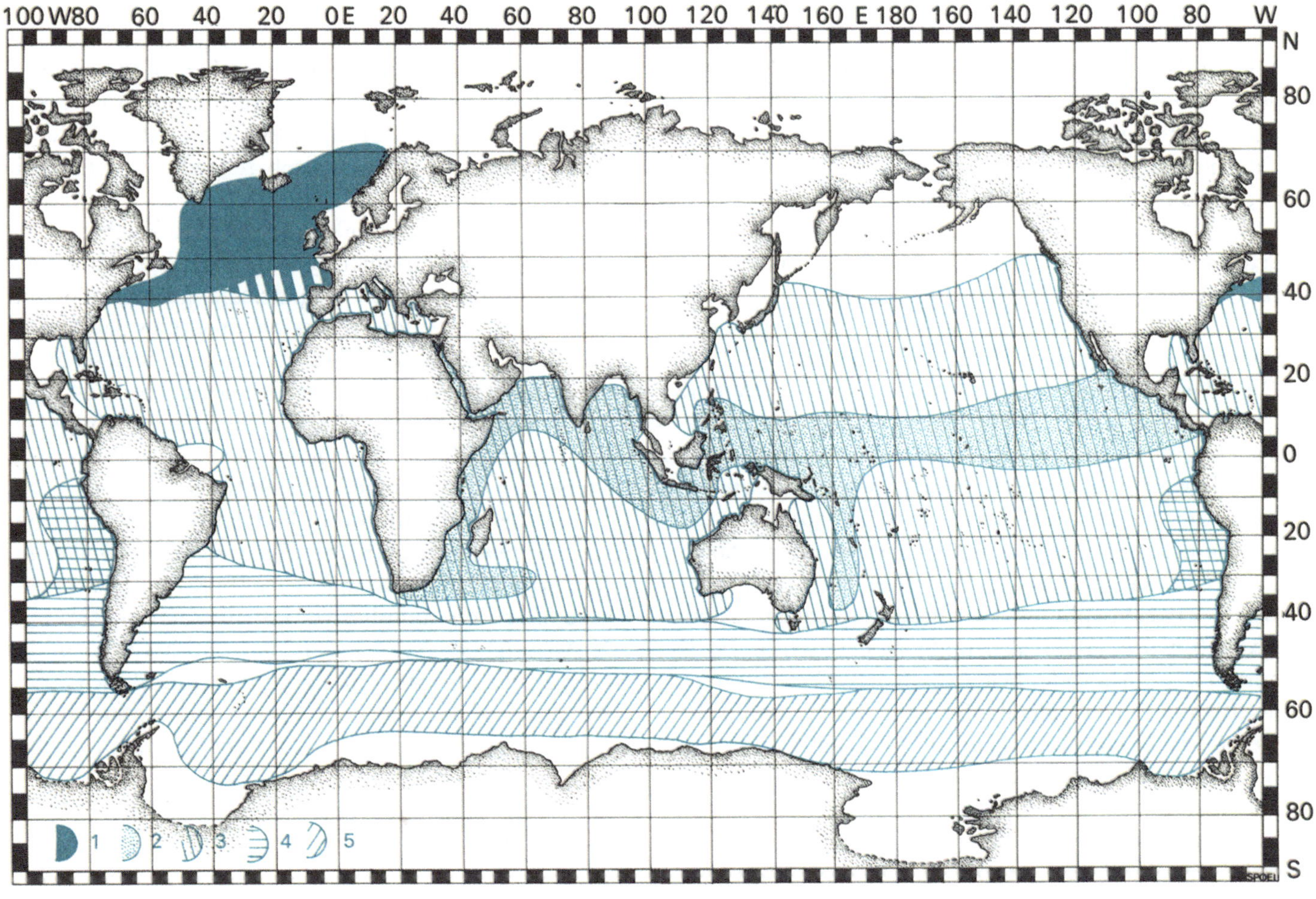
100 W 80 60 40 20 0 E 20 40 60 80 100 120 140 160 E 180 160 140 120 100 80 W
N 80 60 40 20 0 20 40 60 80 S
1 2 3 4 5
SPOEL

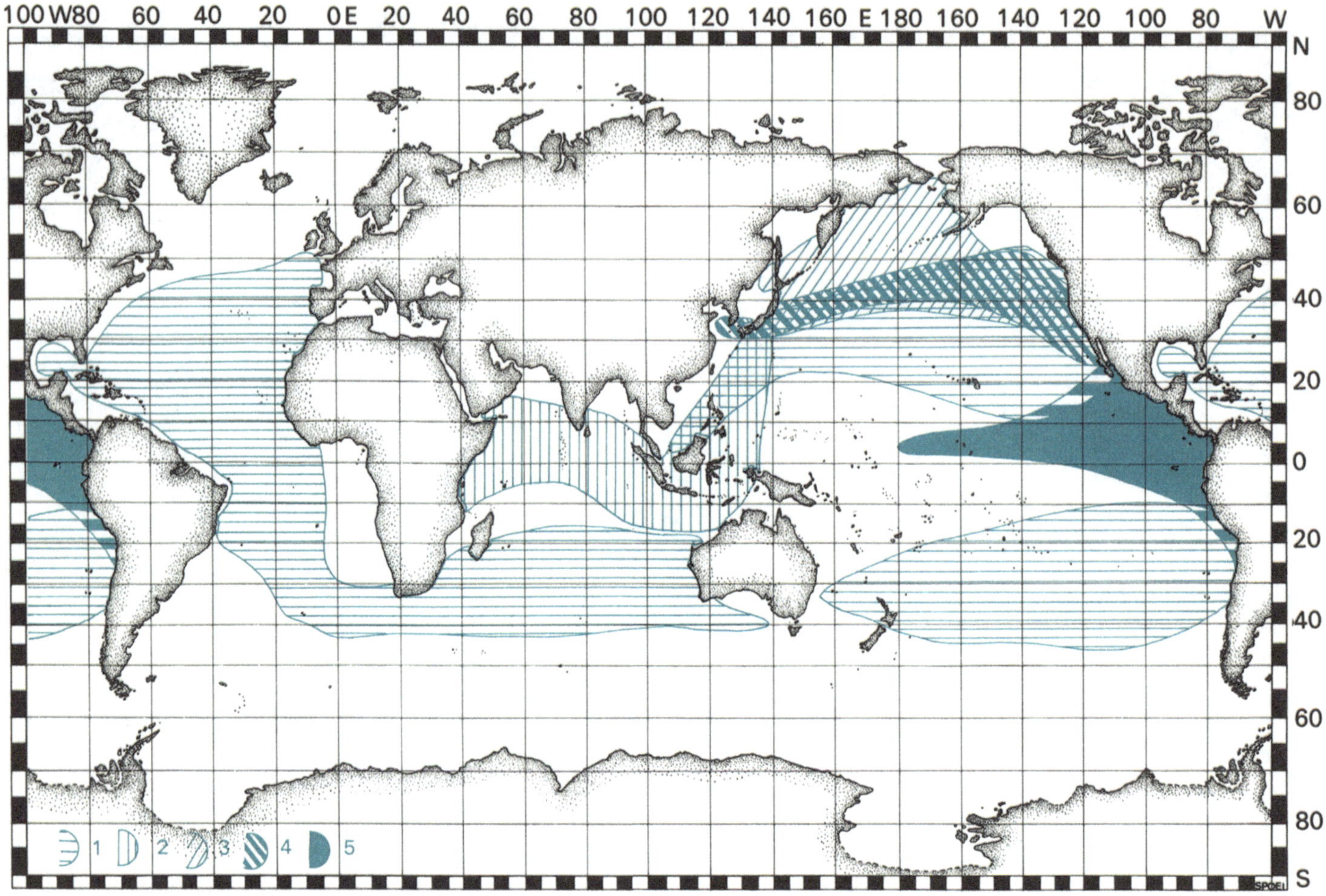

The species-group *Eucalanus hyalinus* (fig. 58) shows a rather complex speciation in the N. and E. Pacific Ocean. The Subarctic *E. bungi bungi* and the Transitional *E. b. californicus* occupy a separate position in the group. *E. hyalinus* is a Central water form, *E. elongatus* is the restricted Indo-Pacific warm-water taxon and probably the E. Pacific *E. inermis* belongs to the *elongatus* group.

Other species-groups mapped in this atlas are the *Pontellina plumata*, the *Diacria trispinosa* and the *Sagitta serratodentata* groups. All the more thoroughly studied species-groups, cold-water taxa excluded, seem to show the following basic zoogeographic principle: Closely related groups of oceanic taxa at about the species level are composed of: a) an Indo-Pacific warm-water taxon living between 30°N and 30°S (or in a narrower belt) which may have split into a Pacific and an Indian Ocean taxon, b) a circum-global taxon living between 40°N and 40°S (or in a wider belt), which may have split into an Atlantic and an Indo-Pacific taxon and c) a group of Central water forms (30°N to 20°N and 20°S to 30°S) or a group of Transitional water forms (50°N to 30°N and 30°S to 50°S).

When, within one species-group, two taxa show both either distribution a) or distribution b), one of the two is less closely related to the other members of the group. When both Central and Transitional water forms are found in one species-group, either the Central or the Transitional taxa are less closely related to the rest of the group.

The type of speciation giving rise to this type of species-groups is frequently found, and one may expect that some phenomenon in a particular geological period caused this speciation. When more complex patterns are found, they are supposed to have been influenced by other phenomena in successive geological periods.

The species-group of *Diacria quadridentata* (p. 99) is not considered here, as it shows some endemic taxa and a tendency to extinction in recent times.

The species-groups of the high latitudes, such as the *Calanus finmarchicus* (fig. 78), the *Calanus helgolandicus* (fig. 79) and the *Limacina helicina* (fig. 82) species-groups, are less complicated. Usually there are two N. Atlantic, two N. Pacific and two Southern Ocean taxa. In groups with high Arctic taxa, one of the two Atlantic, and one of the two Pacific representatives are replaced by one Arctic taxon, *Calanus glacialis* (fig. 78) is such a replacing taxon.

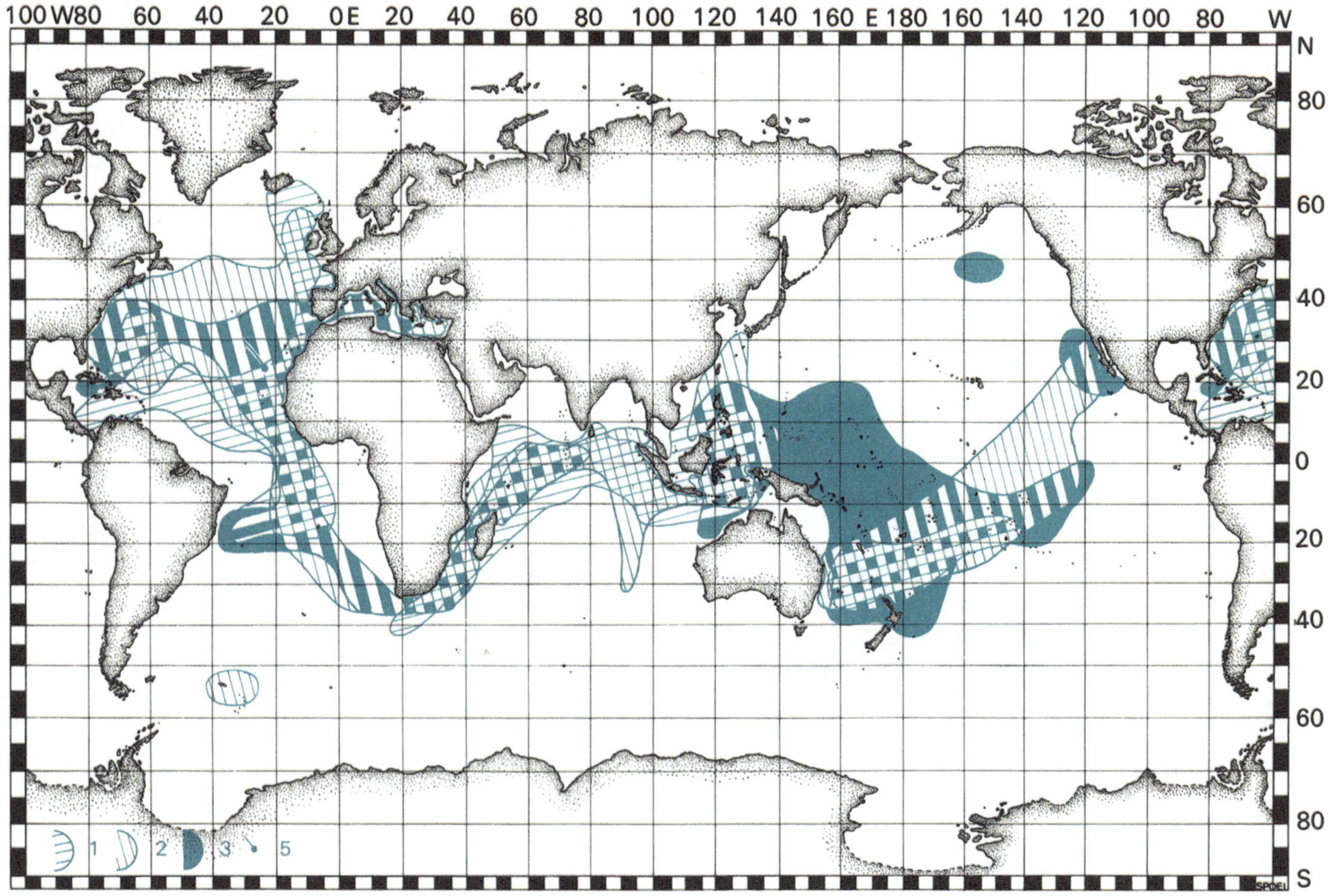

FIG. 58 Distribution of the species-group *Eucalanus hyalinus*. 1 = *E. hyalinus*, 2 = *E. elongatus*, 3 = *E. b. bungi*, 4 = *E. b. californicus*, 5 = *E. inermis*.
FIG. 59 Distribution of the genus *Peraclis* (see also fig. 60). 1 = *P. moluccensis*, 2 = *P. reticulata*, 3 = *P. apicifulva*, 5 = Dakar Centre.

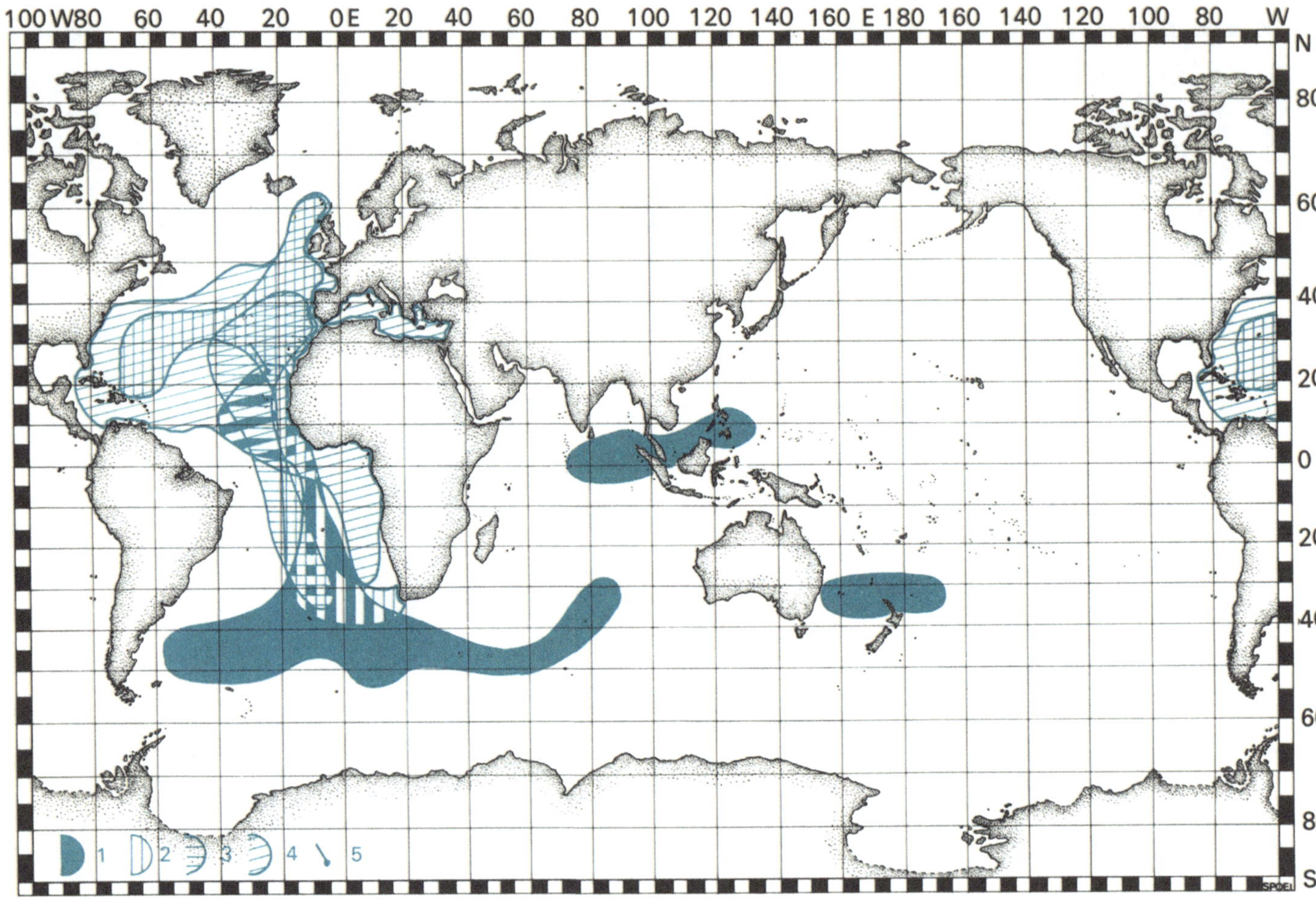

The genus *Peraclis* (figs 59 & 60) provides an example of a group of species spread around one centre (faunal centre 8 in fig. 16). The present genus is chiefly mesoplanktonic but in the epiplanktonic genus *Phronima* exactly the same patterns are found. The *Diacria quadridentata* group seems to have spread from an Indo-Malayan Centre. Zoogeographic patterns of this type point to a mode of distribution different from that of the species and species-groups mentioned above. Speciation in the present genus dates back probably much further than that in species-groups, and the speciation mechanism must have been different also.

The distributions of *Onychia carribaea*, *Brachioteuthis riisei* and *Galiteuthis armata* around a faunal centre (fig. 61) are given to show that nektonic species in some groups are also distributed around the Dakar Centre; their patterns are identical with those of *Peraclis*. For a high number of cephalopod genera and species, centres can be indicated as the places of speciation. Other examples of such patterns in this atlas are *Onychoteuthis banksi* (fig. 74), *Todarodes sagittatus*, *Spirula spirula* (fig. 155) and *Bathyteuthis abyssicolla* (fig. 154).

FIG. 60 Distribution of the genus *Peraclis* (see also fig. 59). 1 = *P. valdiviae*, 2 = *P. triacantha*, 3 = *P. depressa*, 4 = *P. bispinosa*, 5 = Dakar Centre.

FIG. 61 Cephalopod distribution around faunal centre. 1 = *Onychia carribaea*, 2 = *Brachioteuthis riisei*, 3 = *Galiteuthis armata*.

FIG. 62 Distribution of *Euphausia eximia*.

Euphausia eximia (fig. 62), represented by a Tropical E. Pacific and an Atlanto-Mediterranean form, provides an example of speciation by separation and isolation. Though this variation is only of relatively recent date, it is a mechanism which may have been active over extensive geological periods. The disjunction in this species is in all probability the result of the closure of the Panama Isthmus. Remarkable is the split in the E. Pacific population which may be due to the extreme adaptation of the species in this area to waters of very low oxygen concentrations.

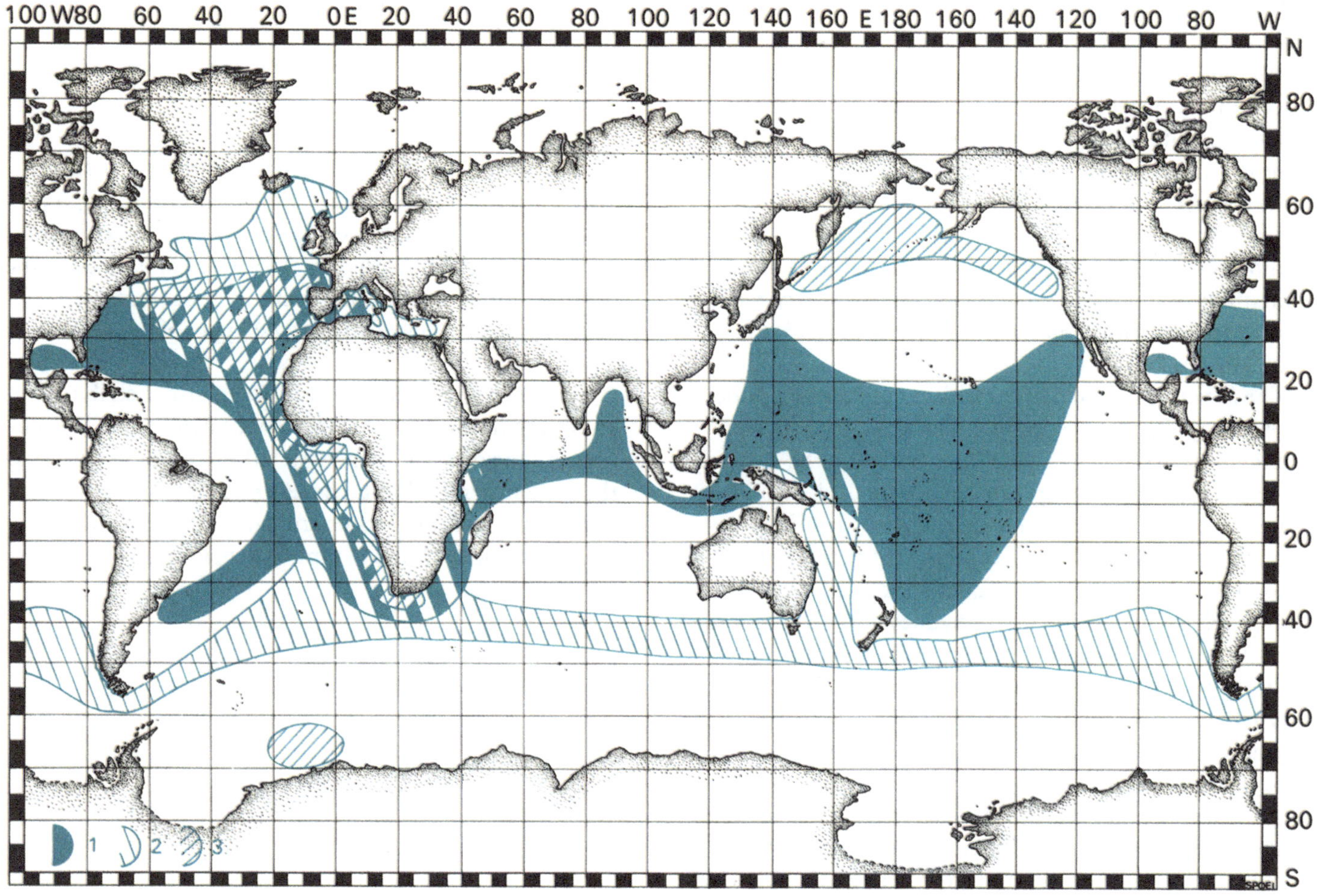

100 W80 60 40 20 0 E 20 40 60 80 100 120 140 160 E 180 160 140 120 100 80 W
N 80 60 40 20 0 20 40 60 80 S
1
2
3
SPOEL

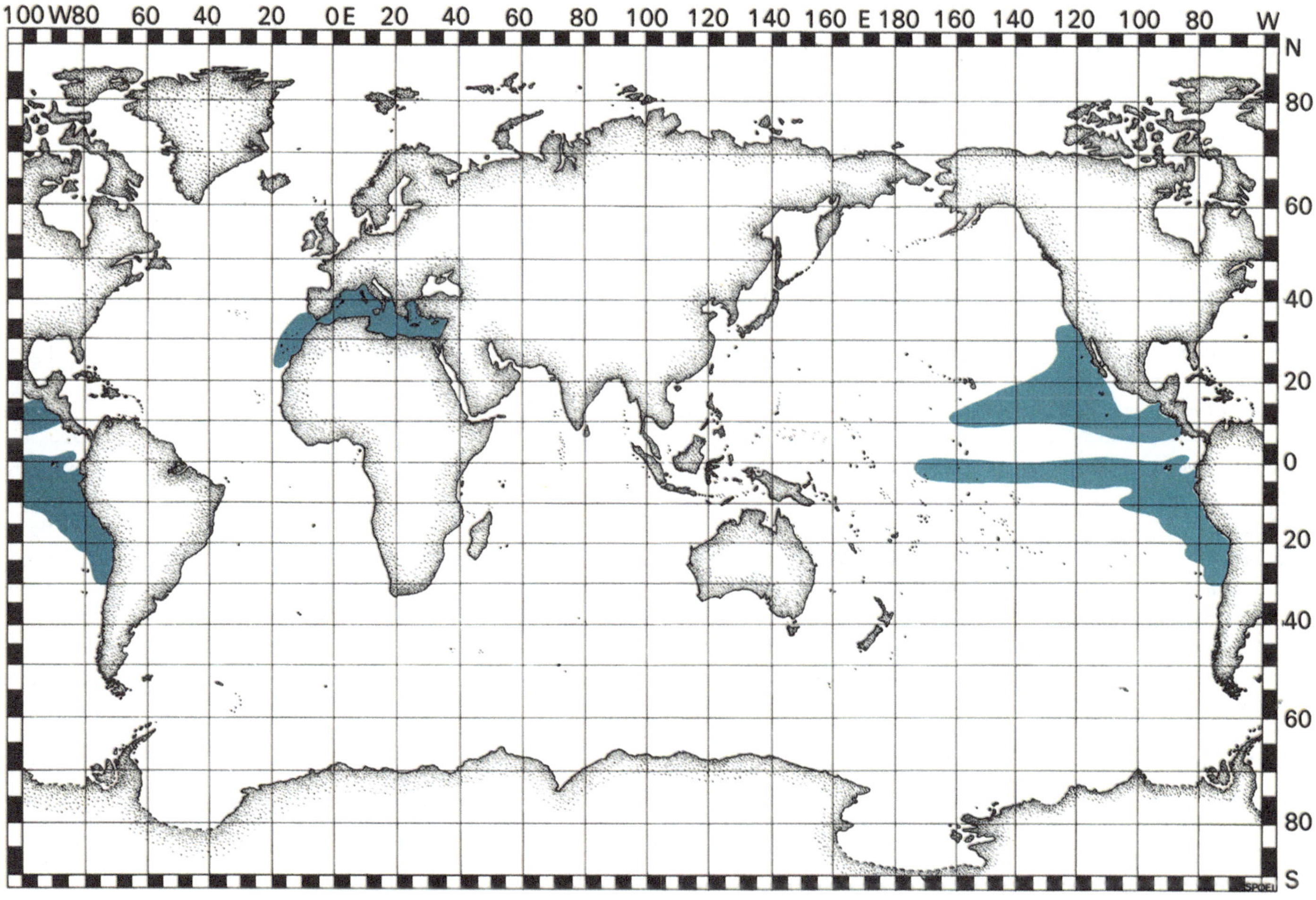

100 W80 60 40 20 0 E 20 40 60 80 100 120 140 160 E 180 160 140 120 100 80 W
N 80 60 40 20 0 20 40 60 80 S
SPOEL

The taxa whose zoogeography has been discussed so far, are to be divided into four large groups: a) the taxa with continuous variation either in north-south or in east-west directions, or in both, b) the taxa which already developed discontinuous variation, which gave rise to mostly allopatric taxa (the species-groups), c) the taxa of the species level grouped around a faunal centre, such as *Peraclis* species and d) the taxa which have been split into disjunct patterns by isolation. Sometimes complex mixtures of the above mentioned patterns and speciation processes are found, as e.g. in the genus *Stylocheiron*.

The *S. maximum* species-group is composed of a wide circum-global species, a more restricted warm-water circum-global species, which also occurs in the Atlantic Ocean, instead of being restricted to the Indo-Pacific, and a Central water species. The species-group *S. longicorne* is much more complicated than most species-groups. With the frequent and total sympatry of different group-members, it becomes hard to accept that really one species-group is concerned. A division into three species-groups seems more logical. Lomakina (1978) proposes the splitting of the *S. longicorne* species-group into the *S. affine* group (with *S. affine*, *S. microphthalma*, *S. suhmi* and *S. insulare*) and the *S. longicorne* group s.s. (with *S. longicorne*, *S. elongatum* and *S. indicus*). This subdivision is, however, not accepted by Mikkelsen (1982) who follows Brinton, but synonymizes under *S. suhmi* the species *S. longicorne*, *S. affine*, *S. microphthalma* and *S. suhmi*. These two subdivisions do not offer an explanation for the zoogeography found.

FIG. 63 Distribution of *Stylocheiron carinatum* (black) and *S. insulare* (hatched).
FIG. 64 Distribution of *Stylocheiron abbreviatum*.

The pattern of *S. insulare* (fig. 63) is of the THALIA SIBOGAOIDES type (fig. 122) and *S. carinatum* (fig. 63) forms the typical example of a 40°N-40°S distribution with a connection south of S. Africa and one through the Strait of Timor, which makes it different from the CLAUSOCALANUS PAUPULOIDES type (fig. 73).

Stylocheiron abbreviatum (fig. 64) is also distributed according to the STYLOCHEIRON CARINATOIDES type (fig. 63), the only difference with the preceding species is the presence of a Mediterranean population. Such relatively small differences are, however, considered to be acceptable within one type of distribution.

TABLE I
Distributions in the genus *Stylocheiron* according to Brinton (1975).

S. carinatum	circum-global (40°N-40°S)	species	fig. 63
S. maximum	circum-global (60°N-60°S)	species-group *S. maximum*	fig. 72
S. abbreviatum	circum-global (40°N-40°S)		fig. 64
S. robustum	Central waters exc. N. Atlantic		fig. 94
S. elongatum	circum-global (60°N-35°S)		fig. 65
S. microphthalma	Central waters		fig. 65
S. suhmi	Central waters		fig. 90
S. insulare	Indo-Malayan + Bay of Bengal		fig. 63
S. indicus	Indo-Malayan + Arabian Sea		fig. 66
S. longicorne			
'long form'	circum-global (50°N-50°S)		fig. 66
'short form'	Central waters	species-group *S. longicorne*	fig. 66
'N. Indian Ocean form'	N. Indian Ocean		fig. 66
S. affine			
'W. Equatorial form'	Equatorial Indo-Pacific		fig. 67
'Indo-Australian form'	Equatorial Indo-W. Pacific		fig. 67
'Central form'	Indo-Pacific Central waters		fig. 91
'E. Equatorial form'	Equatorial E. Pacific waters		fig. 67
'California Curr. form'	E. Pacific Boundary Current		fig. 67

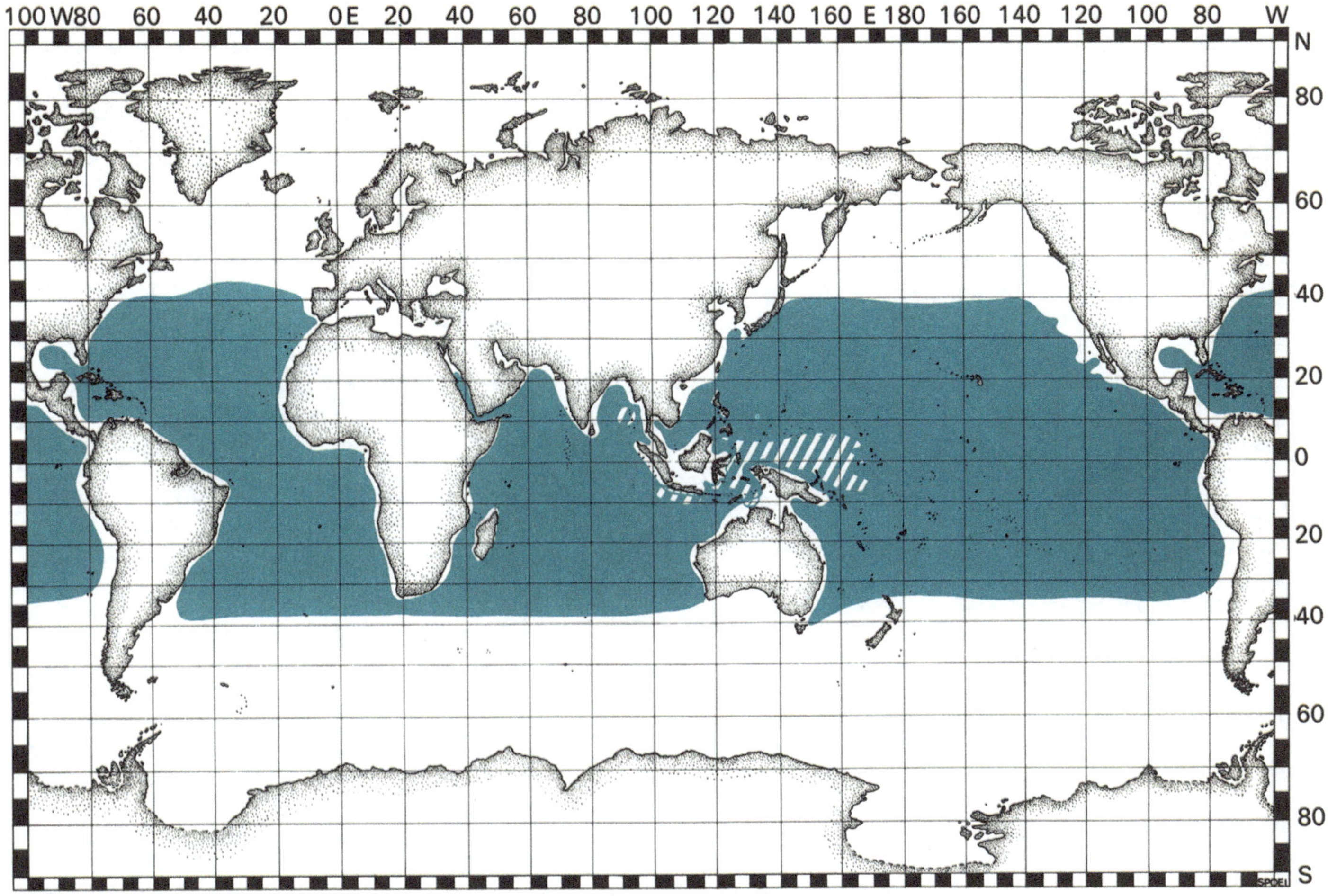
100 W 80 60 40 20 0 E 20 40 60 80 100 120 140 160 E 180 160 140 120 100 80 W
N 80 60 40 20 0 20 40 60 80 S
SPOEI

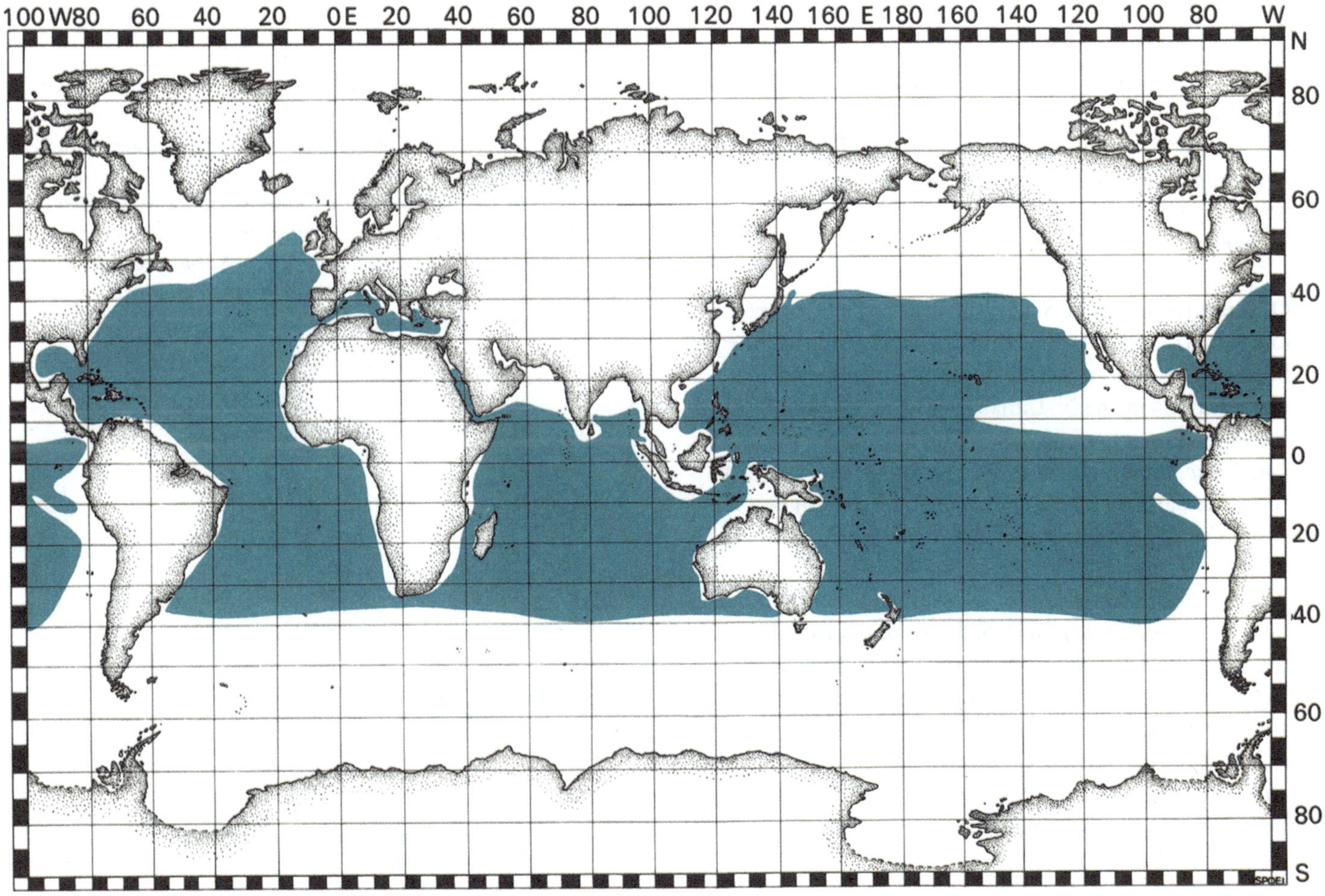
100 W 80 60 40 20 0 E 20 40 60 80 100 120 140 160 E 180 160 140 120 100 80 W
N 80 60 40 20 0 20 40 60 80 S
SPOEI

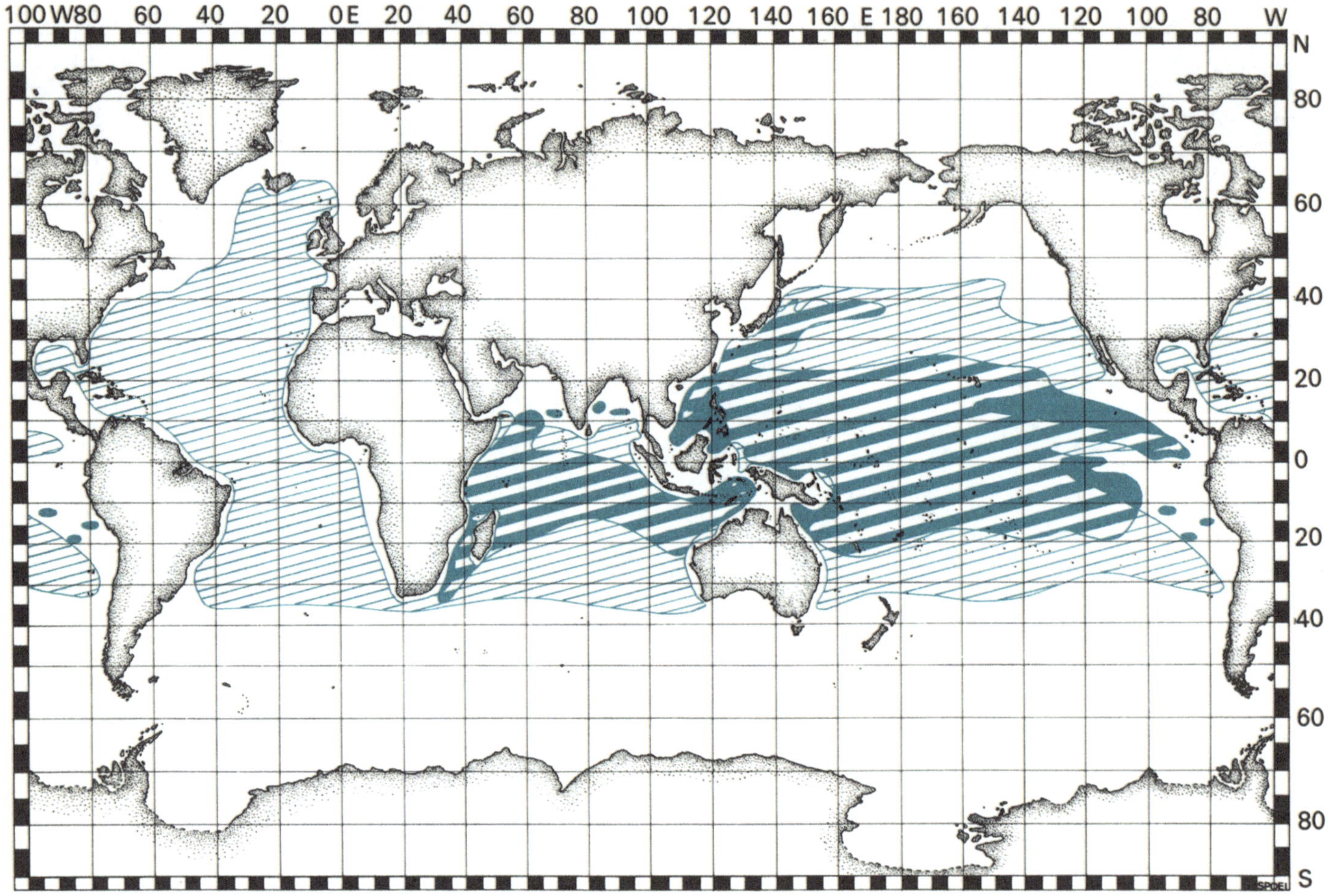

The distribution of *Stylocheiron elongatum* (fig. 65) is of the CLAUSOCALANUS PAUPULOIDES type (fig. 73) (40°N-40°S, without a connection near Timor) and *S. microphthalma* (fig. 65) shows the CAVOLINIA GLOBULOSOIDES type of distribution (fig. 98) with a wide Equatorial Indo-Pacific dispersal.

Of the three forms of *Stylocheiron longicorne* (fig. 66), one is restricted to the N. Indian Ocean, and this one has been chosen as an example of the more endemic distributions in the N. Indian Ocean, as e.g. also shown by *Sergestes semissis* (fig. 102). The *Stylocheiron longicorne* 'long form' and 'short form' shows roughly the same patterns as *S. elongatum* and *S. microphthalma* resp., though the latter is absent from the Atlantic.

Together with fig. 91, fig. 67 gives the distribution of the taxa of *Stylocheiron affine* which are all restricted in distribution. It is especially this species which shows patterns not common for members of species-groups such as the *S. longicorne* one. The frequent sympatry makes it, moreover, very doubtful if the forms in the species *S. affine* are indeed forms instead of genetically isolated species or semispecies.

FIG. 65 Distribution of *Stylocheiron microphthalma* (black) and *S. elongatum* (hatched).
FIG. 66 Distribution of *Stylocheiron longicorne* and *S. indicus*. 1 = *S. longicorne* 'N. Indian Ocean form', 2 = *S. longicorne* 'Long form', 3 = *S. longicorne* 'Short form', 4 = *S. indicus*.
FIG. 67 Distribution of the *Stylocheiron affine* group. 1 = *S. affine* 'W. Equatorial form', 2 = *S. affine* 'California Current form', 3 = *S. affine* 'Indo-Australian form', 4 = *S. affine* 'E. Equatorial form'.

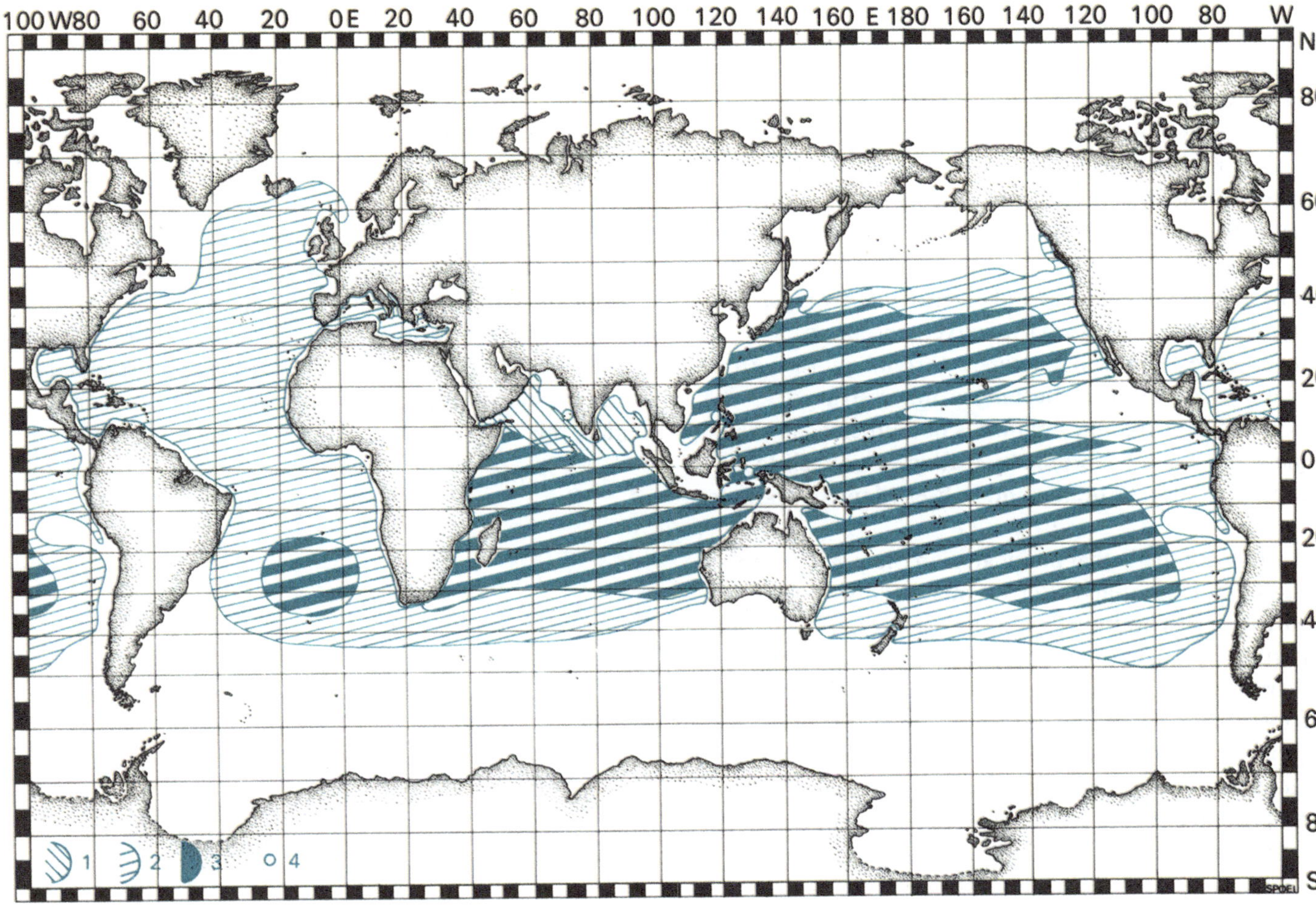
100 W 80 60 40 20 0 E 20 40 60 80 100 120 140 160 E 180 160 140 120 100 80 W
N 80 60 40 20 0 20 40 60 80 S
1 2 3 4

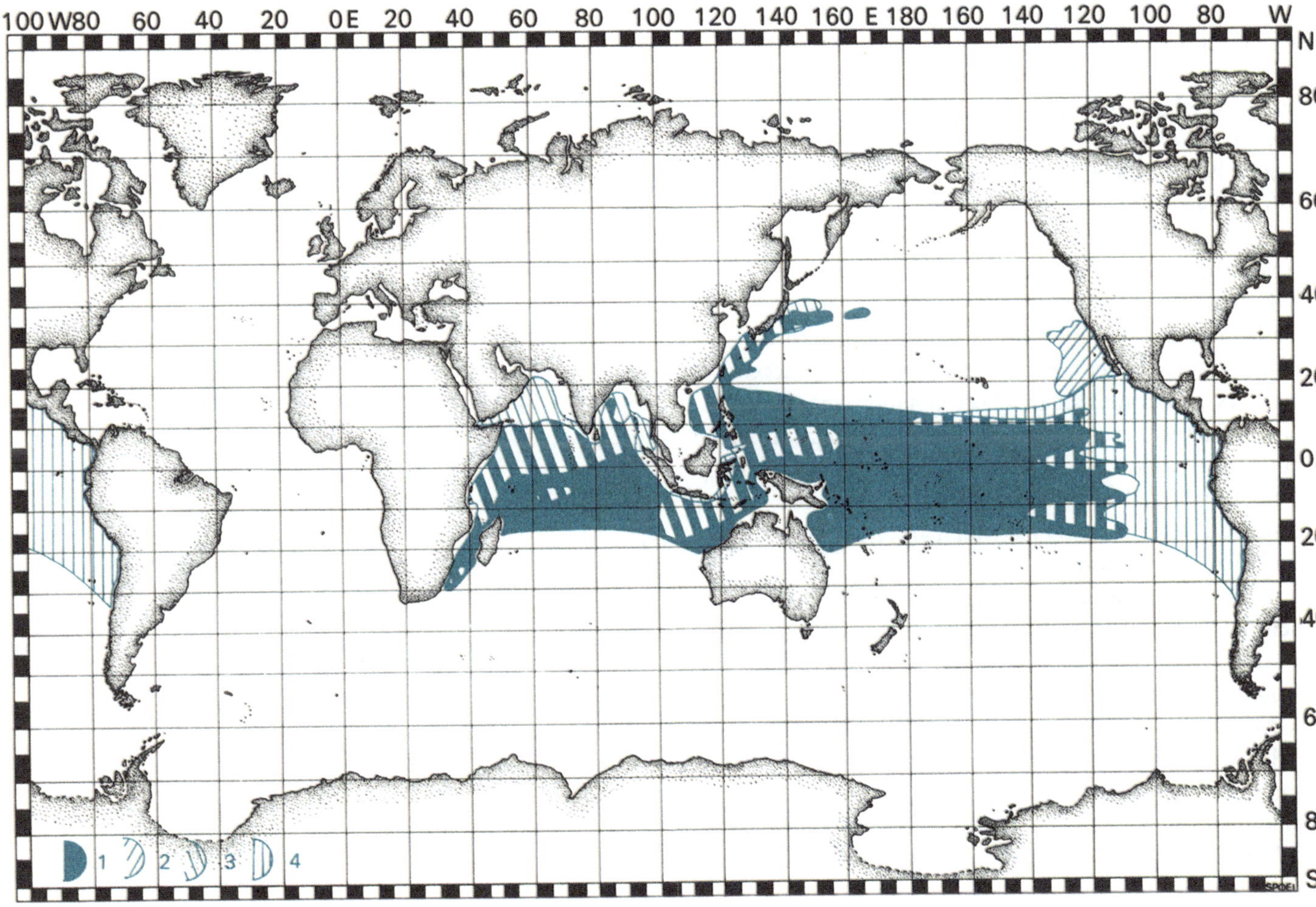
100 W 80 60 40 20 0 E 20 40 60 80 100 120 140 160 E 180 160 140 120 100 80 W
N 80 60 40 20 0 20 40 60 80 S
1 2 3 4

Boundaries and barriers

Though their numbers are few in the oceans, some isolating barriers can be indicated; the continents, divergences and convergences are the most clear-cut ones. Other boundaries and barriers can only be detected and described as places where one or more taxa have their borders of distribution.

The major convergences and divergences are given in fig. 68. Not indicated are the convergences and divergences which do not persist throughout the year, as their zoogeographical meaning is only restricted. Barriers to planktonic organisms not indicated in the map either, can be formed by shallow areas such as the Bering, Timor, Gibraltar and Bab-el-Mandab passages.

A survey of the faunal boundaries in the world ocean (fig. 69) was compiled after various authors by plotting a variety of distributional patterns together. Usually, the borders of ranges do not coïncide exactly, so that a diagrammatical representation has been chosen here. Most areas enclosed by lines in this map are not to be considered provinces or biotopes, as the map is not a summing up of such areas but the representation of ranges and partly of overlapping ranges of different faunal categories.

To get a better impression of barriers and boundaries, and certainly for an explanation of these, one should consider the chemical and physical properties of the oceans (cf. figs 17-35).

In fig. 138 the approximate borders between the major 'provinces' of neritic taxa are given by black markers. Not all borders are given and the actual limits of ranges will not precisely coïncide with the markers. The present map, however, gives a good impression of the areal subdivision of the neritic realm. All the limits indicated, are represented in at least one of the distributional types given in this atlas. The neritic ranges sometimes cover the area between two successive markers, sometimes an area between more markers, disjunct or not, is covered. Most of the borders of neritic ranges can be connected with characteristic isotherms.

Closely linked with the neritic patterns are the distributions of littoral taxa which are, however, excluded from this atlas.

The great resemblance between planktonic and benthic faunas may be due to the fact that both depend on the water masses, as proven e.g. by Buzas & Culver (1980) who stated: 'All published distributional data on recent benthic Foraminifera of the North America Atlantic continental margin were archived into computerized catalogs. Cluster analyses of these data delimited seven large, marginally overlapping provinces exhibiting a congruous relationship with western North Atlantic water masses. The single major latitudinal faunal change occurs at Cape Hatteras.' The borders between the faunal provinces given by these authors coincide with the five markers along the N.E. Atlantic coast in fig. 138. The borders concern shelf provinces and slope-deepwater provinces for benthic organisms, and coastal-neritic and distant-neritic provinces in plankton.

FIG. 68 Continents (1), Convergences (2) and Divergences (3) representing the zoogeographic barriers diagrammatically.

FIG. 69 The total of faunal boundaries in the world ocean. black = Central-Waters and cyclic currents.

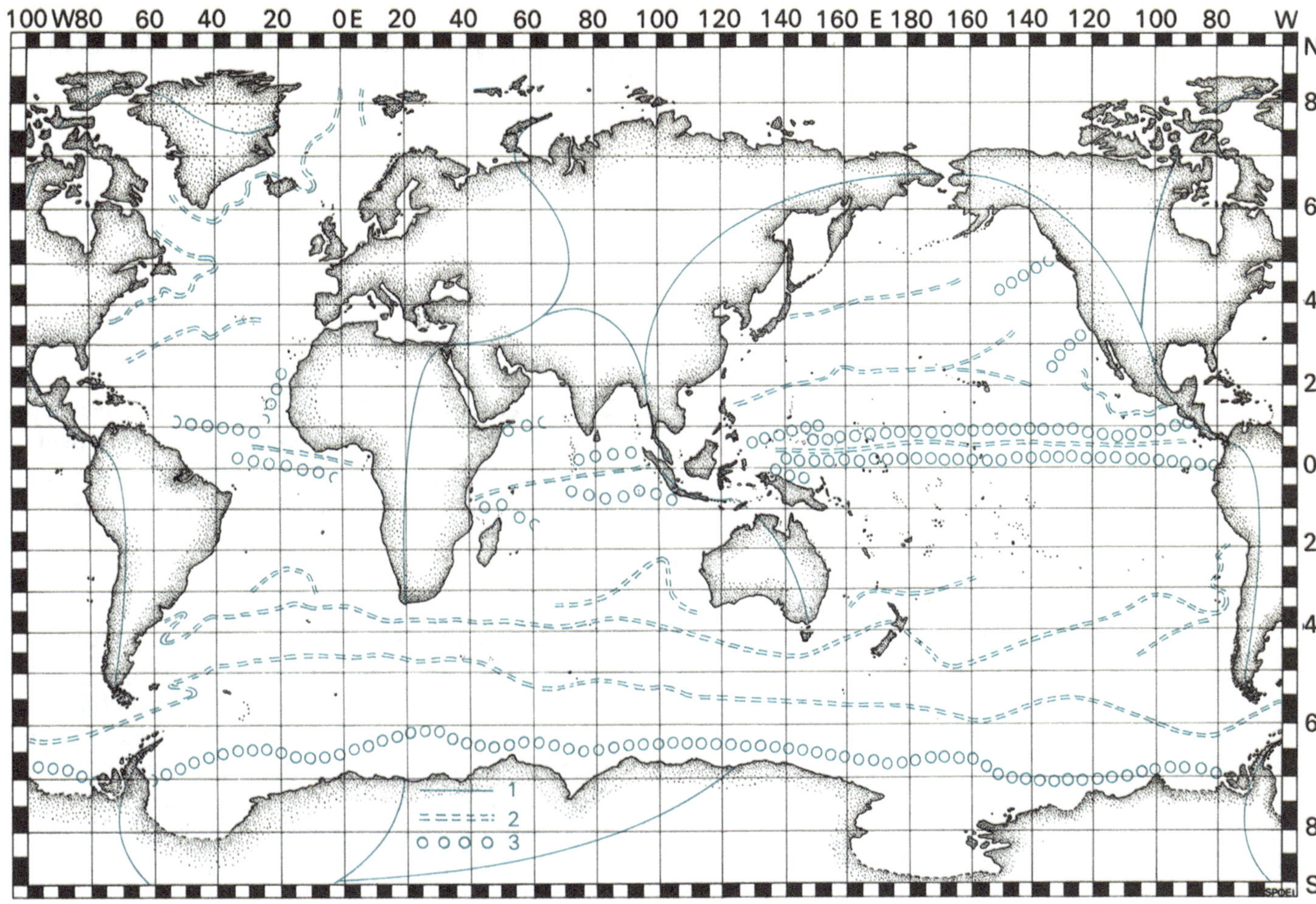
100 W80 60 40 20 0E 20 40 60 80 100 120 140 160 E 180 160 140 120 100 80 W
N 80 60 40 20 0 20 40 60 80 S
1
2
3
SPOEL

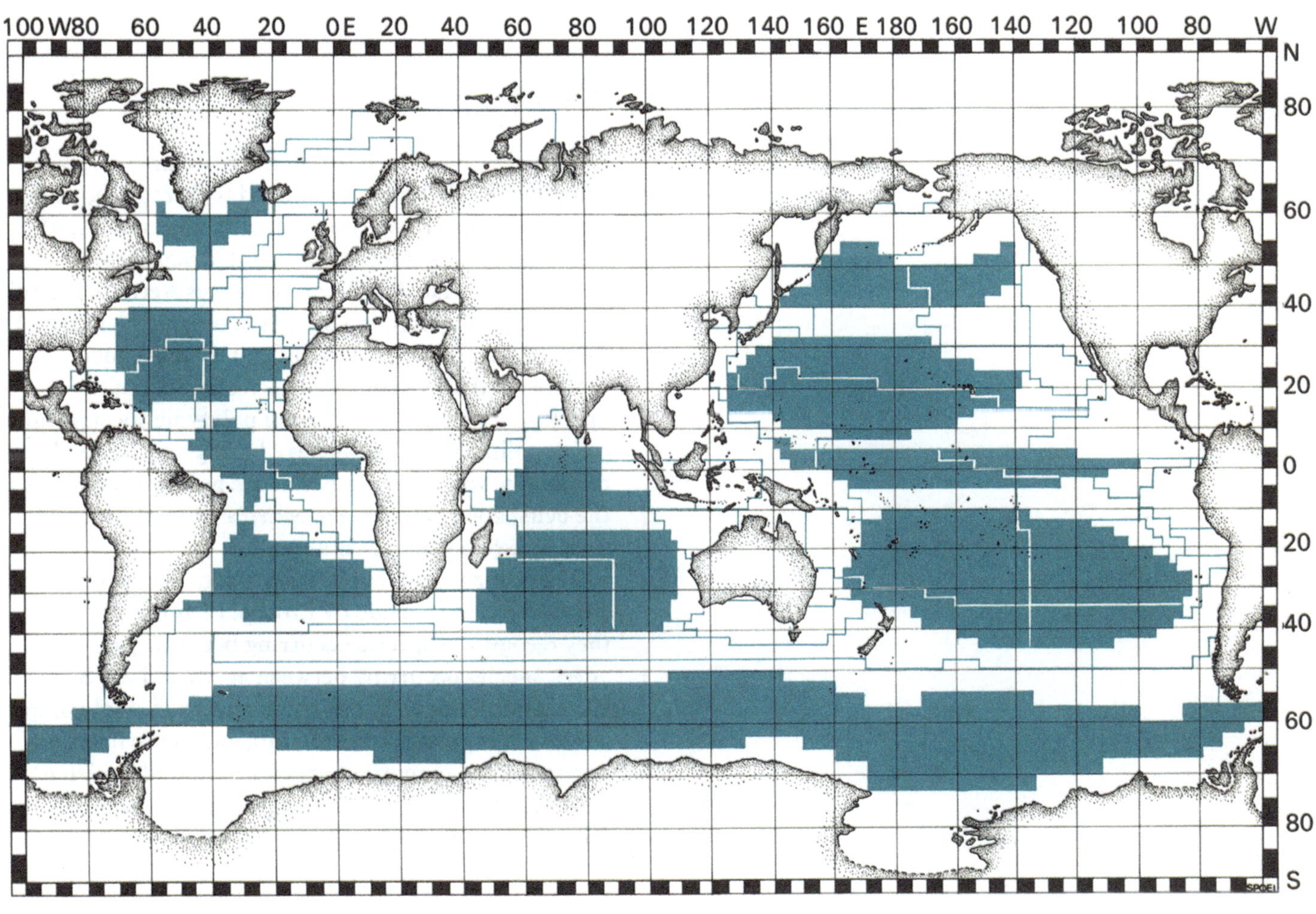
100 W80 60 40 20 0E 20 40 60 80 100 120 140 160 E 180 160 140 120 100 80 W
N 80 60 40 20 0 20 40 60 80 S
SPOEL

3 Types of distribution

Beltshaped patterns—Patterns derived from beltshaped patterns—Interocean connections—Not-beltshaped patterns—Endemics—Neritic distribution patterns—Vertebrate patterns—Mesopelagic and bathypelagic patterns

This chapter gives a survey of distribution patterns. Types of comparable nature have been grouped together, in an effort to achieve a natural structure. A subdivision into epi-, meso- and bathypelagic types is not made, as vertical processes will get attention in Chapter 4, only the patterns occupied exclusively by deep-living organisms are given in the last paragraph.

Beltshaped patterns

All distributions which follow the latitudinal or climatological belts are considered to form one group, which is composed of a) latitudinally wide ranges, b) warm-water ranges and c) cold-water ranges. Two decades ago, wide distributional ranges were considered to be the most common patterns in oceanic plankton. At the moment, however, it is difficult to find a good example of truly circum-global, cosmopolitan, monotypic species. Examples of polytypic species with a cosmopolitan distribution have been given above. The monotypic wide-spread species are here grouped with the beltshaped patterns also when a real indication of their beltshaped nature is absent.

The first zoogeographic approach to beltshaped patterns has been made by Frost & Fleminger (1968); they recognized species occurring between 70°N and 70°S, species occurring between 40°N and 40°S and species occurring between 30°N and 30°S. Only in the last group is provincialism found, and in the first two circum-global monotypic species are found, which is considered to be a rule. In this paragraph essentially the same division is used.

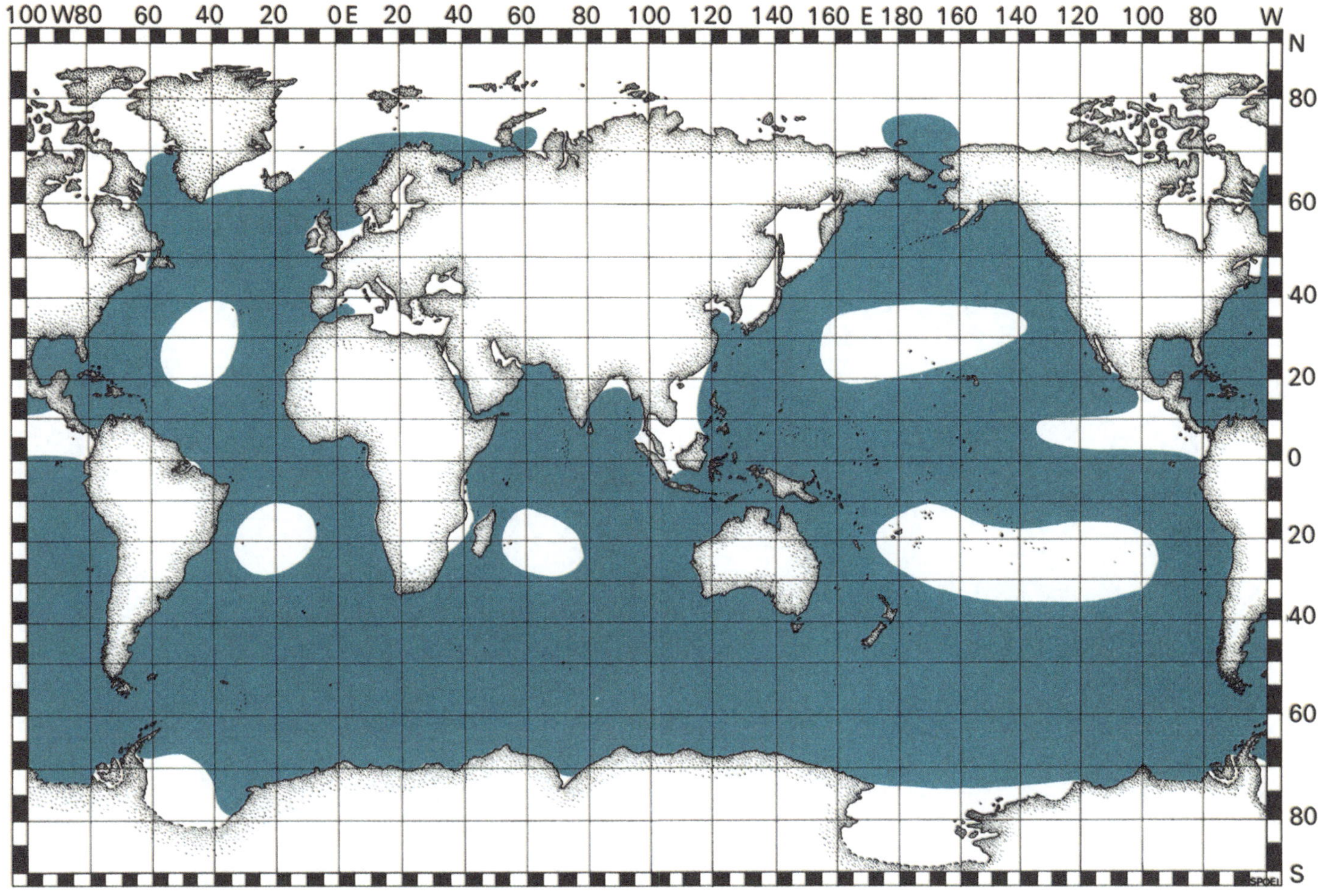

FIG. 70 Distribution of *Rhizosolenia alata*.

LATITUDINALLY WIDE RANGES

RHIZOSOLENIA ALATOIDES type

Taxa of this type (fig. 70) live in all the oceans from pole to pole. The absence from Central waters, so clearly demonstrated in *Rhizosolenia alata* is typical of this pattern and it forms the most clear difference between this type and the next one.

PHRONIMA SEDENTAROIDES type

This type of distribution is considered separately because, as shown in *Phronima sedentaria* (p. 38, fig. 49), it is characterized by a wide range (60° N–60° S) with areas of high abundance in the primary water masses or cyclical waters (cf. fig. 3), which is rather unique and just the reverse of what is found in the RHIZOSOLENIA ALATOIDES type (fig. 70).

SALPA FUSIFORMOIDES type

The distribution of *Salpa fusiformis* (p. 40, fig. 51) is identical with that of the next species and comparable with the preceding pattern, except for the fact that patterns of the present type show clear north-south variation, usually of the clinal type.

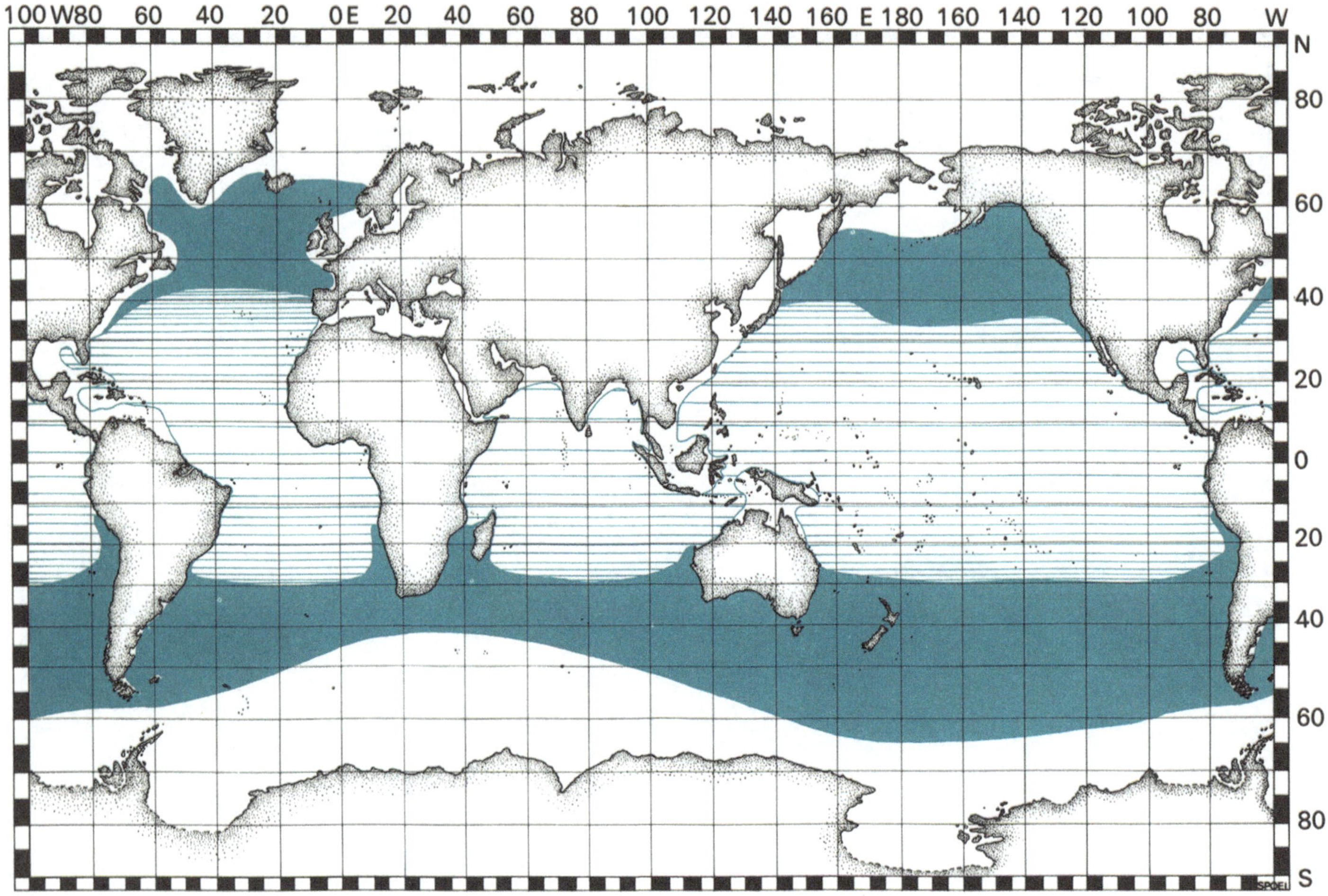

SAGITTA ZETESIOIDES type

This type of wide distribution (fig. 71) is associated with bathymetric variation. *Sagitta planctonis* forma *zetesios* is found between 65°N and 60°S, but it prefers cold water, in contrast with *S.p.* forma *planctonis*. The forma *zetesios* lives in the Tropics at greater depth, as indicated in the map by hatching. In the warmer surface waters between 40°N and 30°S the forma *planctonis* is found. This bathymetric variation shows parallels with the clinal variations in *Salpa fusiformis* and *Clio pyramidata*, where adaptations to warmer waters are achieved by the development of special forms instead of by going deeper with the colder waters.

STYLOCHEIRON MAXIMOIDES type

A cosmopolitan distribution only avoiding Arctic and Antarctic waters, is shown by *Stylocheiron maximum* (fig. 72), which occurs between 55°N and 55°S.

CAVOLINIIOIDES type

This type of distribution resembles the preceding type but the taxon concerned shows provincialism; in the example selected, *Cavolinia tridentata*, nine allopatric forms are distinguished (p. 42 & 99, fig. 53).

FIG. 71 Distribution of *Sagitta planctonis* forma *zetesios*, with tropical submergence represented by hatching.
FIG. 72 Distribution of *Stylocheiron maximum*.
FIG. 73 Distribution of *Clausocalanus paupulus*.

CLAUSOCALANUS PAUPULOIDES type

The species of this type are in reality warm-water taxa which are here recognized as the narrow type of 'wide distributions' (fig. 73). This type can manifest, as *Clausocalanus paupulus* does, a circum-global distribution, as it occurs still south of S. Africa. In the Indo-Malayan Archipelago its distribution is discontinuous, however, which distinguishes it from the next type.

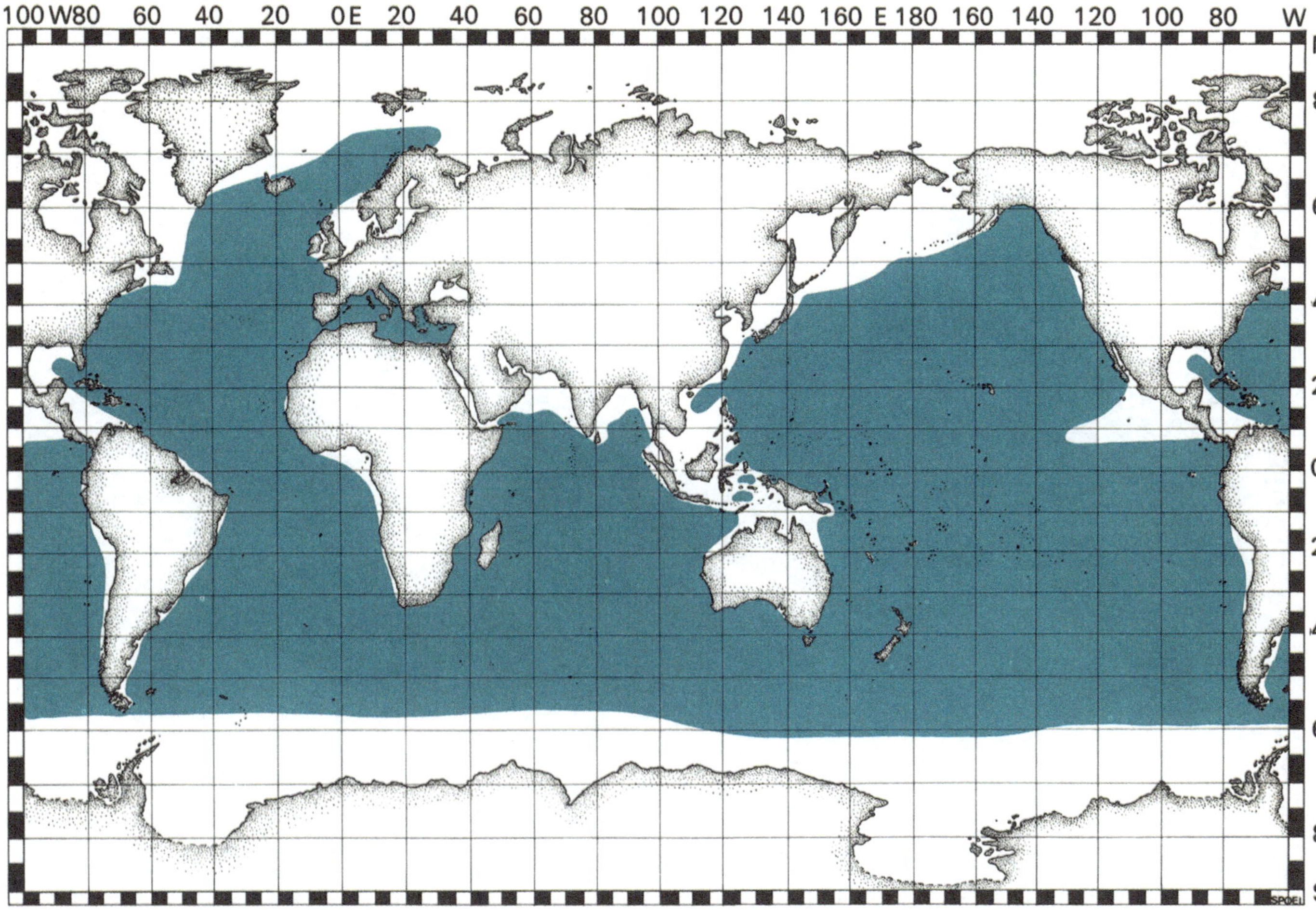
100 W80 60 40 20 0 E 20 40 60 80 100 120 140 160 E 180 160 140 120 100 80 W
N
80
60
40
20
0
20
40
60
80
S

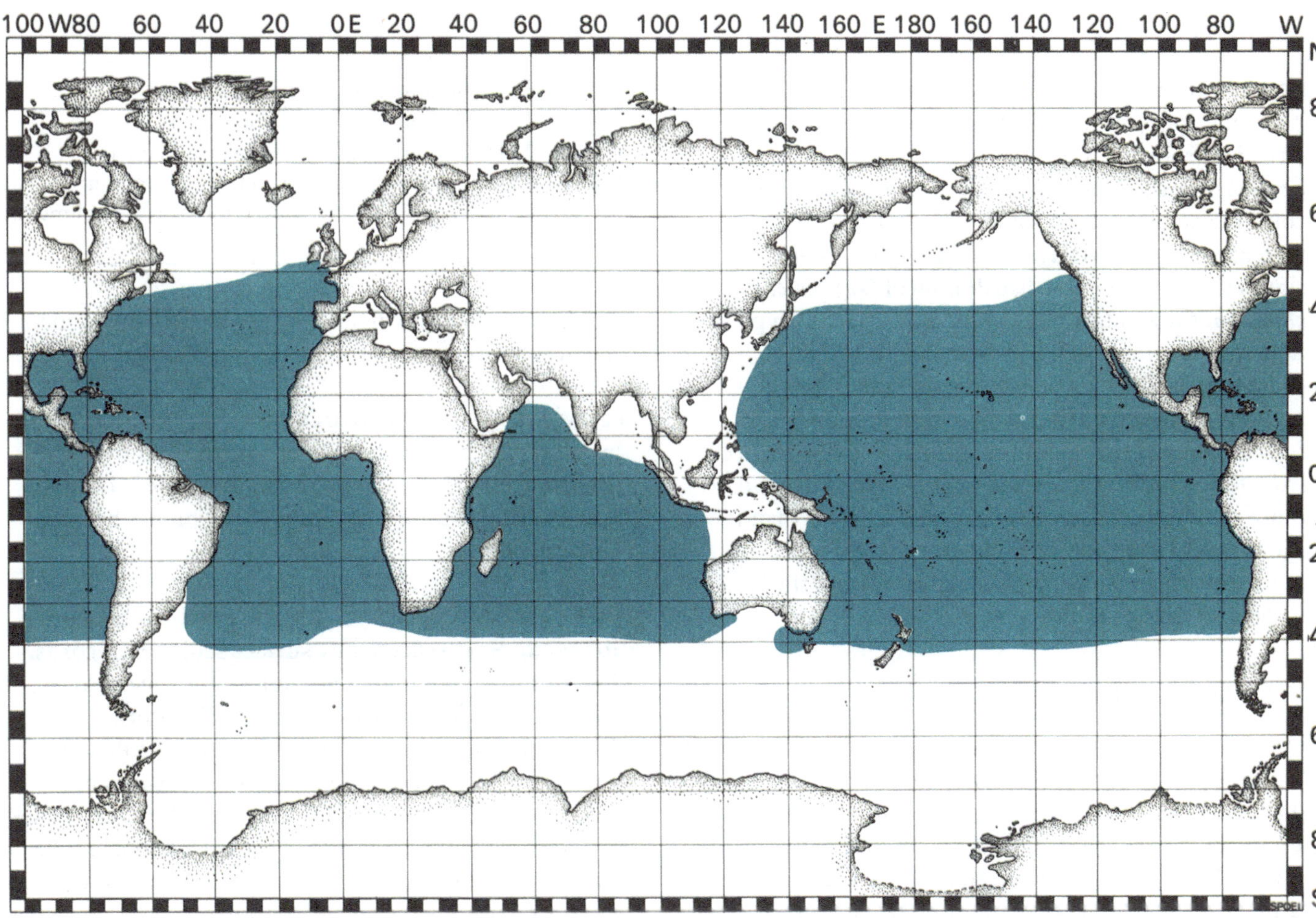
100 W80 60 40 20 0 E 20 40 60 80 100 120 140 160 E 180 160 140 120 100 80 W
N
80
60
40
20
0
20
40
60
80
S

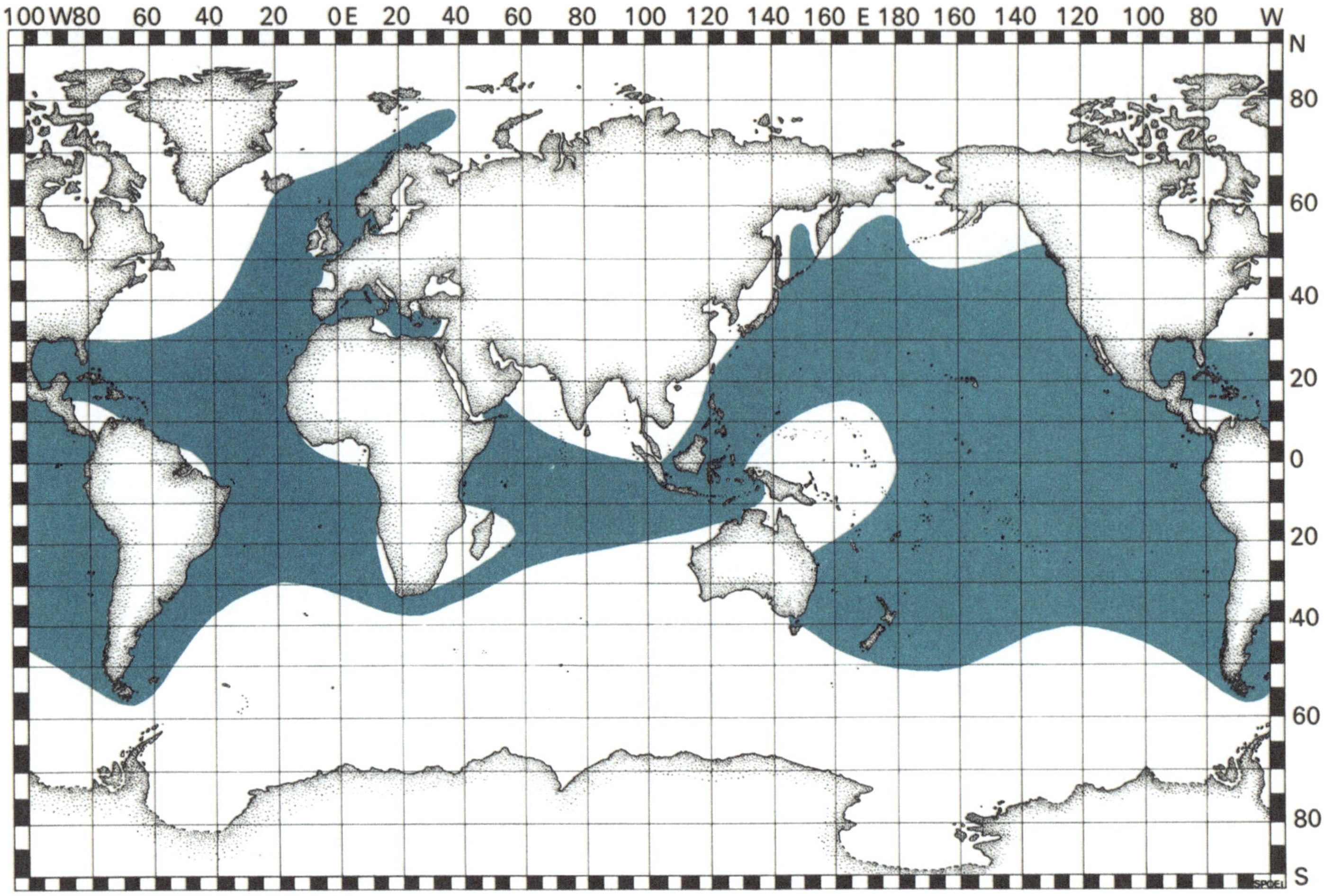

STYLOCHEIRON CARINATOIDES type

The range of this type, like that of the preceding one, stretches from 40°N to 40°S but, as shown by *Stylocheiron carinatum* (p. 50, fig. 63), in this type a continuous distribution south of S. Africa as well as all through the Indo-Malayan Archipelago is found. Fig. 74 shows the cosmopolitan distribution of *Onychoteuthis banksi*, of the STYLOCHEIRON CARINATOIDES type. The species concerned, is, however, a nektonic cephalopod, which demonstrates the resemblance between nektonic and planktonic distributions. The typical shape of the present range, marked by absence from the S. Indian Ocean and north of Australia and a more common occurrence in the East Atlantic than in the West Atlantic, points to the fact that this species, like many other cephalopods, originates from the Dakar Centre (cf. figs 59 & 60).

WARM-WATER RANGES

EUPHAUSIA TENEROIDES type

This is still a circum-global type, though the distribution in a 30°N-30°S belt, as illustrated by *Euphausia tenera* (fig. 75) is usually discontinuous or incomplete, and as a rule this induces variation and provincialism (cf. figs 52 & 53). The absence of variation in this type can be explained by the narrow but evident contacts between the populations of the different oceans.

FIG. 74 Distribution of *Onychoteuthis banksi*.
FIG. 75 Distribution of *Euphausia tenera*.
FIG. 76 Distribution of *Pulleniatina obliquiloculata*.
black = high abundance, double hatched = common occurrence, hatched = only recorded from surface sediments.

EUCALANUS SUBTENUOIDES type

In *Eucalanus subtenuis* (p. 40, fig. 52) the first traces of east-west variation and speciation developed. In most taxa showing this type of range, variation has become much stronger.

Pulleniatina obliquiloculata (fig. 76) also lives between 30°N-30°S, as reflected by surface sediment records. The pattern of common and high abundances proves, however, that this is a case of a more Equatorial distribution, of the EUPHAUSIA PSEUDOGIBBOIDES type (fig. 77). The possibility of a passage south of S. Africa, only reflected in the surface sediments, makes it acceptable that for some taxa a circum-global range may be found, though the normal boundaries of the taxon are found between 30°N and 30°S.

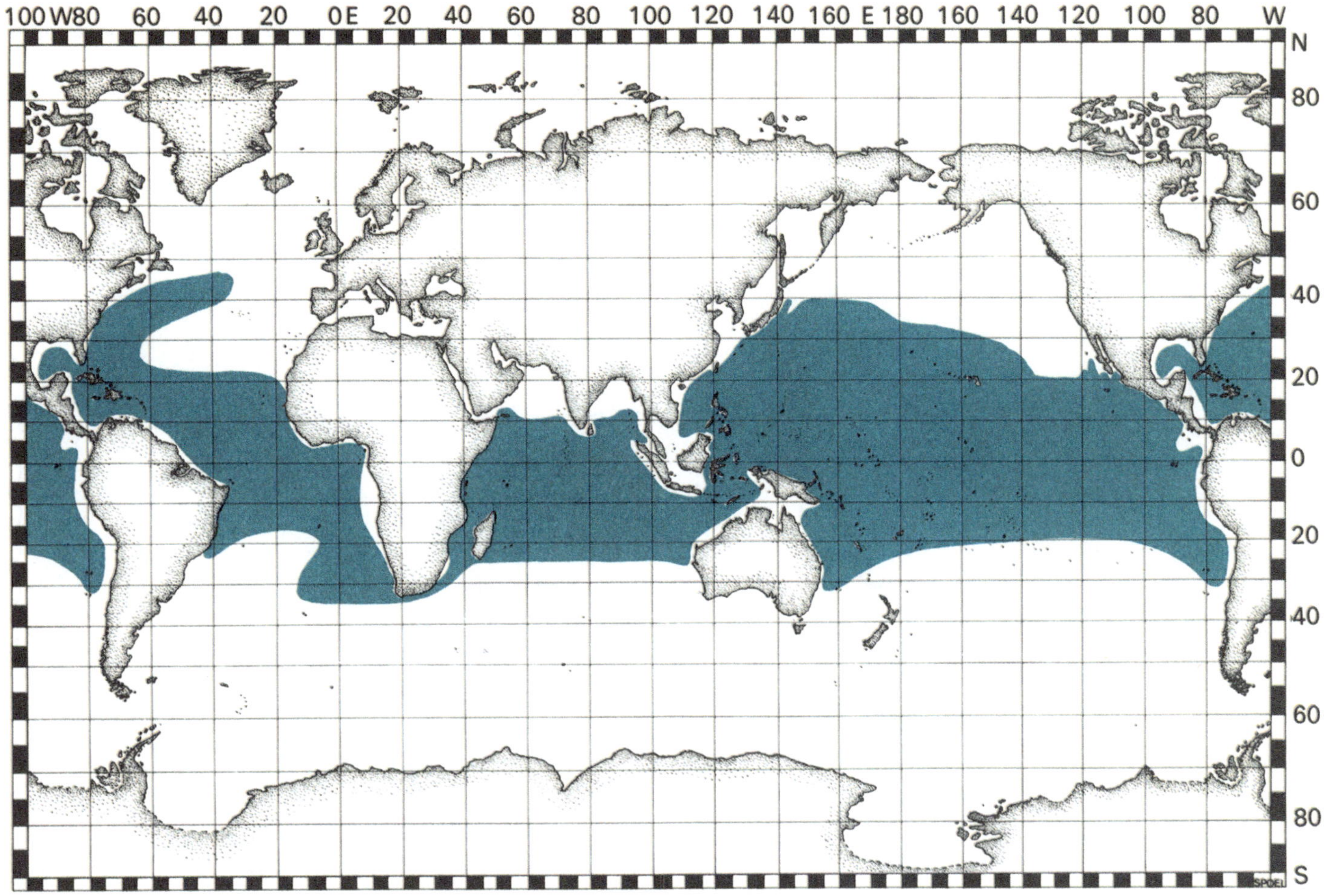
100 W80 60 40 20 0 E 20 40 60 80 100 120 140 160 E 180 160 140 120 100 80 W
N 80 60 40 20 0 20 40 60 80 S
SPOEL

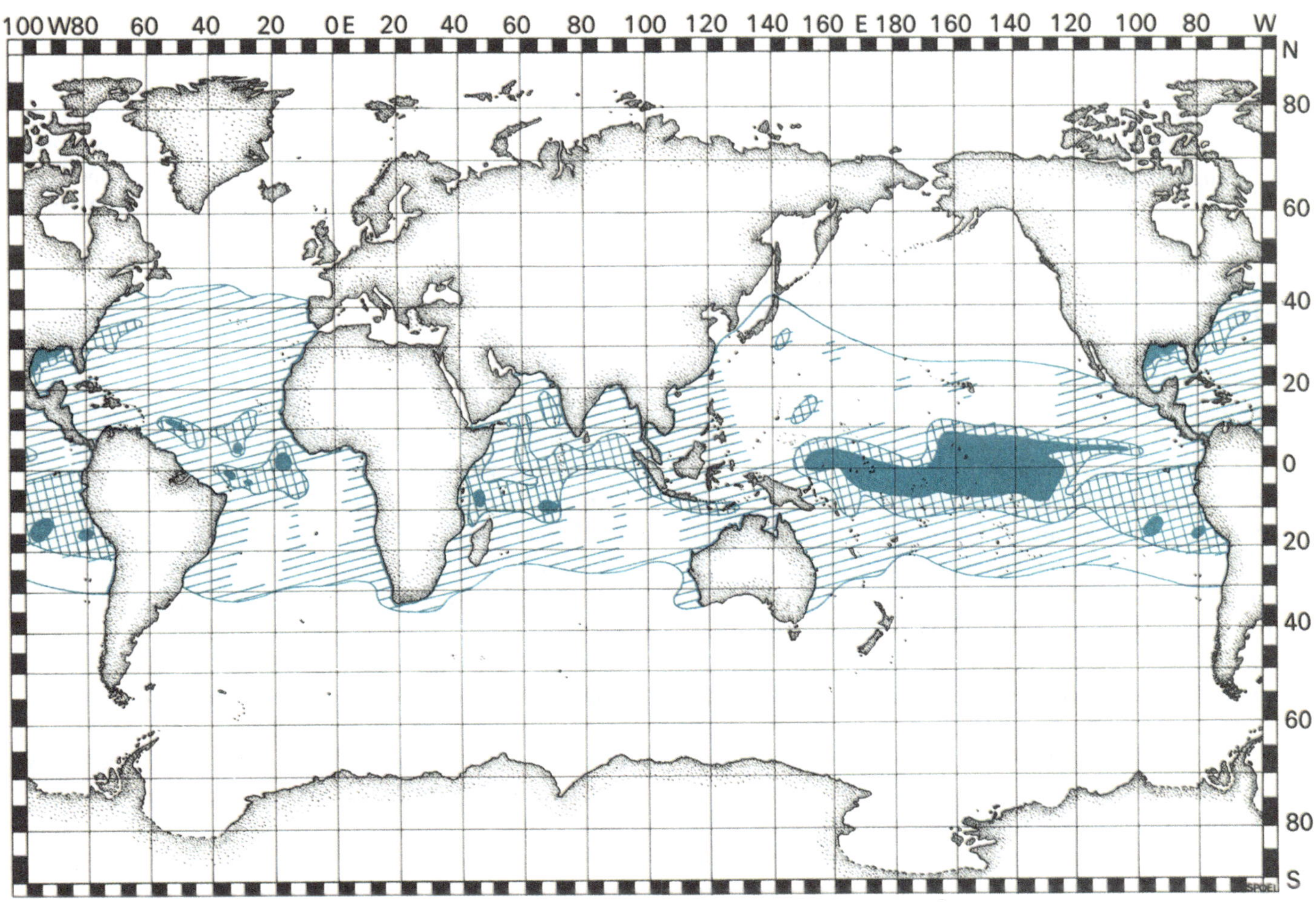
100 W80 60 40 20 0 E 20 40 60 80 100 120 140 160 E 180 160 140 120 100 80 W
N 80 60 40 20 0 20 40 60 80 S
SPOEL

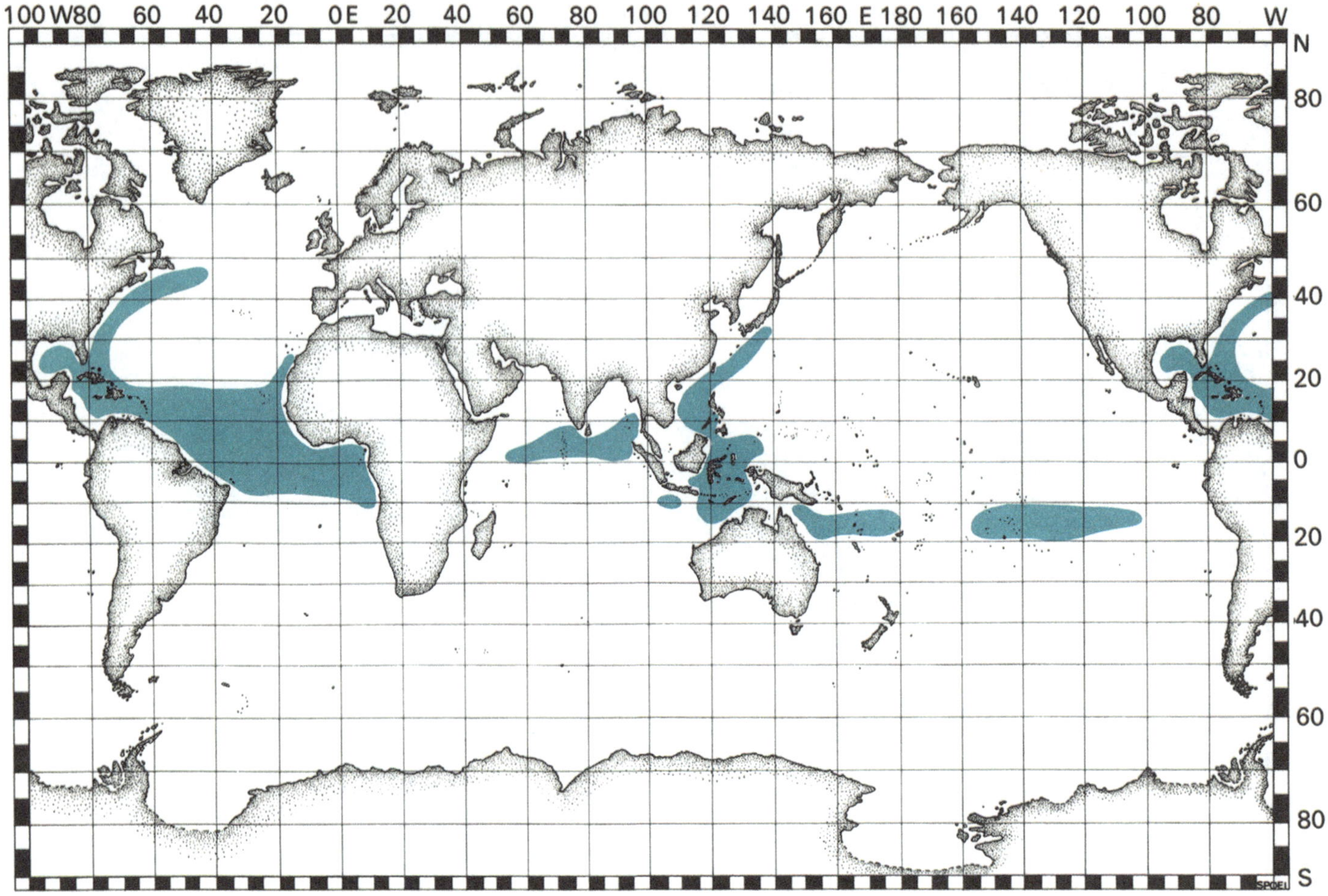

EUPHAUSIA PSEUDOGIBBOIDES type

Euphausia pseudogibba (fig. 77) is an example of the unexpected distribution between 20°N and 20°S which is, though discontinuous, still circum-global. The expatriation shown in the preceding map may also occur in the present species, which would explain the monotypy. Monotypical species usually have only one reproductive range (Beklemishev, 1971), but when the disjunct reproduction ranges are not too far apart, expatriation may still keep the species monotypic, as shown here.

CLIO CONVEXOIDES type

Species such as *Clio convexa* (p. 44, fig. 57) are distributed throughout the Equatorial belt of the Indo-Pacific Oceans between 20°N and 20°S, and are absent from the Atlantic. This pattern is much more common than the preceding one. This type differs from the HYDROMYLES GLOBULOSOIDES type (fig. 112), which is not a beltshaped pattern, in reaching the W. American coasts.

The distribution of Central water species such as *Euphausia brevis* (fig. 89) and related patterns may be considered to be beltshaped but they are given separately as distinctly different types of distribution (p. 70).

FIG. 77 Distribution of *Euphausia pseudogibba*.

FIG. 78 Distribution of the *Calanus finmarchicus* species-group. 1 = *C. glacialis*, 2 = *C. finmarchicus*, 3 = *C. marshallae*.

FIG. 79 Distribution of the *Calanus helgolandicus* species-group. 1 = *C. helgolandicus*, 2 = *C. australis*, 3 = *C. pacificus*, 4 = *C. chilensis*, 5 = *C. sinicus*.

COLD-WATER RANGES

CALANUS GLACIALOIDES type

Calanus glacialis (fig. 78), member of the Arctic-Subarctic species-group *C. finmarchicus* depicted here, provides a perfect example of Arctic distribution. The Subarctic taxa are not considered to belong to the real beltshaped patterns, as they are supposed to have developed separately in the two oceans (p. 93).

The *Calanus helgolandicus* group (fig. 79) is closely related to the *C. finmarchicus* group and does not show beltshaped patterns either, as the members clearly developed in separate basins. The distributional types represented by the five species are treated on pp 105-112.

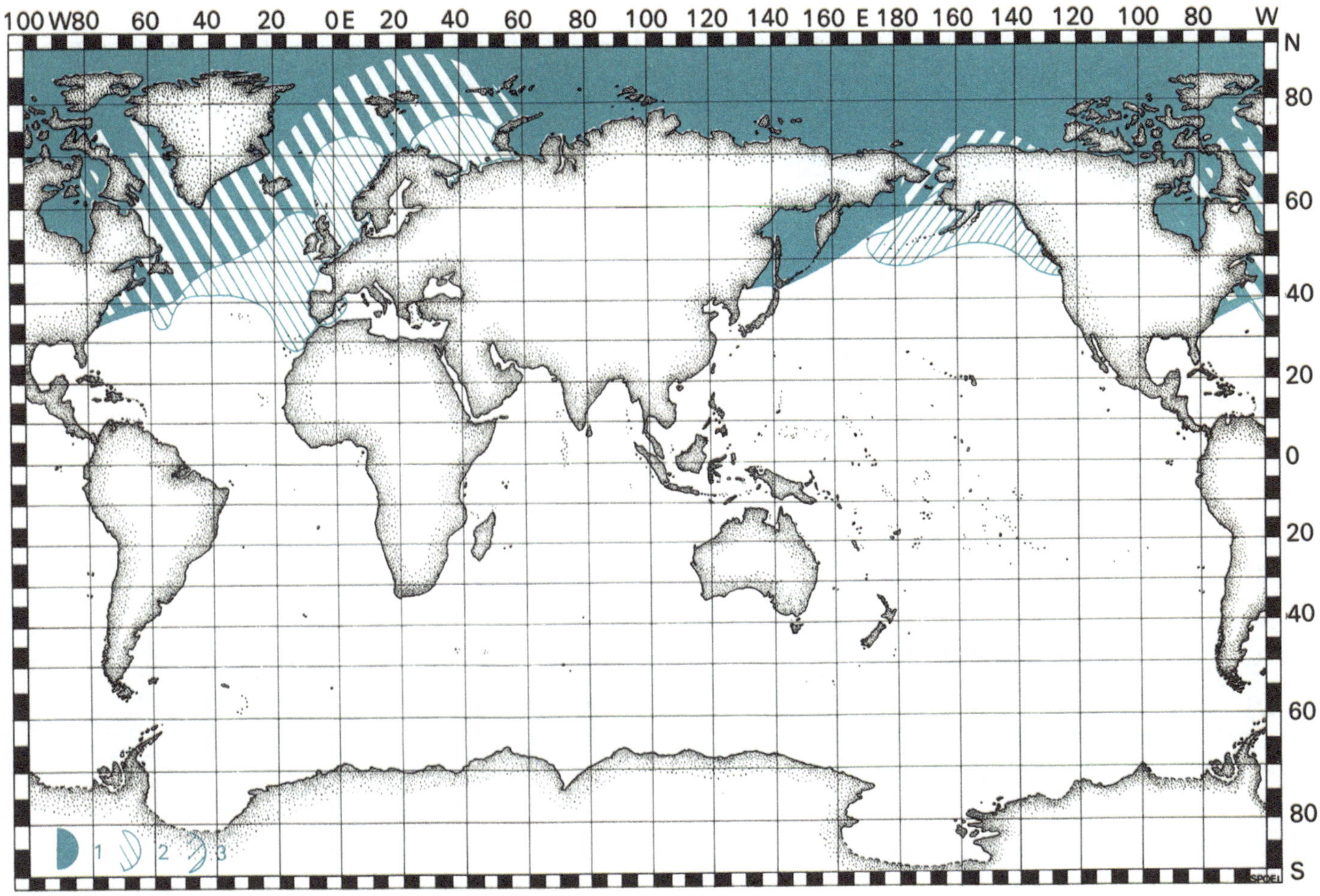
100 W80 60 40 20 0E 20 40 60 80 100 120 140 160 E 180 160 140 120 100 80 W
N
80
60
40
20
0
20
40
60
80
S
1
2
3

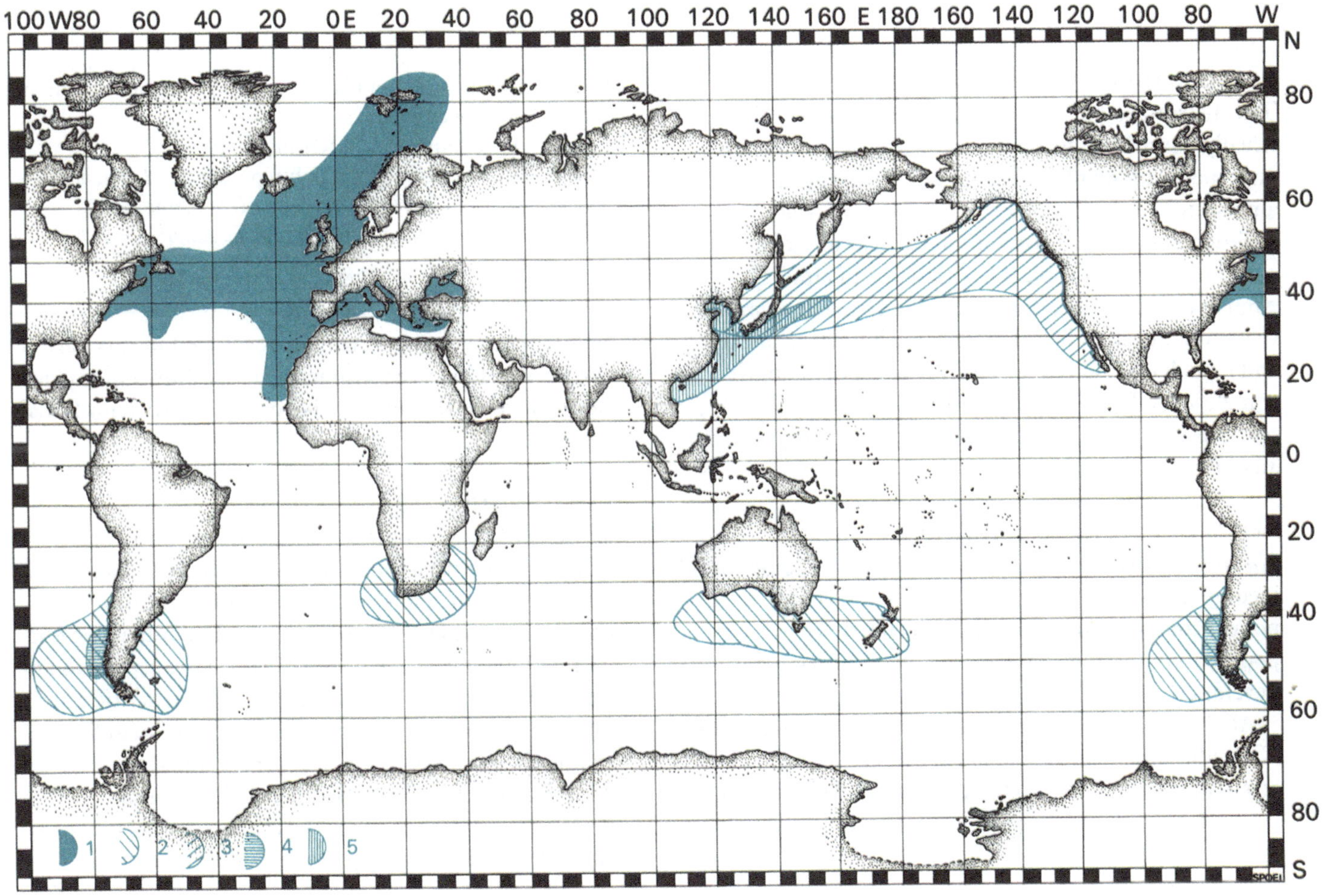
100 W80 60 40 20 0E 20 40 60 80 100 120 140 160 E 180 160 140 120 100 80 W
N
80
60
40
20
0
20
40
60
80
S
1
2
3
4
5

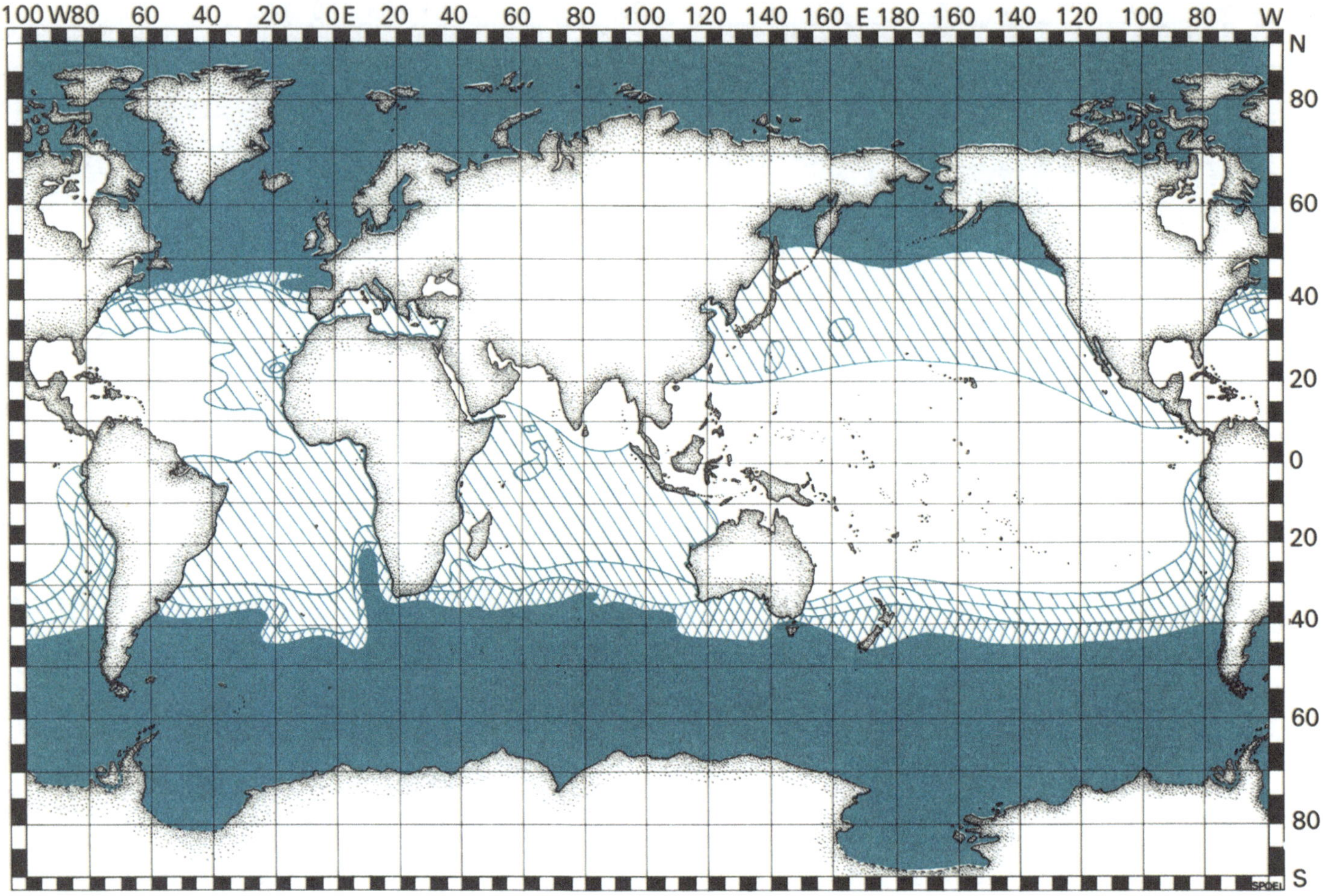

GLOBOQUADRINA PACHYDERMOIDES type

The distribution of *Globoquadrina pachyderma* (fig. 80) is usually considered to exemplify bipolarity and when only the living plankton is studied, it emerges as an almost monotypic bipolar species. In the surface sediments, however, distribution with a full north-south continuity in the Tropical E. Atlantic and almost a connection in the Tropical E. Pacific Ocean is found. It remains a question if real monotypic bipolar taxa exist and if such taxa do indeed entirely lack north-south connections. For the nearly discontinuous bipolar patterns the present type of distribution is proposed as an example.

NITZSCHIA CYLINDRICOIDES type

Nitzschia cylindrica (fig. 81) is an example of some sort of incomplete bisubpolarity. Usually the southern populations are more polar than the northern populations and the N. Pacific populations have not developed.

As there is no connection between the N. Atlantic and N. Pacific and as the north-south contact in the Pacific was always more interrupted than in the Atlantic, the N. Pacific occupies a separate zoogeographic position.

FIG. 80 Distribution of *Globoquadrina pachyderma*, shading intensity increases with abundance.

FIG. 81 Distribution of *Nitzschia cylindrica*.

FIG. 82 Distribution of the polytypic species *Limacina helicina*. 1 = *L.h.* forma *helicina*, 2 = *L.h.* forma *acuta*, 3 = *L.h.* forma *pacifica*, 4 = *L.h.* forma *rangi*, 5 = *L.h.* forma *antarctica*.

Sometimes even separate taxa developed in this part of the Pacific. However, also patterns as shown in the next figure develop through evolutionary and distributional processes.

LIMACINOIDES type

Limacina helicina (fig. 82) consists of a group of cold-water forms with a more or less beltshaped distribution. This pattern of semi-bipolarity and variation is considered to form a separate type. The Subarctic and Transitional taxa clearly demonstrate the separation of the N. Atlantic and N. Pacific. In the N. Atlantic only one taxon is found, which is not an exceptional feature, though in a few species also separate Subarctic and Transitional taxa developed here. The present distribution type concerns the entire species, the separate forms are grouped under different types (pp 68–94).

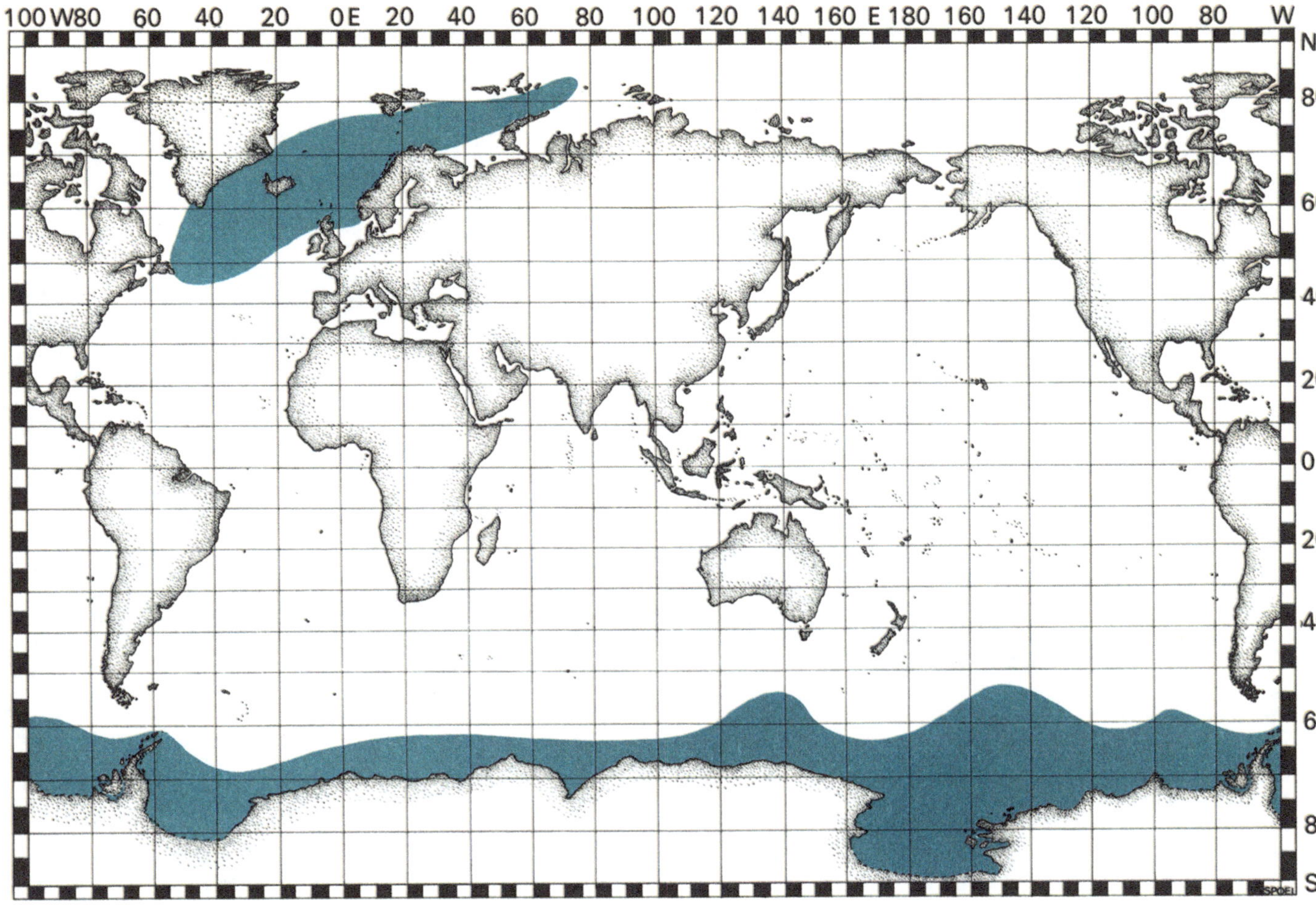
100 W80 60 40 20 0 E 20 40 60 80 100 120 140 160 E 180 160 140 120 100 80 W
N 80 60 40 20 0 20 40 60 80 S
SPOEL

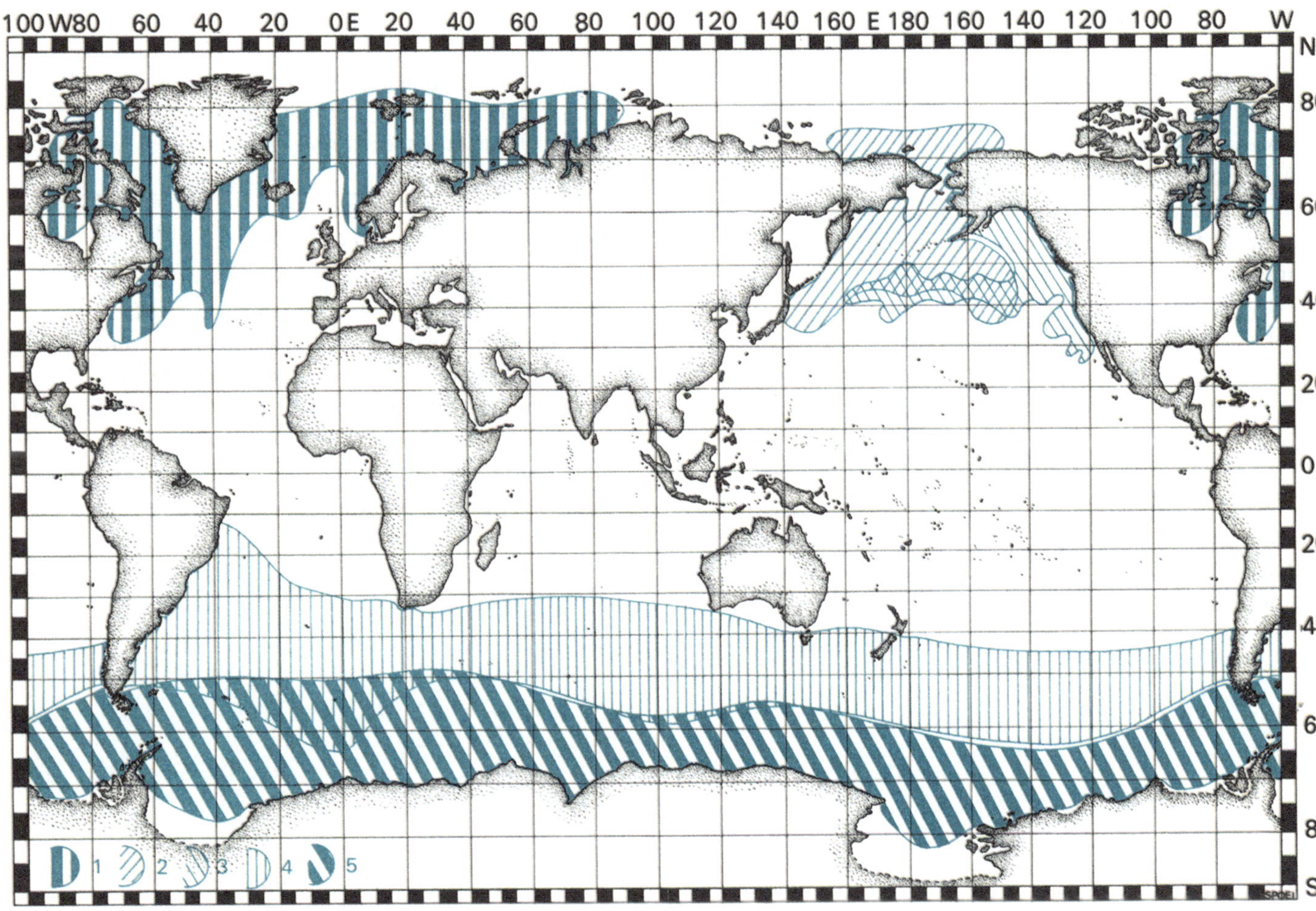
100 W80 60 40 20 0 E 20 40 60 80 100 120 140 160 E 180 160 140 120 100 80 W
N 80 60 40 20 0 20 40 60 80 S
1 2 3 4 5
SPOEL

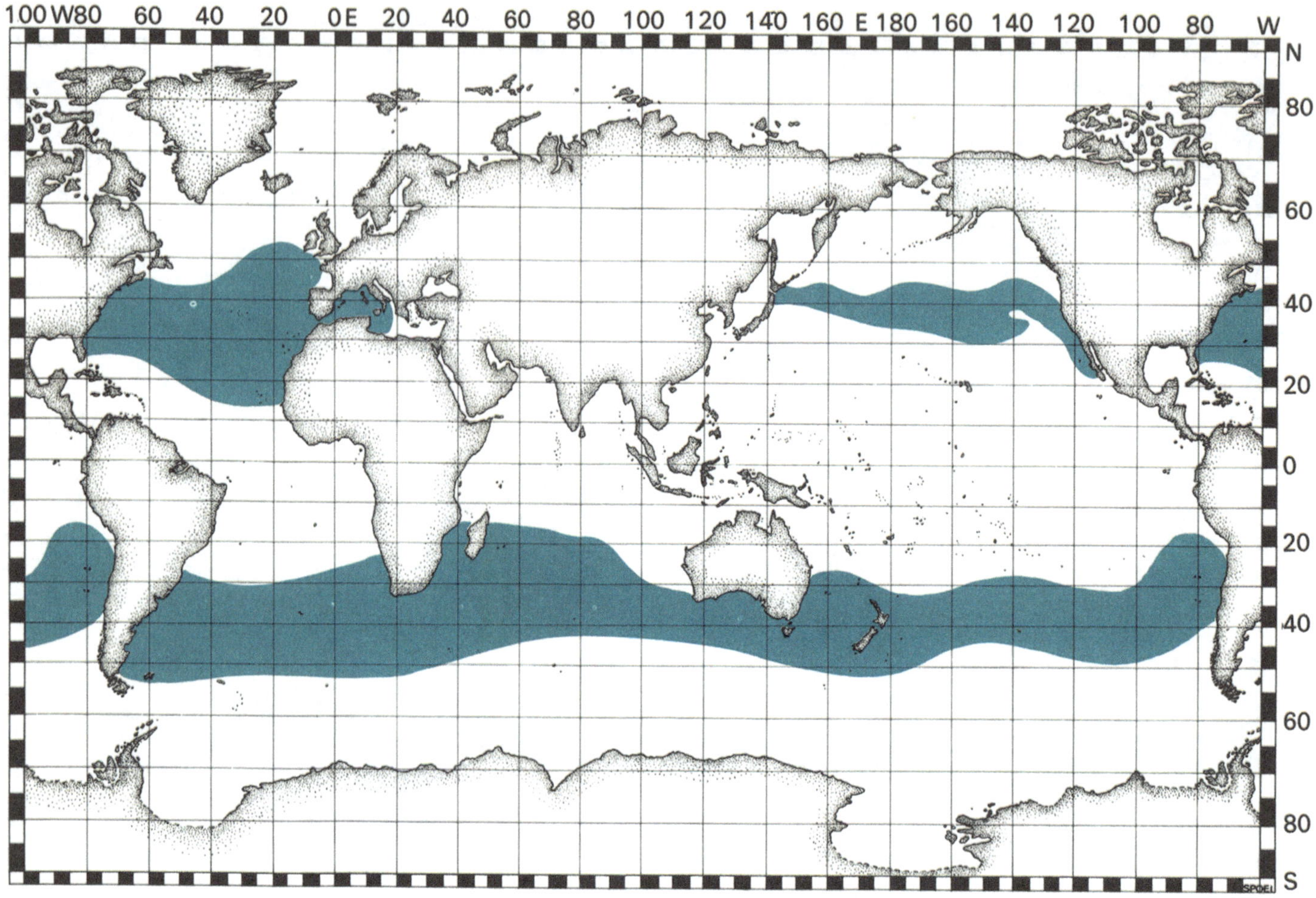

FIG. 83 Distribution of *Thysanoessa gregaria*.
FIG. 84 Distribution of *Euphausia lucens*.
FIG. 85 Distribution of *Euphausia similis*.

SAGITTA ELEGANSOIDES type

Sagitta elegans (fig. 124) is an example of the pattern covering the Arctic and Subarctic waters in both the Atlantic and Pacific Oceans (for a more detailed discussion of 'Subarctic' see p. 146). This pattern shows a connection between the two oceans north of Canada and probably north of Siberia.

THYSANOESSA GREGARIOIDES type

Thysanoessa gregaria (fig. 83) shows a rather exceptional bitransitional pattern.

EUPHAUSIA LUCENSOIDES type

There are only a few species, e.g. *Euphausia lucens* (fig. 84), which are endemic to the southern Transitional waters. Usually the range extends into Central waters or into Boundary and Equatorial areas.

EUPHAUSIA SIMILOIDES type

This rare type of distribution, represented by *Euphausia similis* (fig. 85) concerns Transitional water taxa with large populations in the S. Subtropical Terminal, partly Equatorial, waters of the Indian Ocean and in W. Boundary Subtropical waters of the Pacific Ocean. It is questionable whether an actual distribution is concerned here, but as more ranges are known to cover completely different water masses, the present pattern is considered a valid type.

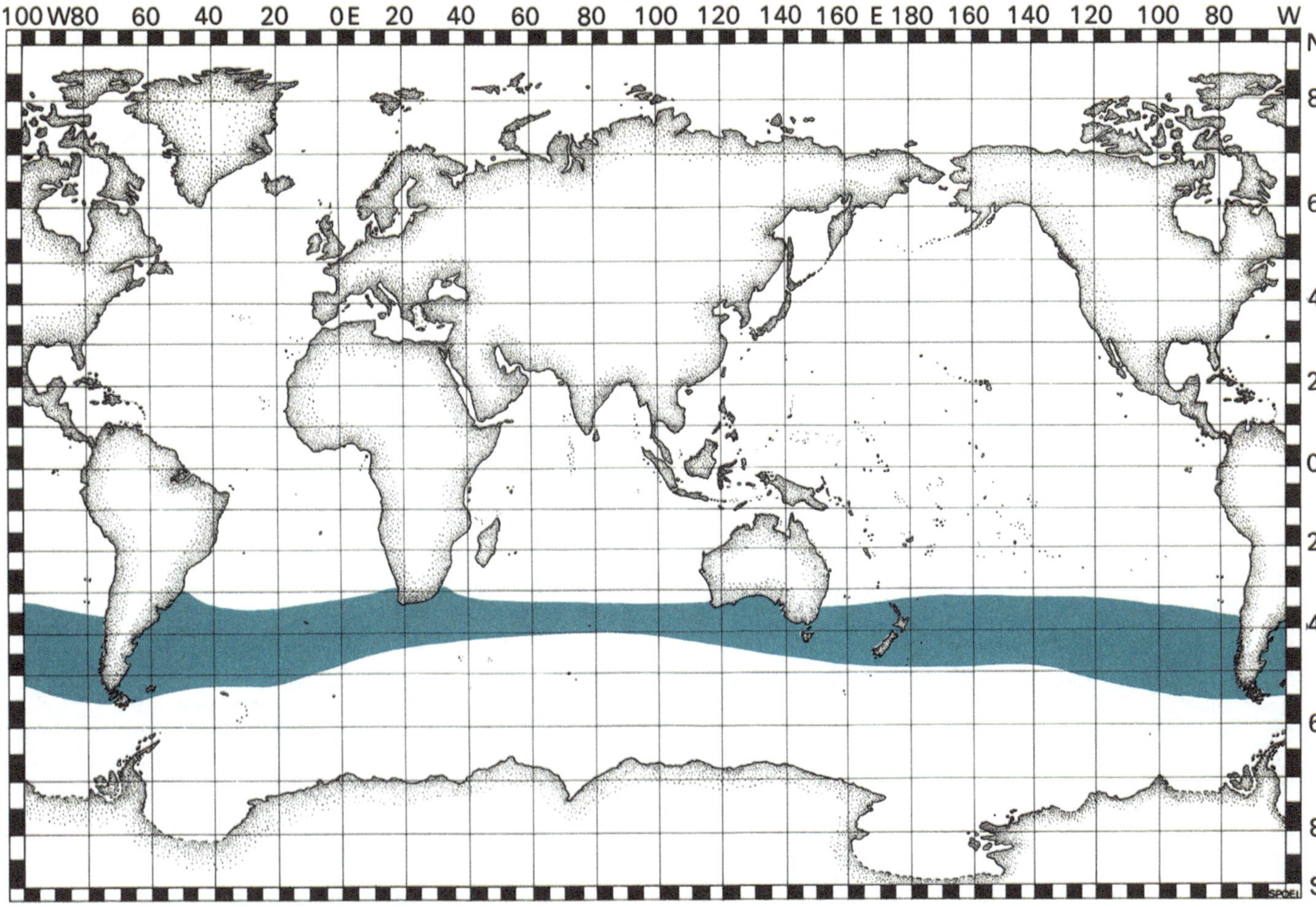
100 W80 60 40 20 0E 20 40 60 80 100 120 140 160 E 180 160 140 120 100 80 W
N 80 60 40 20 0 20 40 60 80 S

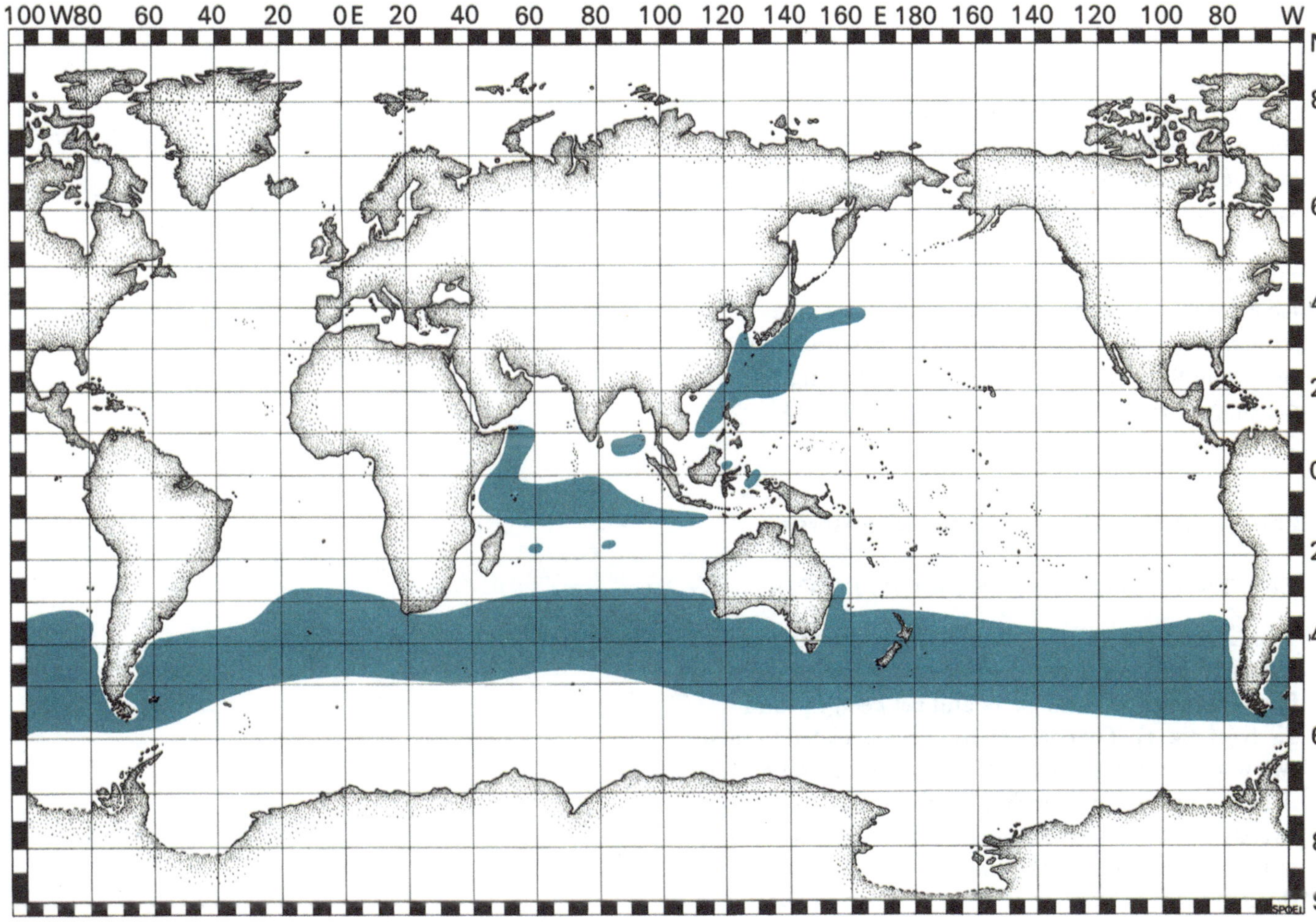
100 W80 60 40 20 0E 20 40 60 80 100 120 140 160 E 180 160 140 120 100 80 W
N 80 60 40 20 0 20 40 60 80 S

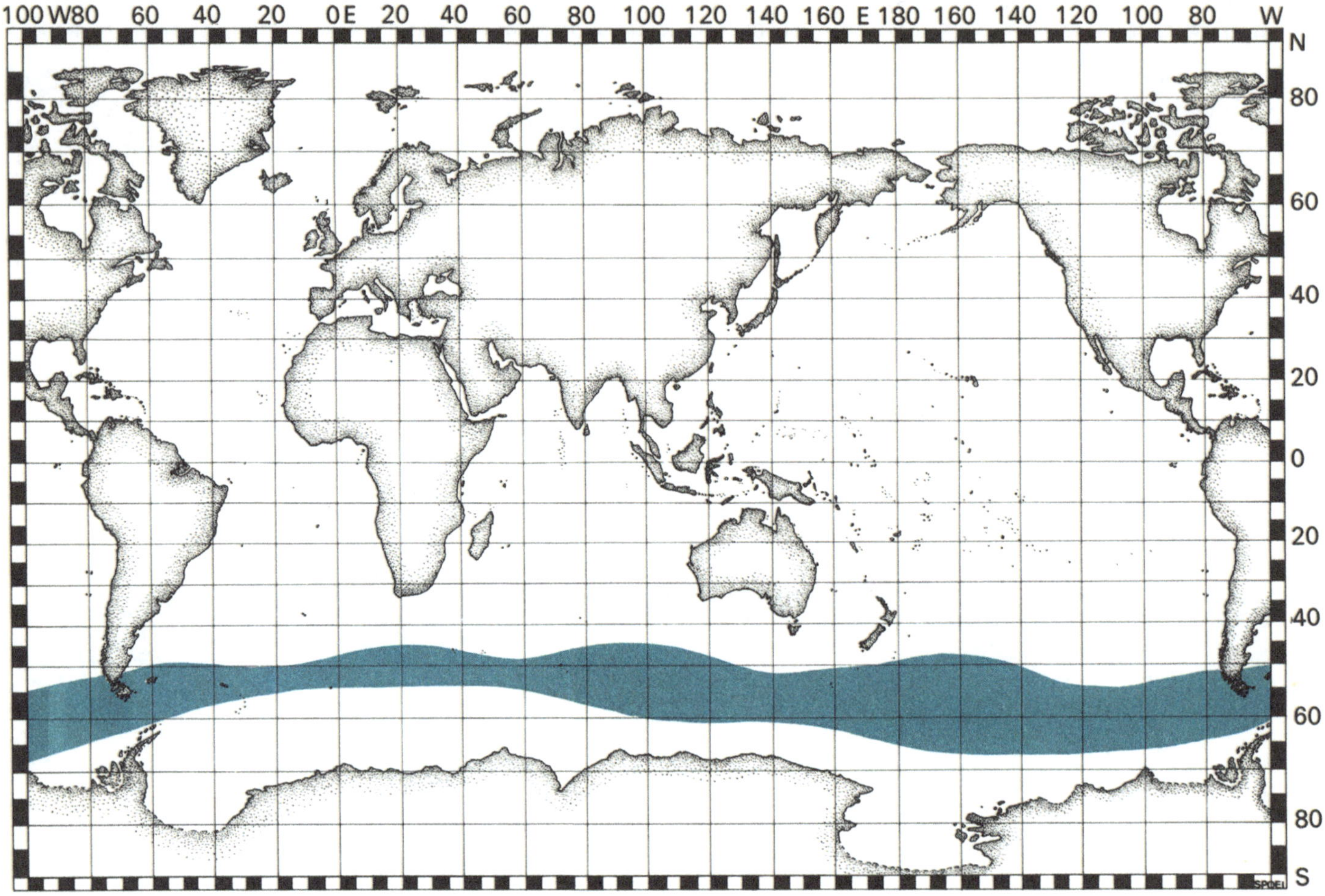

EUPHAUSIA TRIACANTHOIDES type
Subantarctic distributions are very variable in occupying, depending on the taxon, wider or smaller belts. *Euphausia triacantha* gives a good example of the average Subantarctic range (fig. 86).

THYSANOESSA MACRUROIDES type
Like the Subantarctic range, the Antarctic range strongly varies in width. *Thysanoessa macrura*, however, well exemplifies oceanic-Antarctic patterns (fig. 87). The neritic pattern has been termed the EUPHAUSIA CRISTALLOROPHIOIDES type (fig. 135).

Patterns derived from beltshaped patterns

CENTRAL WATER DISTRIBUTIONS

The Central waters of the Pacific Ocean are the most regularly and completely developed ones; they are found between 38° and 15°N and between 12° and 40°S (cf. fig. 1). There are many species related to these two water bodies, some of which also live in Indian and Atlantic waters. Van der Spoel & Pierrot-Bults (1979) already mentioned some regular patterns in the Central water taxa. Species of the Central water masses have small expatriation areas and species of mixed water

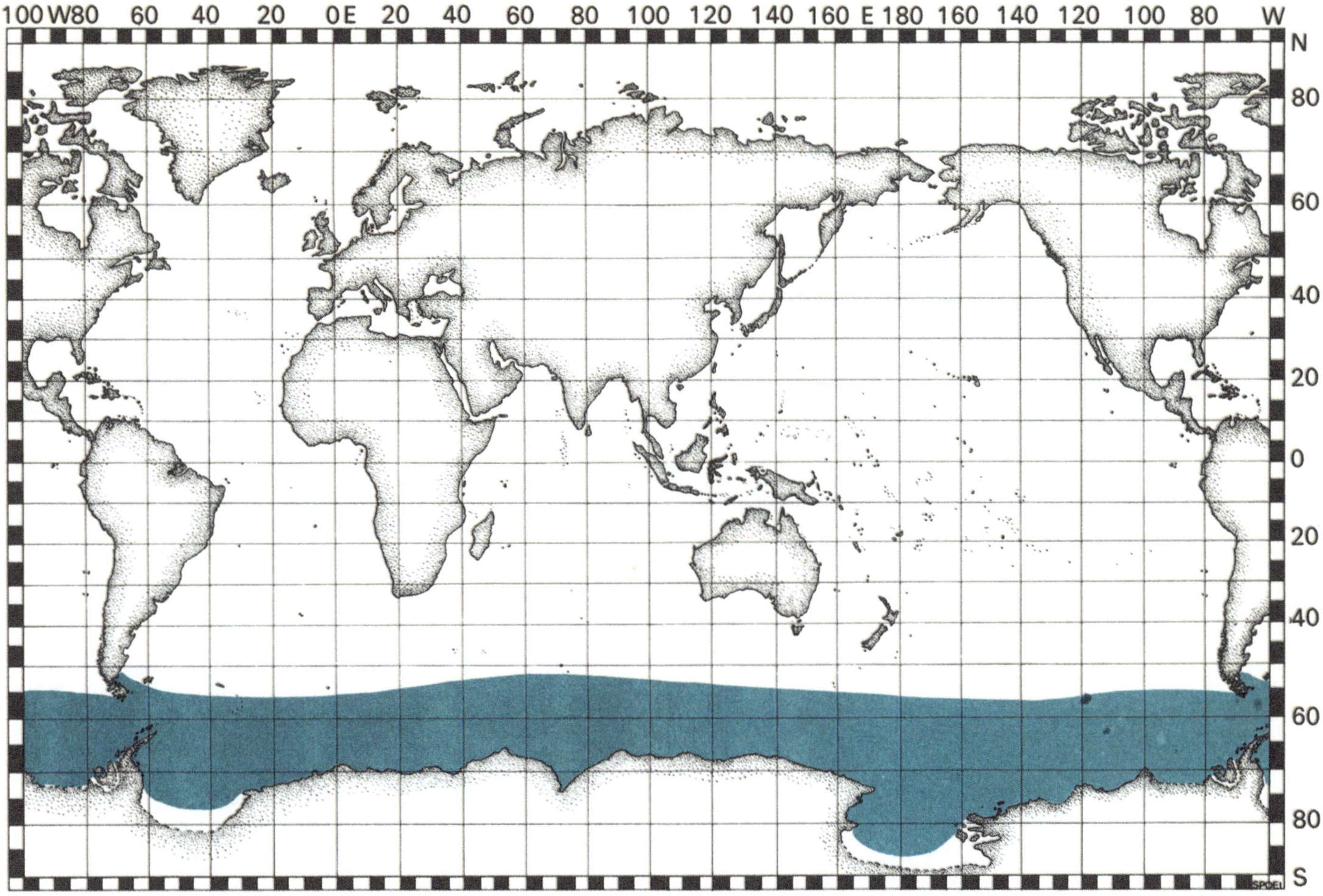

FIG. 86 Distribution of *Euphausia triacantha*.
FIG. 87 Distribution of *Thysanoessa macrura*.

Diagram II

Patterns of Central water distributions

C = Central water; N = north; S = south; T = tropical; W = west;
// = represented by a related taxon; —— = continuous distribution; - - - - - = discontinuous distribution.

PACIFIC INDIAN ATLANTIC	representative TAXON:
/SC/ - - - - - - - NC - - - - - - - - - SC - - - - SC - - - NC	*/Euphausia gibba/ – E.hemigibba*
SC - - - - - - - - - NC - - - - - - - - - SC absent	*Stylocheiron affine* 'Central form'
SC - - - - - - - - - NC - - - - - - - - - SC - - - - - - - - - - - NC	*Diacria major*
SC - - - - - - - - - NC - - - - - - - - - SC - - - - SC	no good example available
SC - - - - - - - - - NC - - - - - - - - - SC - - - - SC - - - NC	*Euphausia brevis*
SC - - - - - - - - - NC - - - - - - - - - SC - - - - SC— NC	*Stylocheiron suhmi*
SC—TW— NC - - - - T— SC absent	no good example available
SC—TW— NC - - - - T— SC - - - - SC	*Stylocheiron robustum*
SC—TW— NC - - - - T— SC - - - - SC— NC	*Thysanopoda obtusifrons*
SC—TW— NC—— T— SC—— SC— NC	*Diacria trispinosa*

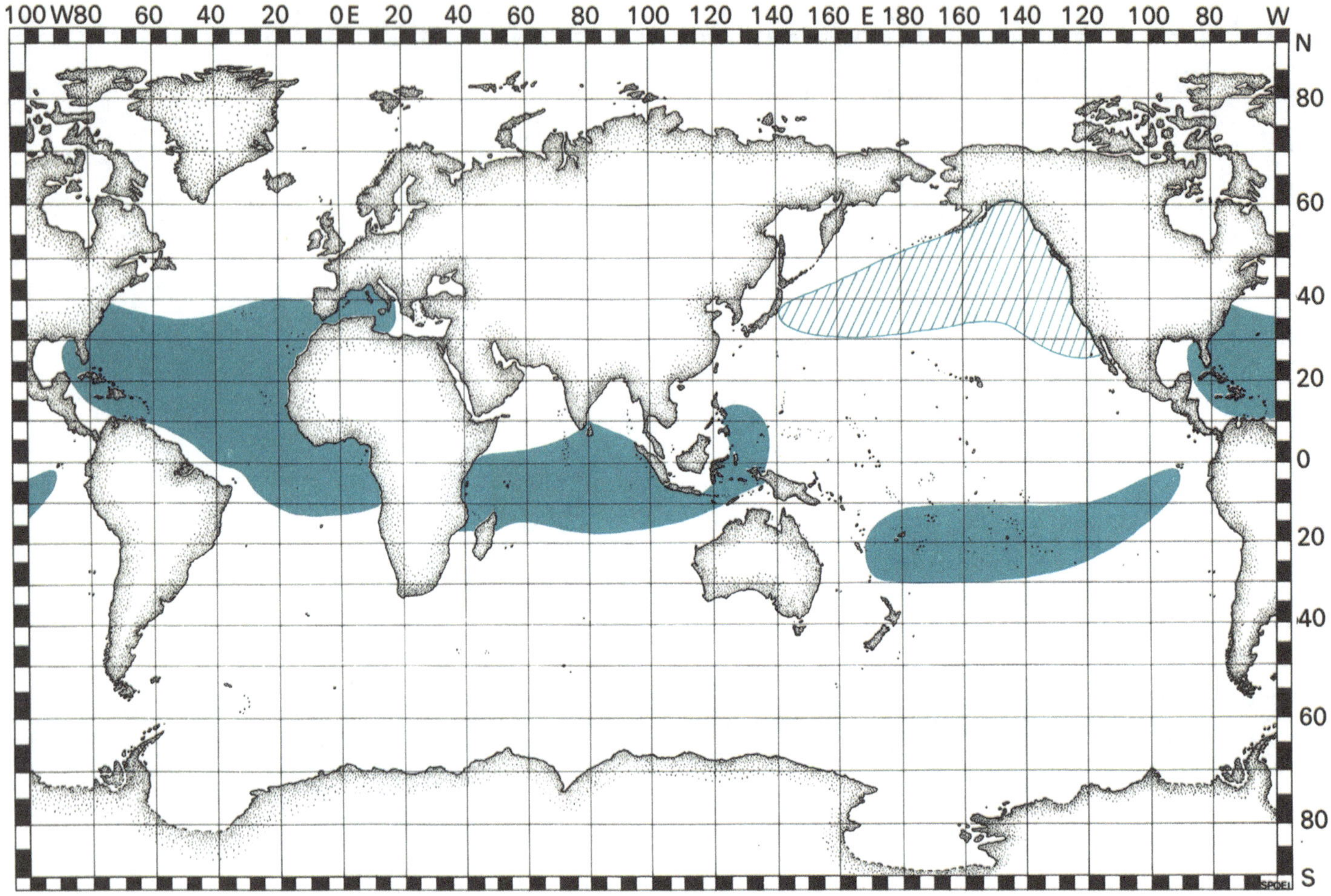

masses have large expatriation areas and small, usually neritic, reproductive ranges (Beklemishev, 1971). This restricted expatriation makes it possible to give a rigid outline (diagram II) of the Central water patterns. The distribution of Central water taxa represents a unique phenomenon and is probably the oldest pattern still recognisable. Real Central water distributions can date back to the Eocene. The knowledge of geological and climatological influences since the Eocene makes it possible to explain different Central water patterns, and to predict the possible and impossible patterns within the Central water masses of the world (cf. Van der Spoel, 1983). The Central water patterns are given in the following maps. Relations of Central water taxa with Transitional water masses are frequently found and they may be very complex as shown by *Pneumoderma atlantica* (fig. 88) and *Sagitta serratodentata* (fig. 118). In addition, there are taxa occurring in Central as well as Transitional waters (fig. 99).

EUPHAUSIA BREVOIDES type

Euphausia brevis (fig. 89) is a species which shows the most common distribution in Central waters; it is found in the North and South Central waters of all the oceans. This type of distribution was already mentioned in relation with *L. lesueuri* (p. 36).

FIG. 88 Distribution of *Pneumoderma atlanticum*. black = *P.a.* forma *atlanticum*, hatched = *P.a.* forma *pacificum*.

FIG. 89 Distribution of *Euphausia brevis*.

FIG. 90 Distribution of *Stylocheiron suhmi*.

STYLOCHEIRON SUHMOIDES type

Stylocheiron suhmi (fig. 90) is a representative of the common type of Central water distribution in which, however, the populations of the Atlantic Ocean expanded, so that the species occupies the entire Atlantic between 40°N and 40°S. This modification of the Central water pattern may be due to influences of the Ice-Ages (see p. 144). The specific characteristics of some of the following types of Central water distribution have probably also been effected by these influences.

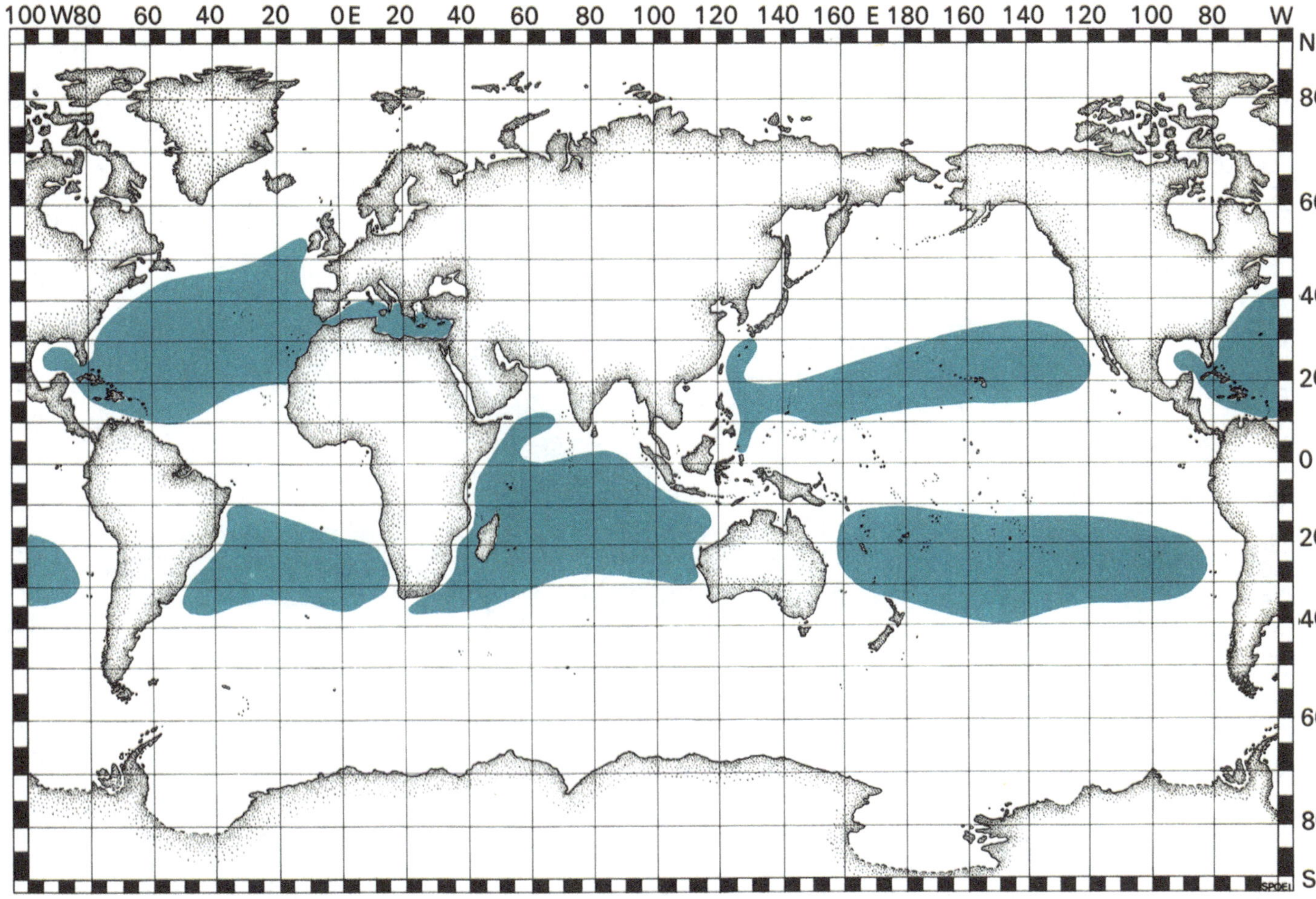

100 W80 60 40 20 0 E 20 40 60 80 100 120 140 160 E 180 160 140 120 100 80 W
N
80
60
40
20
0
20
40
60
80
S
SPOEL

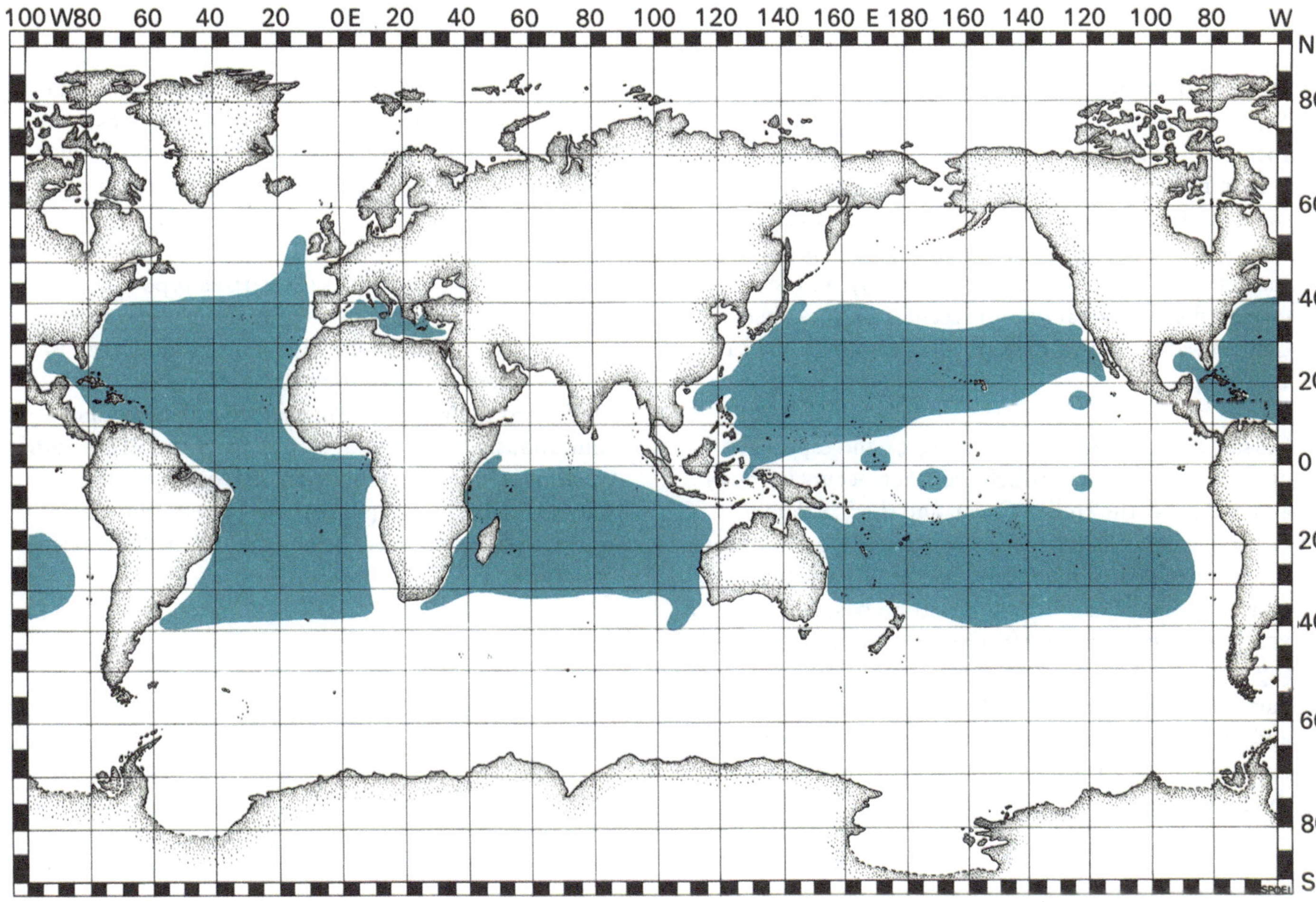

100 W80 60 40 20 0 E 20 40 60 80 100 120 140 160 E 180 160 140 120 100 80 W
N
80
60
40
20
0
20
40
60
80
S
SPOEL

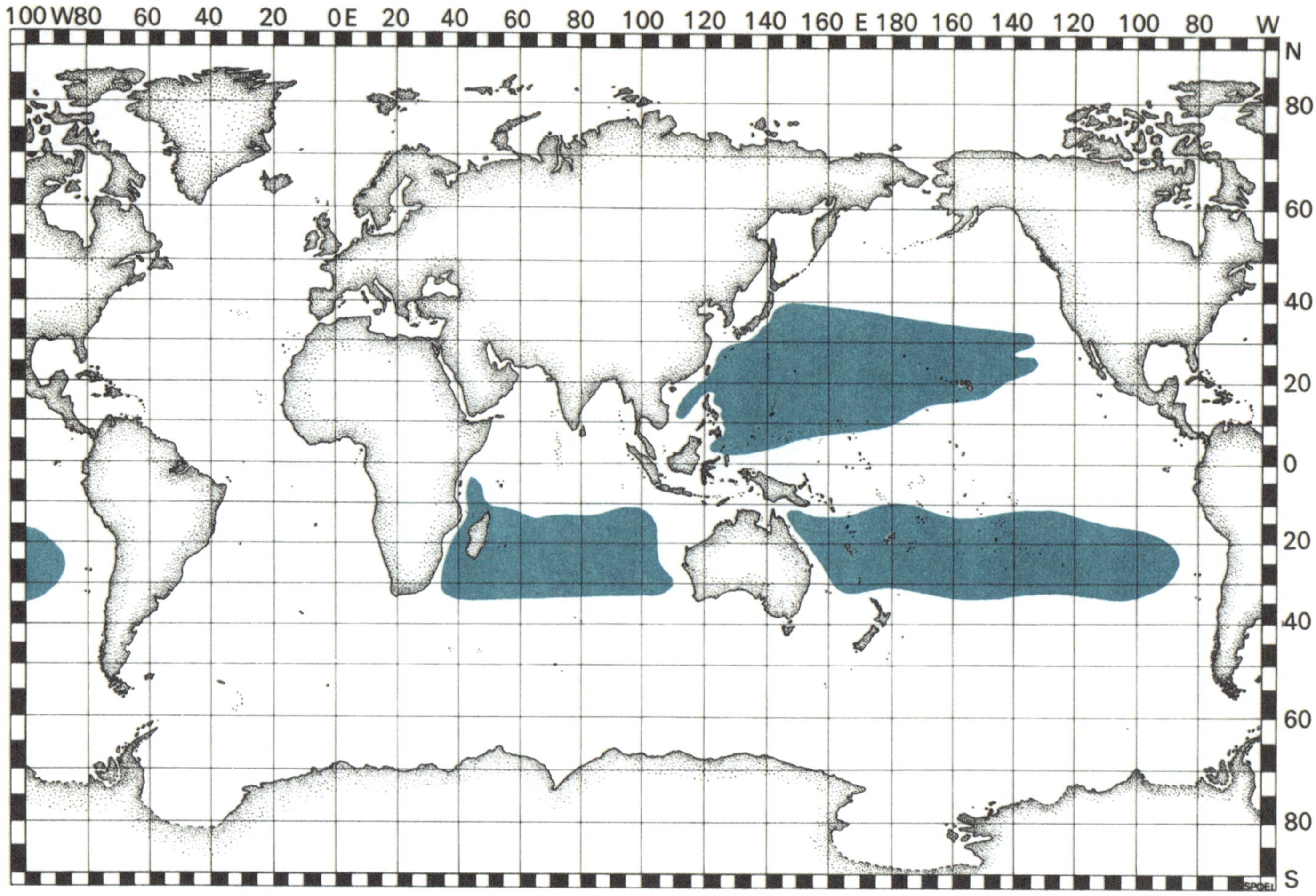

STYLOCHEIRON AFFINOIDES type

Stylocheiron affine 'Central form' (fig. 91) shows another aberration in the Atlantic Ocean: it is completely absent. It is an Indo-Pacific form which probably did not succeed in reaching the Atlantic. Phylogenetically this form is much younger than for example the preceding species and those of the figs 93, 94 and 97, and so it forms a distributional type different from the other Central water types, which are all of an old geological age (p. 144).

Closely related to the pattern of fig. 91 and especially fig. 90 is the pattern in which the taxon occurs in the N. and S. Pacific, the Indian Ocean and the S. Atlantic Central waters. As can also be concluded from some of the following patterns, absence in the N. Atlantic is logical. A good example of this pattern has, however, not yet been found. The EUPHAUSIA RECURVOIDES type (fig. 99) resembles this hypothetical type, but major parts of its range are situated in Transitional waters; and, in addition, more penetration into the Tropics is expected, so that the hypothetical figure 92 must do for the time being.

FIG. 91 Distribution of *Stylocheiron affine* 'Central form' representing the STYLOCHEIRON AFFINOIDES type.
FIG. 92 Hypothetical Central-water type of distribution.
FIG. 93 Distribution of *Thysanopoda obtusifrons*.

THYSANOPODA OBTUSIFRONSOIDES type

This type, represented by *Thysanopoda obtusifrons* (fig. 93), seems to be more widely distributed than in Central waters only. In the Pacific it lives in W. Equatorial waters and in the Atlantic it occurs continuously between 40°N and 30°S and in the Indian Ocean it is found also slightly north of the Central waters. As discussed (p. 144), this penetration into warmer waters may also be an effect of the Pleistocene Ice-Ages.

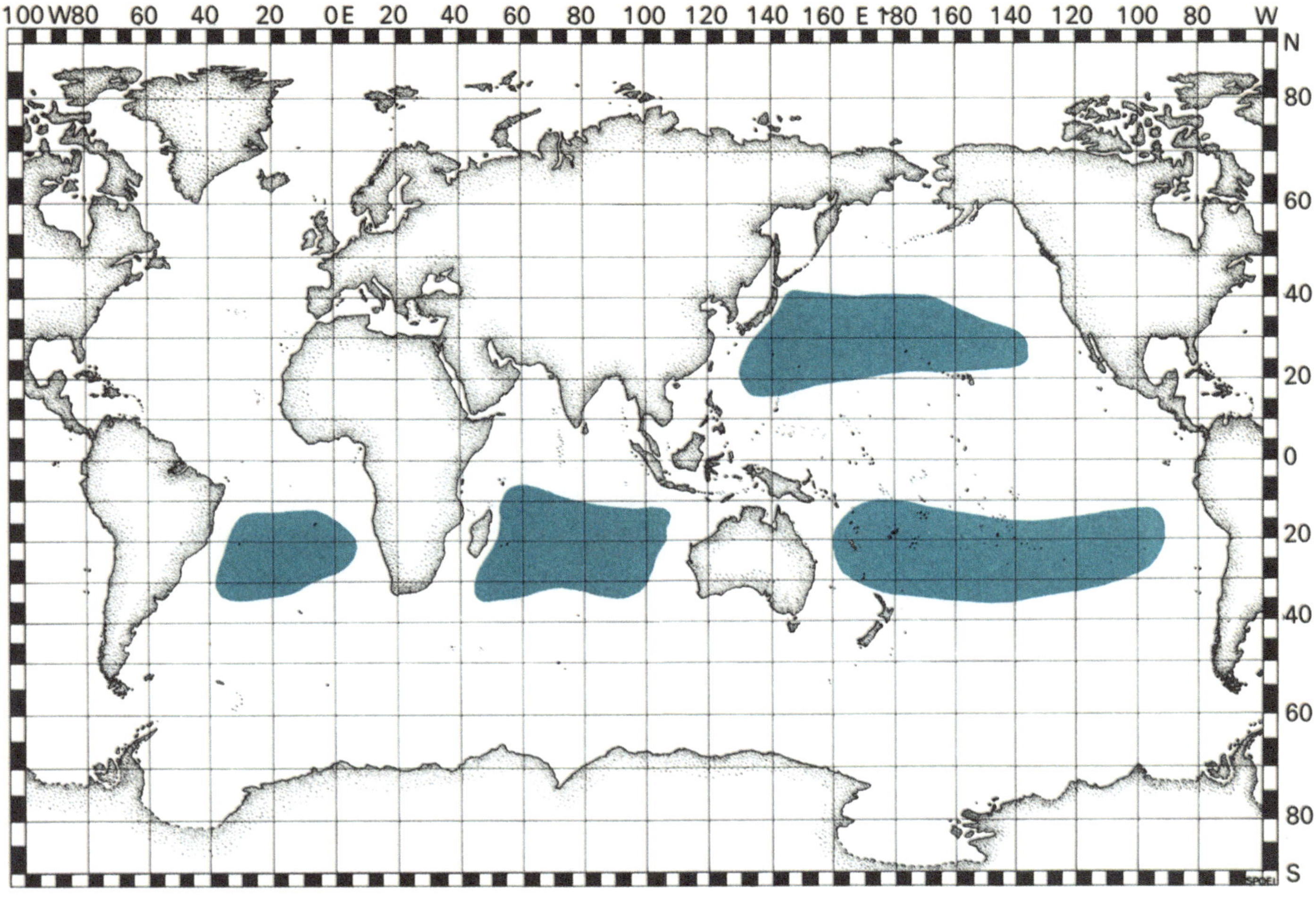
100 W80 60 40 20 0E 20 40 60 80 100 120 140 160 E 180 160 140 120 100 80 W
N
80
60
40
20
0
20
40
60
80
S

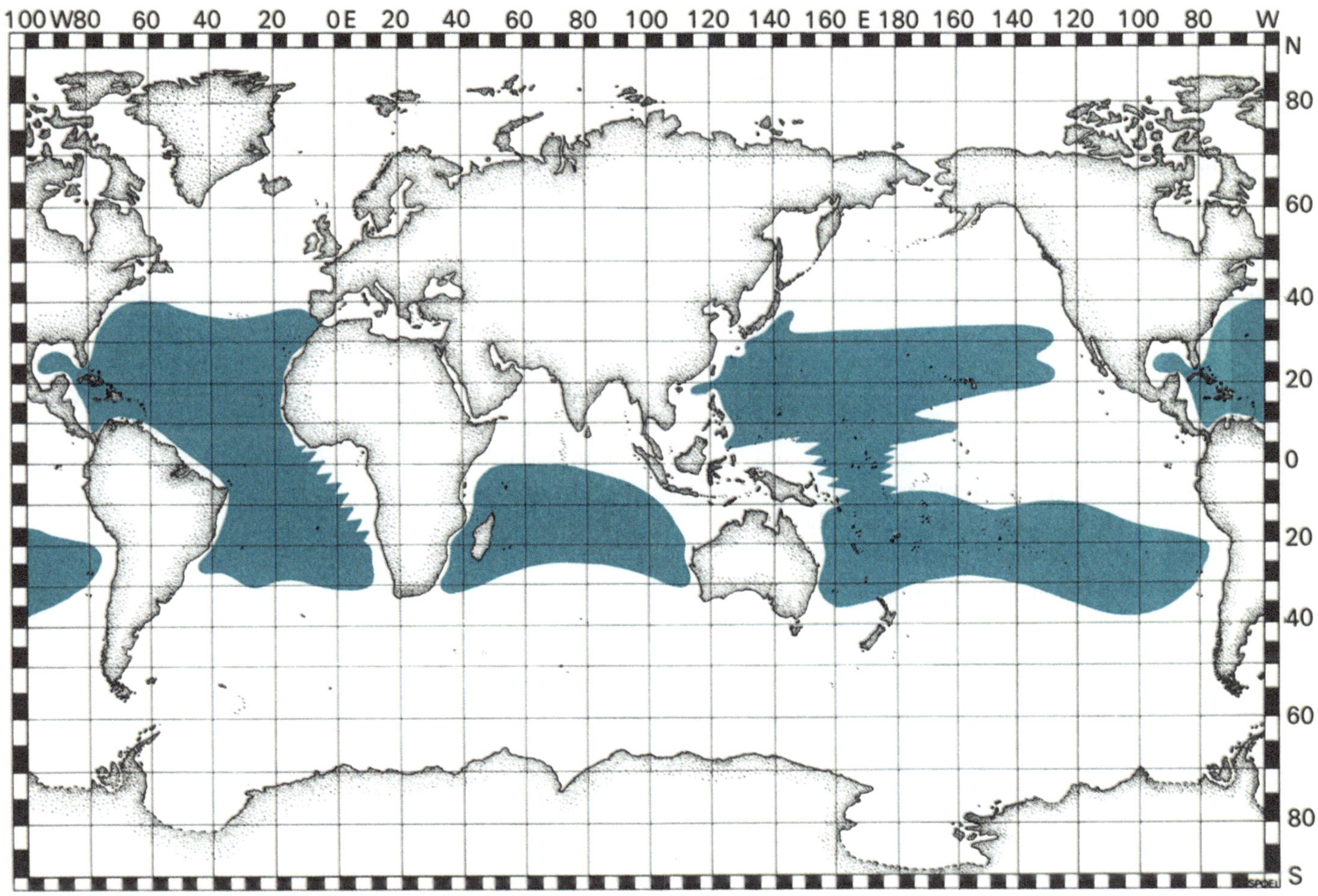
100 W80 60 40 20 0E 20 40 60 80 100 120 140 160 E 180 160 140 120 100 80 W
N
80
60
40
20
0
20
40
60
80
S

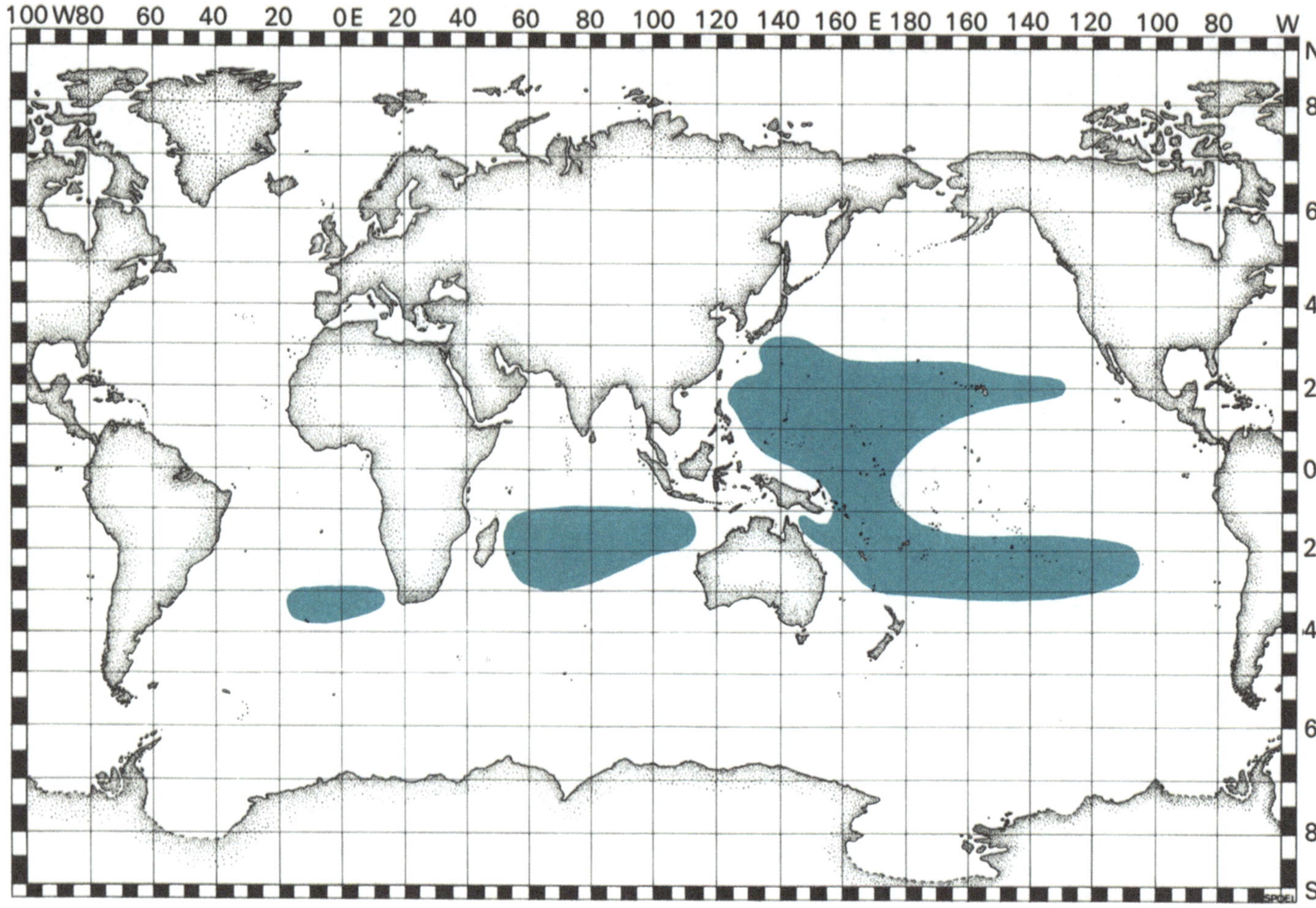

STYLOCHEIRON ROBUSTOIDES type
This pattern resembles most the hypothetical type (fig. 92). The present pattern, represented by *Stylocheiron robustum* (fig. 94), occurs in the W. Equatorial waters of the Pacific which indicates an Ice-Ages influence absent in the pattern of fig. 92. So, if the patterns of figs 92 and 94 both exist, the one in fig. 92 is in all probability much younger than the STYLOCHEIRON ROBUSTOIDES type.

DIACRIA TRISPINOSOIDES type
Diacria trispinosa trispinosa (fig. 95) shows a Central water type of distribution occurring in the N. and S. Central and the Equatorial waters of the Pacific, and throughout the Indian and Atlantic Ocean between 40°N and 40°S. This species forms one species-group together with the two species in fig. 96.

DIACRIA ATLANTICOIDES type
The form *Diacria trispinosa atlantica* (fig. 95) is restricted to the N.E. Atlantic, a pattern resembling that of e.g. *Cavolinia tridentata* forma *dakarensis* (fig. 53).

FIG. 94 Distribution of *Stylocheiron robustum*.
FIG. 95 Distribution of *Diacria trispinosa*. hatched = *D.t. atlantica*, black = *D.t. trispinosa*.
FIG. 96 Distribution of *Diacria rampali* (black) and *D. major* (hatched).

DIACRIA MAJOROIDES type
Diacria major (fig. 96) represents the type of distribution covering the Central waters except for those in the S. Atlantic Ocean. *Diacria rampali*, of the EUPHAUSIA PSEUDOGIBBOIDES type (fig. 77) with wide extension into the S. Atlantic, is given here to complete the survey of the species-group.

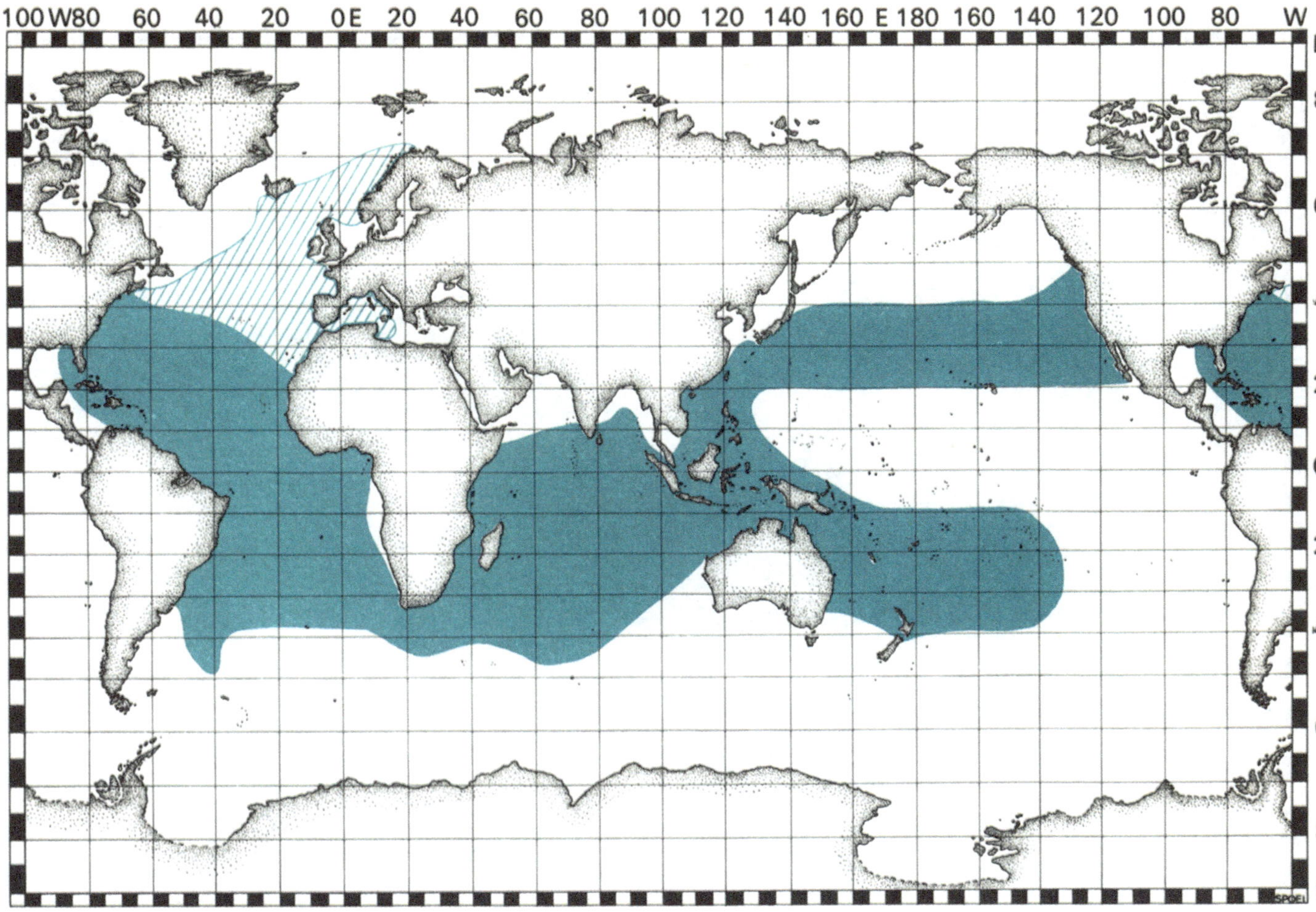
100 W 80 60 40 20 0 E 20 40 60 80 100 120 140 160 E 180 160 140 120 100 80 W
N 80 60 40 20 0 20 40 60 80 S

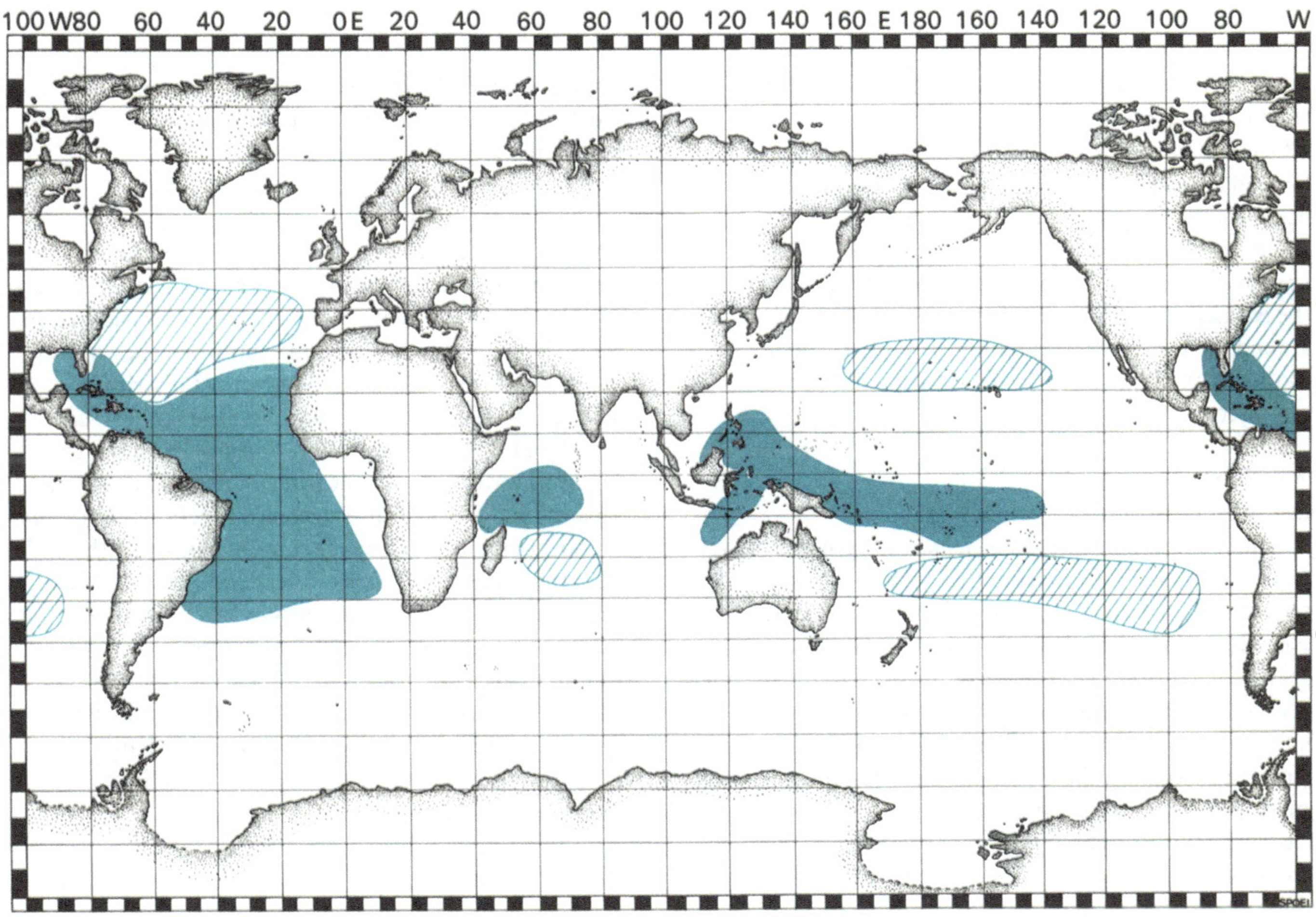
100 W 80 60 40 20 0 E 20 40 60 80 100 120 140 160 E 180 160 140 120 100 80 W
N 80 60 40 20 0 20 40 60 80 S

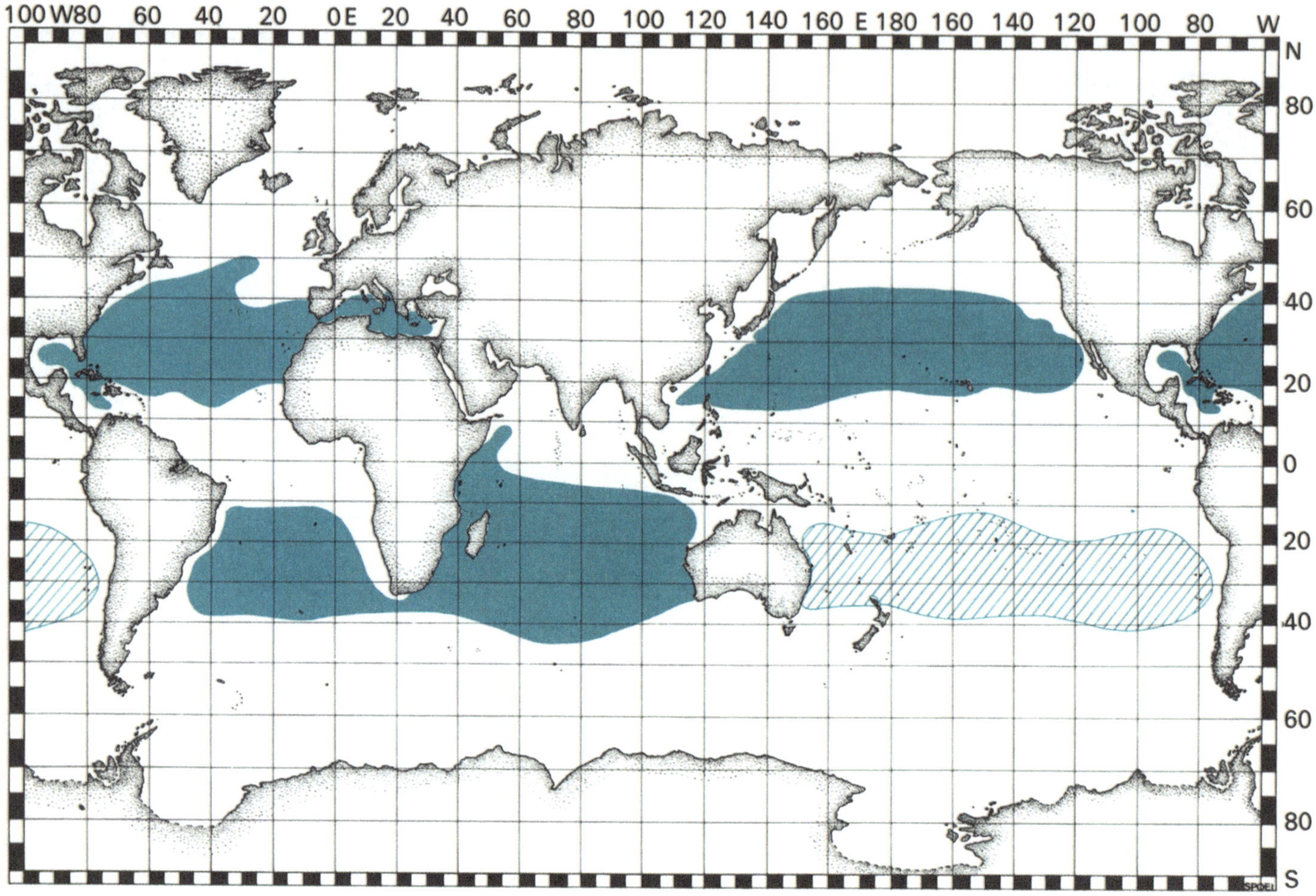

EUPHAUSIA HEMIGIBBOIDES type and EUPHAUSIA GIBBOIDES type

The more complicated Central water patterns are formed by pairs of taxa, as in fig. 97. *Euphausia hemigibba* occupies the Central waters except in the S. Pacific, where *E. gibba* lives. The two taxa are closely related and speciation in the S. Pacific is explained by the fact that the S. Pacific Central waters occupy an isolated position among the Central waters. In the STYLOCHEIRON AFFINOIDES type (fig. 91), this isolation did not have any effect.

TROPICAL PATTERNS

CAVOLINIA GLOBULOSOIDES type

The distribution of *Cavolinia globulosa* (fig. 98) is closely related to that of *Clio convexa* (fig. 57) and also somewhat to that of *Euphausia tenera* (fig. 75), which are both considered to be beltshaped (cf. also *Melamphaes spinifer*, fig. 116). There seems to be no relation with the HYDROMYLES GLOBULOSOIDES type, a pattern situated around the Indo-Malayan Archipelago (fig. 112), although there do exist some resemblances. In principle, the present pattern is a 30°N-30°S distribution throughout the Indo-Pacific Oceans except for a hiatus near 140°W. *Pontellina morii* (fig. 98) is another typical representative of this pattern. Characteristic of this type is a strong penetration into the Agulhas Current and a very low abundance in the W. Pacific populations.

CLAUSOCALANUS FARRANOIDES type

Taxa living between 30°N and 30°S meet the African continent as a barrier and therefore they cannot disperse into the Atlantic Ocean. They cannot be considered to have a beltshaped distribution even when they occupy the entire Indo-Pacific belt. A group of distribution patterns, well-exemplified by *Clausocalanus farrani* (fig. 56), also shows a complete absence from the Indo-Malayan waters. This type is characterized by a great isolating influence of interocean connections: *C. farrani* cannot pass south of Cape Town, nor through the Indo-Malayan waters. *Eucalanus attenuatus* (fig. 54) is also considered to belong to this group as it fails to penetrate the Atlantic, though populations are present in the Agulhas Current. The SAGITTA PACIFICOIDES type (fig. 117) is another example of restricted Indo-Pacific distributions; it is, however, much wider than the present type and there are indications that it has a different origin (cf. p. 92).

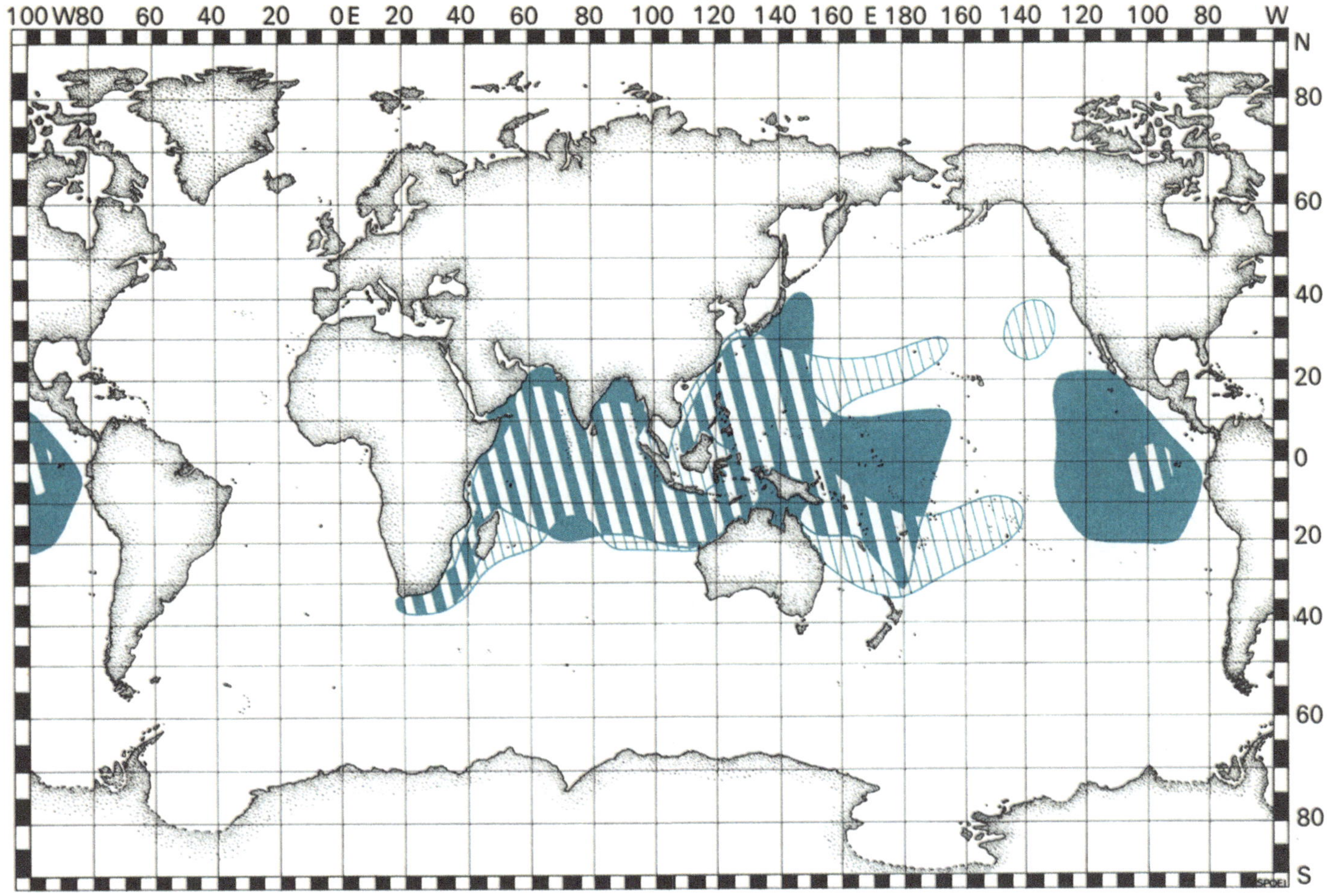

CLAUSOCALANUS MINOROIDES type

This type, represented by *Clausocalanus minor* (fig. 55) is roughly identical with the preceding type and differs only essentially in its total absence from the Indo-Malayan waters, and in a further penetration into the boundary currents. It appears to be a more oceanic species.

FIG. 97 Distribution of *Euphausia hemigibba* (black) and *E. gibba* (hatched).

FIG. 98 Distribution of *Pontellina morii* (black) and *Cavolinia globulosa* (hatched).

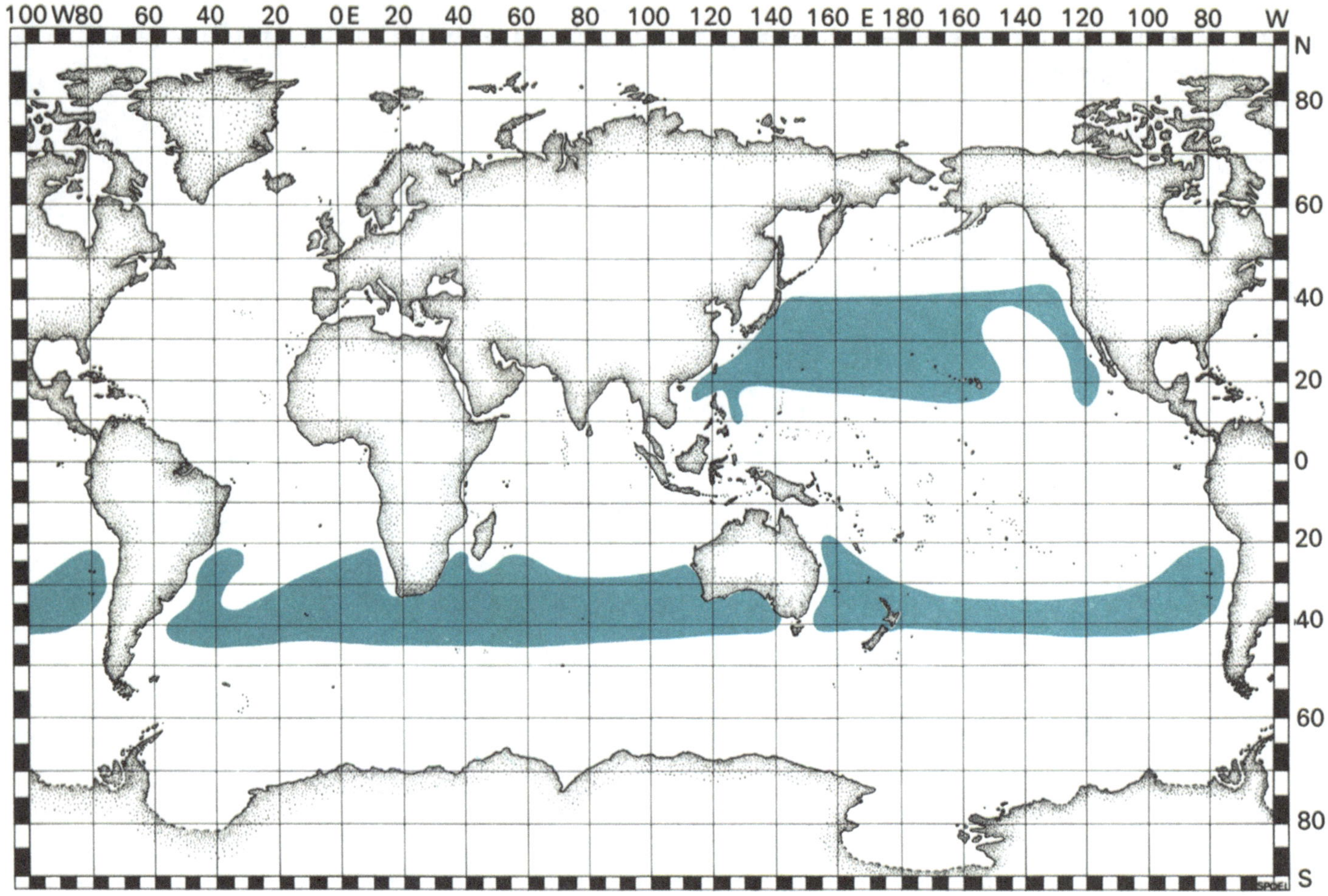

COLD-WATER PATTERNS

EUPHAUSIA RECURVOIDES type

The rather strange distribution shown by *Euphausia recurva* (fig. 99) and other taxa such as *Winteria telescopa* (fig. 100), can be considered to have derived from either Central waters or Transitional waters of the beltshaped type. This type covers the southern Transitional belt in all the oceans and the Central waters of the N. Pacific Ocean. Some resemblance exists between this pattern and that of the EUPHAUSIA SIMILOIDES type (fig. 85).

Most of the other cold-water patterns which are more or less beltshaped are occupied by neritic species (p. 100).

Interocean connections

Taxa living in the warm-water belt between 30°N and 30°S meet the continents as isolating elements, and when they are not related to taxa distributed over the cold-water belts, there must have been connections for these taxa with relatives in other oceans by pathways now blocked. There have been three typical pathways for warm-water plankton, viz.: a) via the Panama area before closure, b) via the now completely vanished

FIG. 99 Distribution of *Euphausia recurva*.
FIG. 100 Distribution of *Winteria telescopa*.
FIG. 101 Distribution of the *Sergestes edwardsi* species-group (see also fig. 102). 1 = *S. orientalis*, 2 = *S. tantillus*, 3 = *S. brevispinatus*, 4 = *S. edwardsi*.

inland seas of Peru, Columbia and Brazil, and c) via the Strait of Timor, and probably around S. Australia.

The geological connection between the Atlantic and Pacific Oceans via the Panama area is best shown by the *Sergestes edwardsi* group.

In the *Sergestes edwardsi* group (figs 101 & 102), *S. brevispinatus* and *S. edwardsi* are more closely related to each other than to the other representatives of the group. Two pairs of species are nearly sympatric, viz.: *S. geminus* with *S. brevispinatus* and *S. gibbilobatus* with *S. tantillus*. According to Judkins (1978) the affinity matrix proves that there are subgroups within the species-group and the sympatric taxa belong to different subgroups.

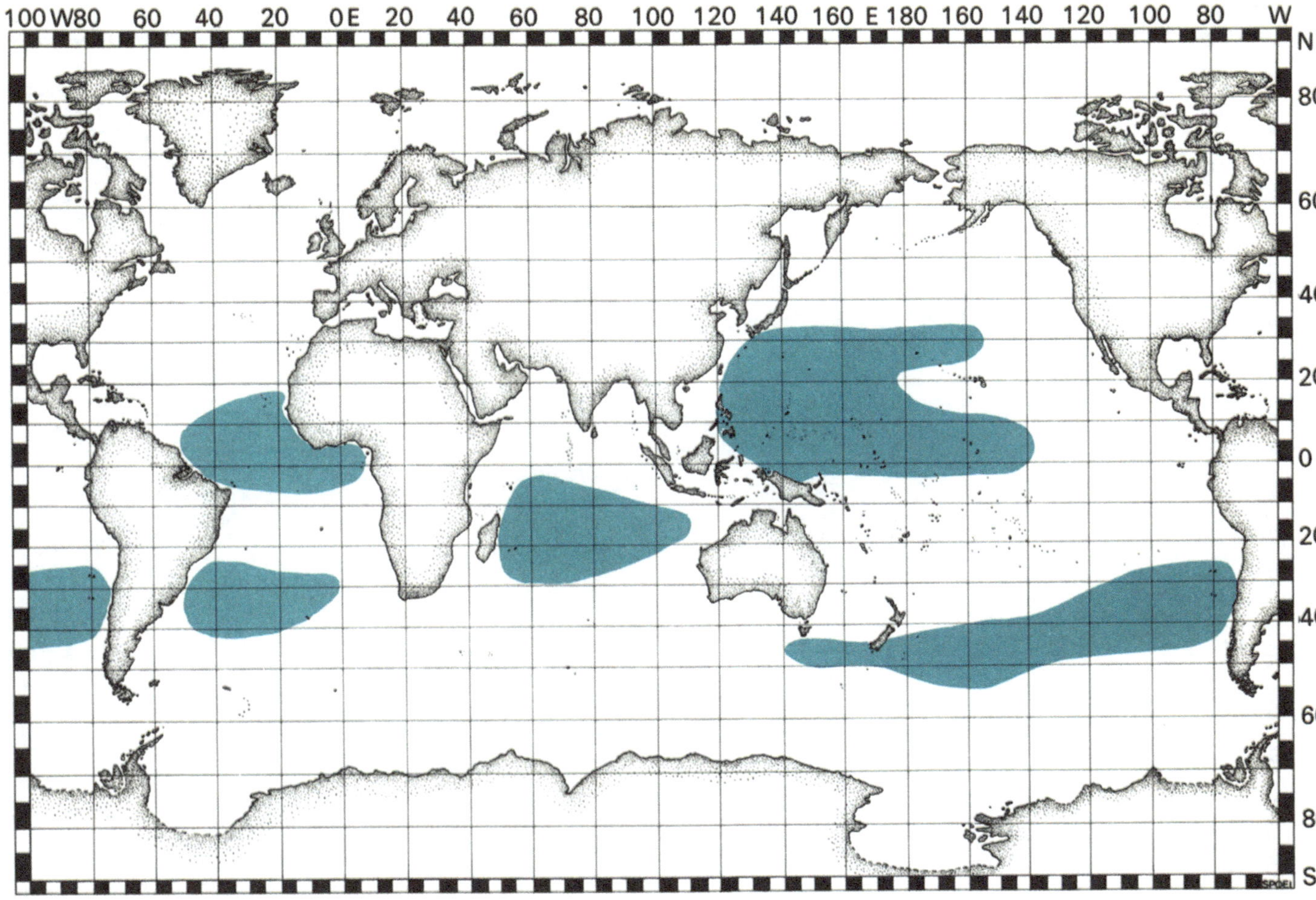
100 W80 60 40 20 0 E 20 40 60 80 100 120 140 160 E 180 160 140 120 100 80 W
N 80 60 40 20 0 20 40 60 80 S

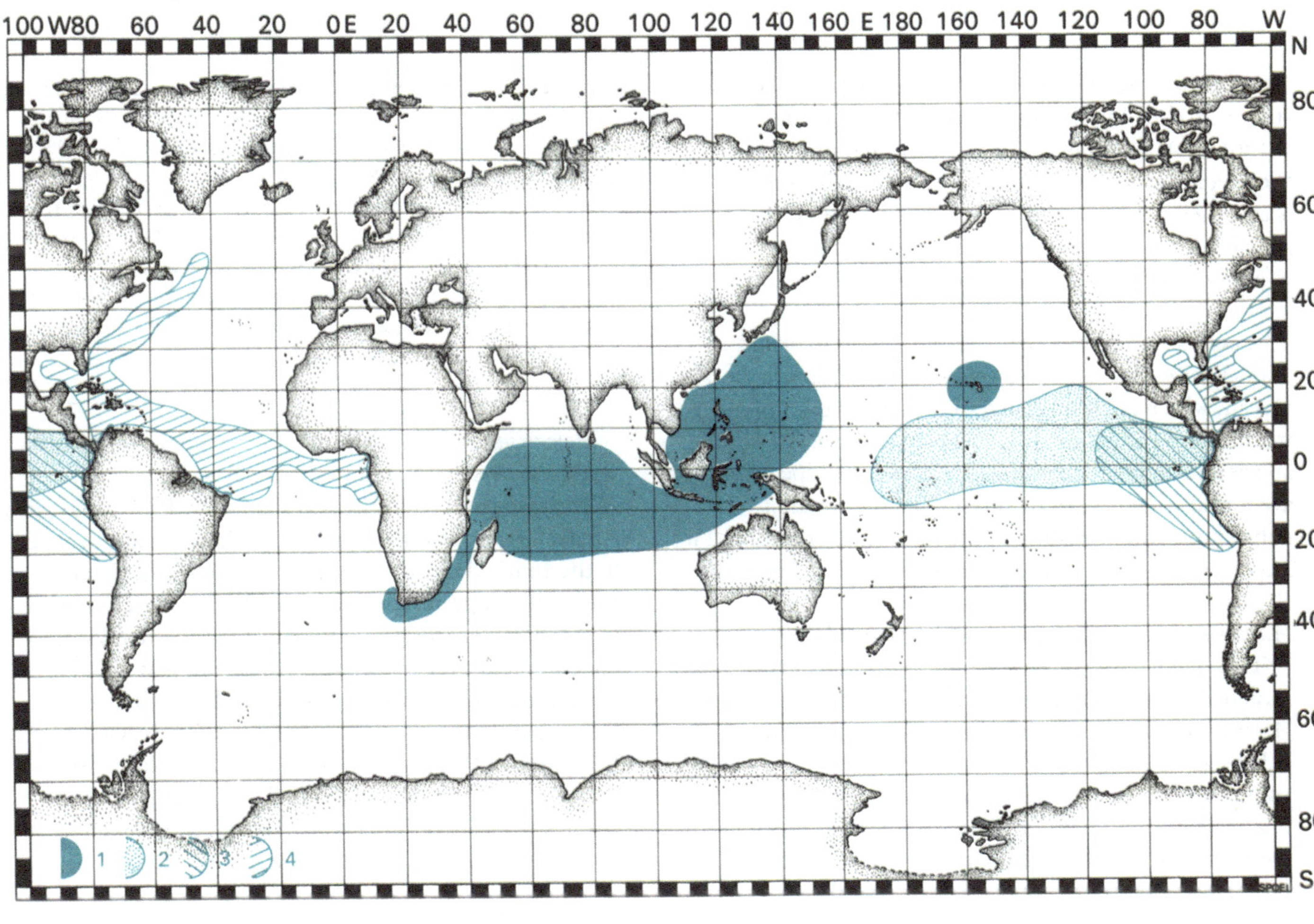
100 W80 60 40 20 0 E 20 40 60 80 100 120 140 160 E 180 160 140 120 100 80 W
N 80 60 40 20 0 20 40 60 80 S
1 2 3 4

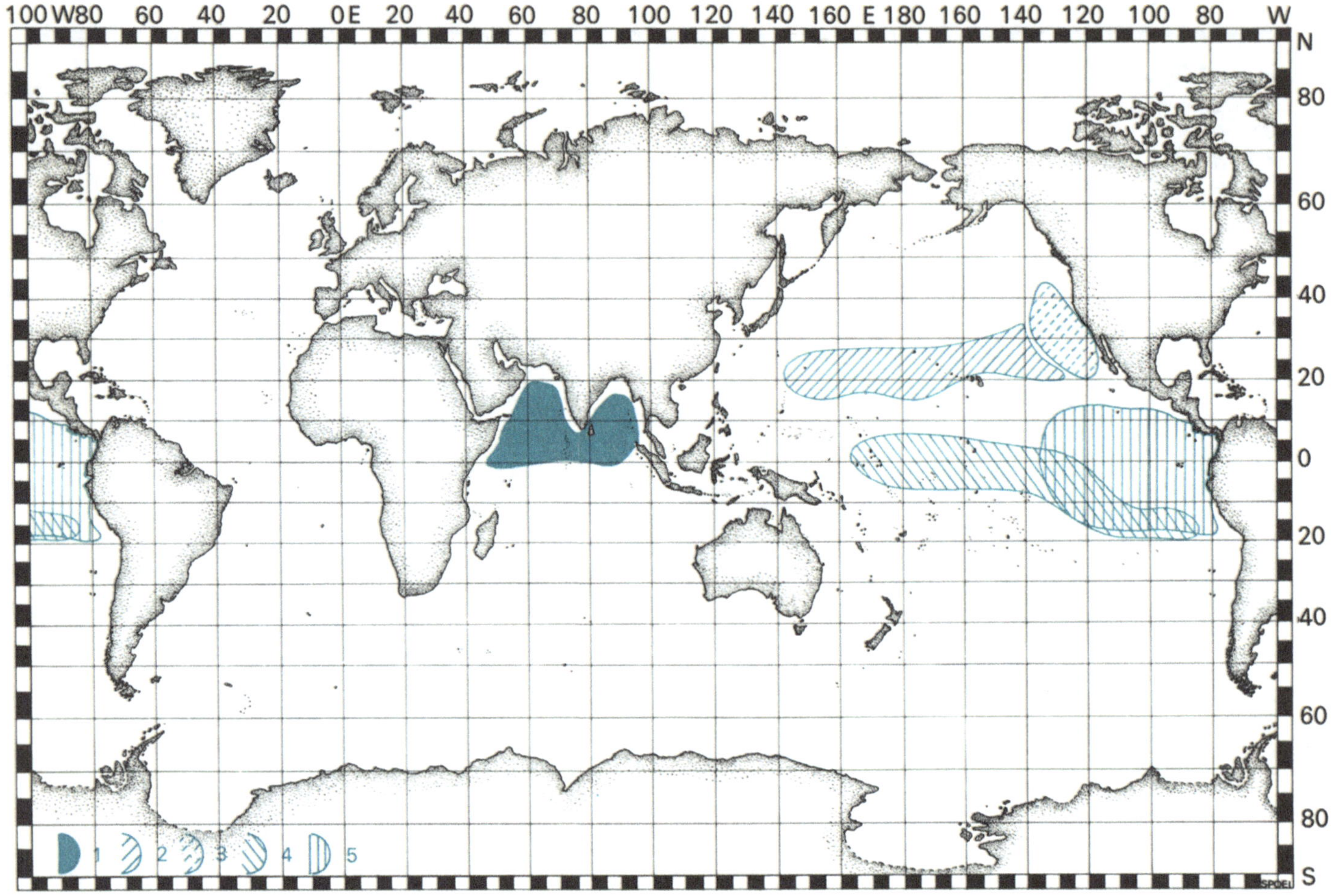

S.edwardsi, *S.brevispinatus* and *S.consobrinus* form one subgroup showing the distribution of the EUPHAUSIA GIBBOIDOIDES type (fig. 107). The remaining taxa of the species-group are Indo-Pacific in distribution and it is accepted that the Indo-Pacific group, and more specifically *S.brevispinatus*, split off the Atlantic *S.edwardsi* before the closure of the Panama Isthmus.

The genus *Acetes* (figs 103, 104 and 105) is in origin Indo-Pacific with only two species in the Atlantic Ocean, which are less related to each other than to other members of the genus. It is accepted that the Atlantic species developed separately. In *Acetes*, two species-groups are found which are to be subdivided into smaller subgroups, which results in the following structure: The *A.erythraeus* species-group composed of the *A.marinus* + *A.paraguayensis* subgroup and the *A.erythraeus* + *A.intermedius* + *A.sibogae* + *A.vulgaris* subgroup.

The *A.japonicus* species-group composed of the *A.indicus* subgroup, the *A.serrulatus* + *A.johni* + *A.natalensis* subgroup, the *A.japonicus* + *A.chinensis* subgroup and the *A.americanus* + *A.binghami* subgroup. The close relation of *A.americanus* and *A.binghami*, and their distributions, point to a former connection through the Panama area. The Atlantic populations are therefore of a Pacific origin.

FIG. 102 Distribution of the *Sergestes edwardsi* species-group (see also fig. 101). 1 = *S.semissis*, 2 = *S.consobrinus* 'Central form', 3 = *S.consobrinus* 'California Current form', 4 = *S.gibbilobatus*, 5 = *S.geminus*.

FIG. 103 Distribution of the genus *Acetes* (see also figs 104 & 105). 1 = *A.binghami*, 2 = *A.a.carolinae*, 3 = *A.a.americanus*, 4 = *A.chinensis*, 5 = *A.indicus*, 6 = *A.japonicus*.

FIG. 104 Distribution of the genus *Acetes* (see also figs 103 & 105). 1 = *A.serrulatus*, 2 = *A.s.sibogalis*, 3 = *A.s.sibogae*, 4 = *A.s.australis*.

A.marinus, with a high tolerance of slightly brackish waters, is closely related to the fresh-water species *A.paraguayensis* from the Rio Paraguay, Rio Nunay, Rio Amazonas and Rio Paraná, and the latter is most related to the Indo-Malayan *A.erythraeus* subgroup. In all probability a population of some Pacific representative of the *A.erythraeus* group penetrated the inland sea, once situated over the western Brazilian plate. This inland sea and the Amazonas drained into the Pacific Ocean before the rise of the Andes. After the closure of this outflow, this population became an isolated fresh-water taxon, later on splitting off the Atlantic Ocean species *A.marinus*. This is an example of the second Pacific-Atlantic interocean connection.

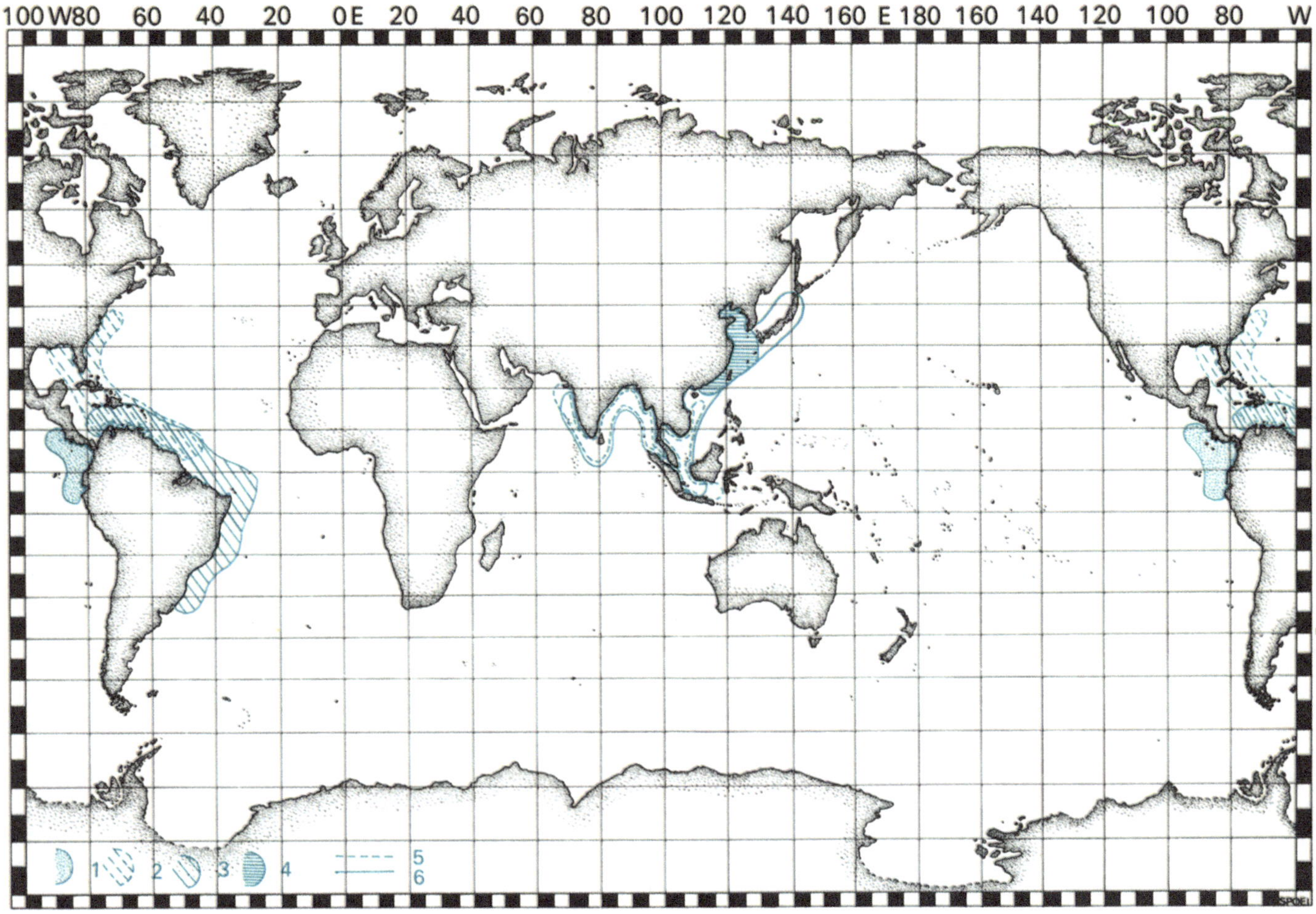
100 W 80 60 40 20 0 E 20 40 60 80 100 120 140 160 E 180 160 140 120 100 80 W
N 80 60 40 20 0 20 40 60 80 S
1 2 3 4
5
6

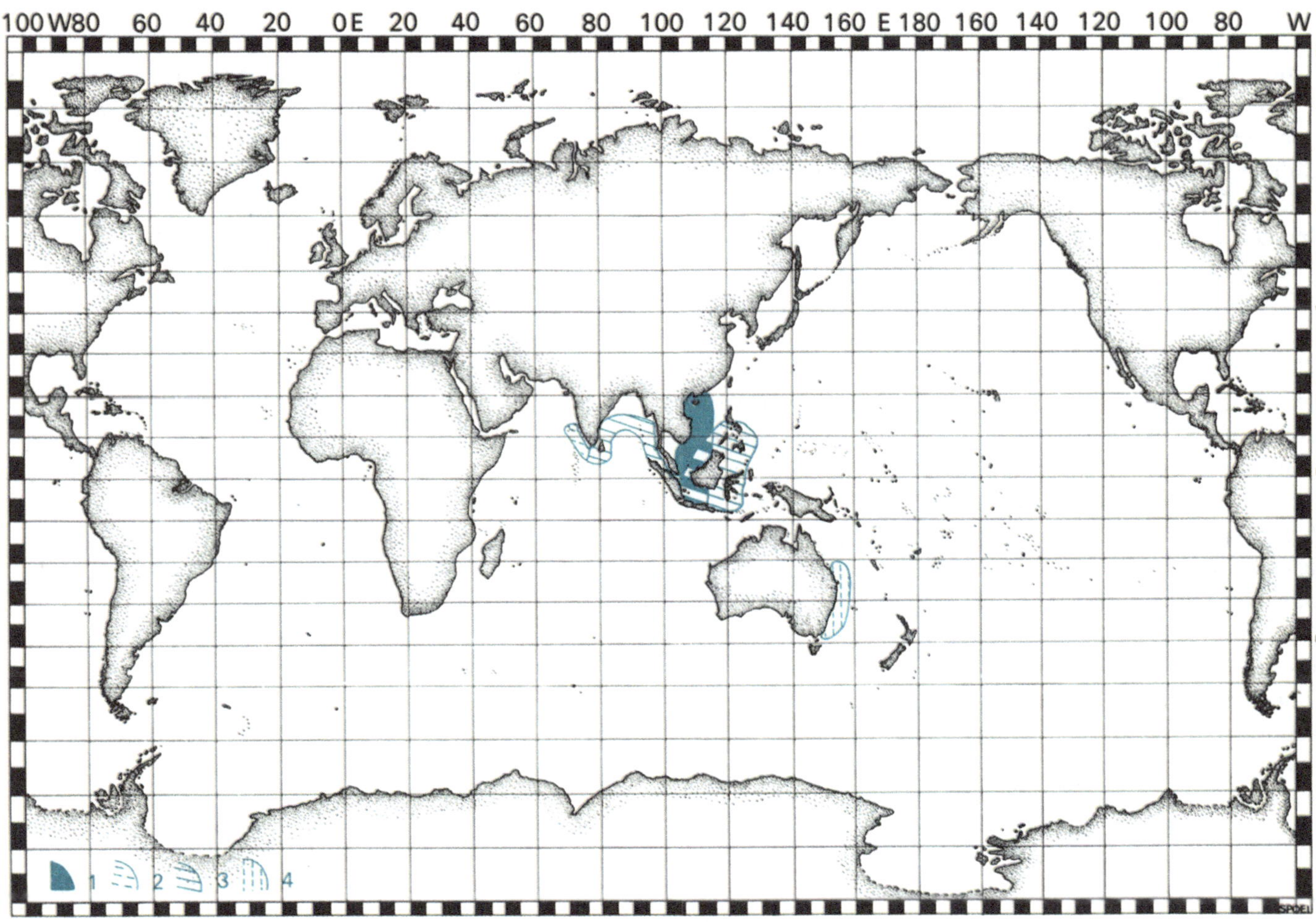
100 W 80 60 40 20 0 E 20 40 60 80 100 120 140 160 E 180 160 140 120 100 80 W
N 80 60 40 20 0 20 40 60 80 S
1 2 3 4

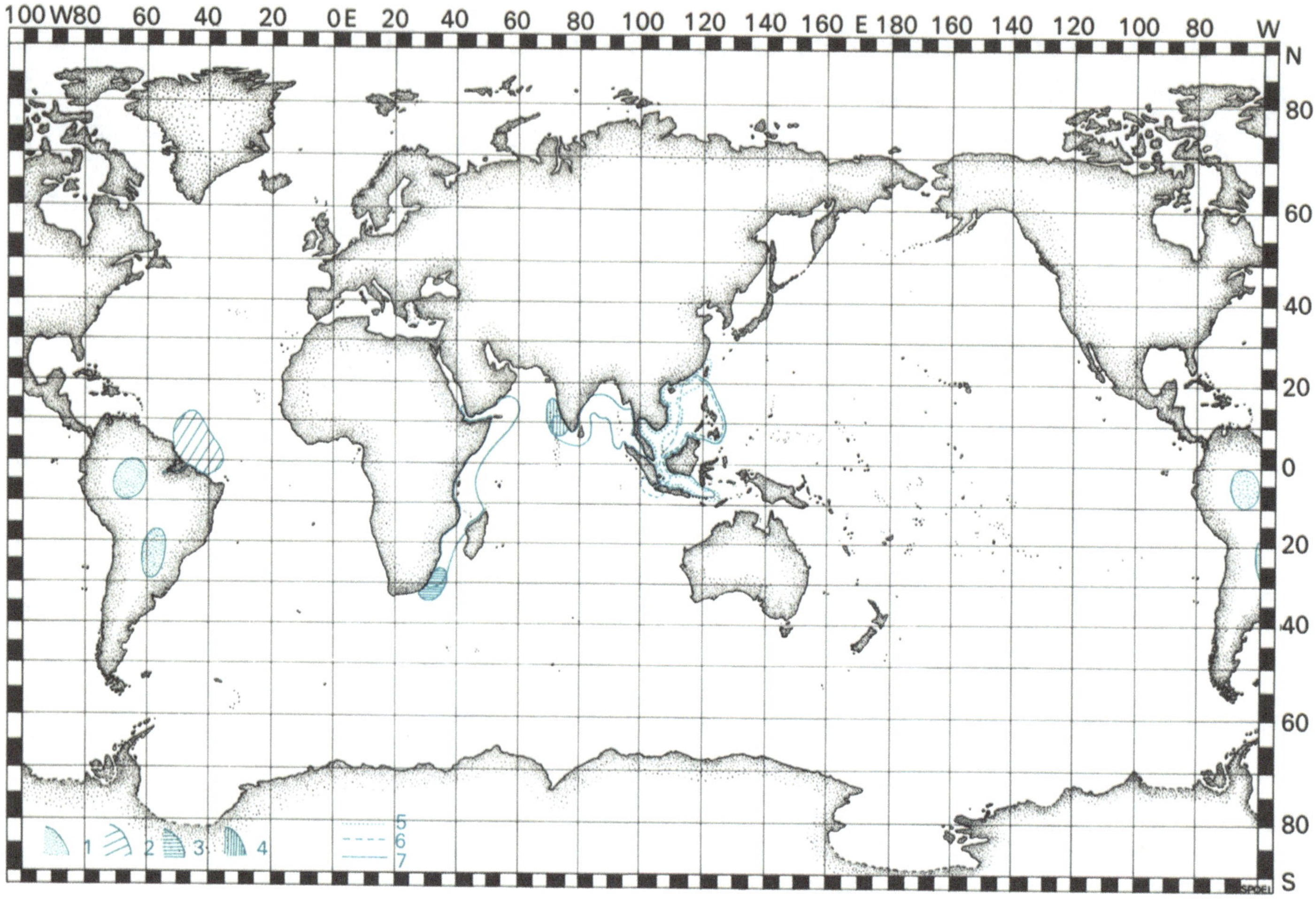

COROLLA SPECTABILOIDES type

Figure 106 illustrates that the former Panama connection has not always resulted in speciation into two different taxa in the two oceans and that different closely related Panamerican taxa are not always to be explained by the Panama phenomenon.

Corolla spectabilis has not (yet) developed taxonomically different populations at both sides of the isthmus, after it was split. *Pontellina platychella* and *P. sobrina* are closely related but they have in all probability never connected via the Panama area; they are members of the *P. plumata* species-group (cf. figs 98, 106 & 115), a group comparable with the *Clausocalanus arcuicornis* (figs 55 & 56) and the *Eucalanus attenuatus* (fig. 54) groups. In such groups circum-global species usually provide the possibility to develop different offshoots in the different oceans.

EUPHAUSIA GIBBOIDOIDES type

The distribution of *Euphausia gibboides* (fig. 107) represents a special type of Panama connection as this species occurs in the Atlantic and E. Pacific Tropical waters but also in the Pacific N. Transitional waters.

FIG. 105 Distribution of the genus *Acetes* (see also figs 104 & 105). 1 = *A. paraguayensis*, 2 = *A. marinus*, 3 = *A. natalensis*, 4 = *A. johni*, 5 = *A. vulgaris*, 6 = *A. intermedius*, 7 = *A. erythraeus*.

FIG. 106 Distribution of *Corolla spectabilis* (1), *Pontellina platychela* (2) and *P. sobrina* (3).

FIG. 107 Distribution of *Euphausia gibboides*.

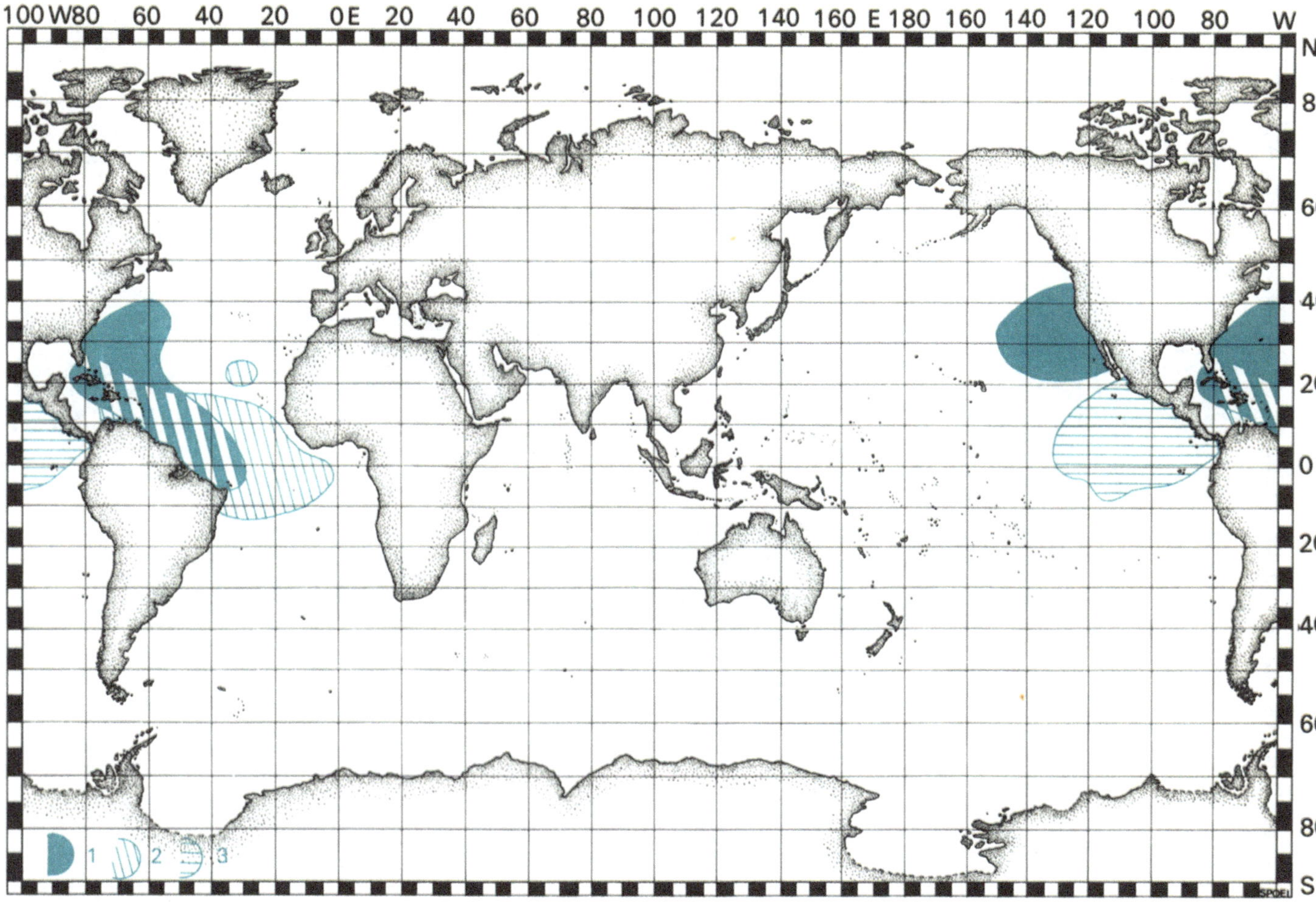
100 W80 60 40 20 0 E 20 40 60 80 100 120 140 160 E 180 160 140 120 100 80 W
N
80
60
40
20
0
20
40
60
80
S
1
2
3

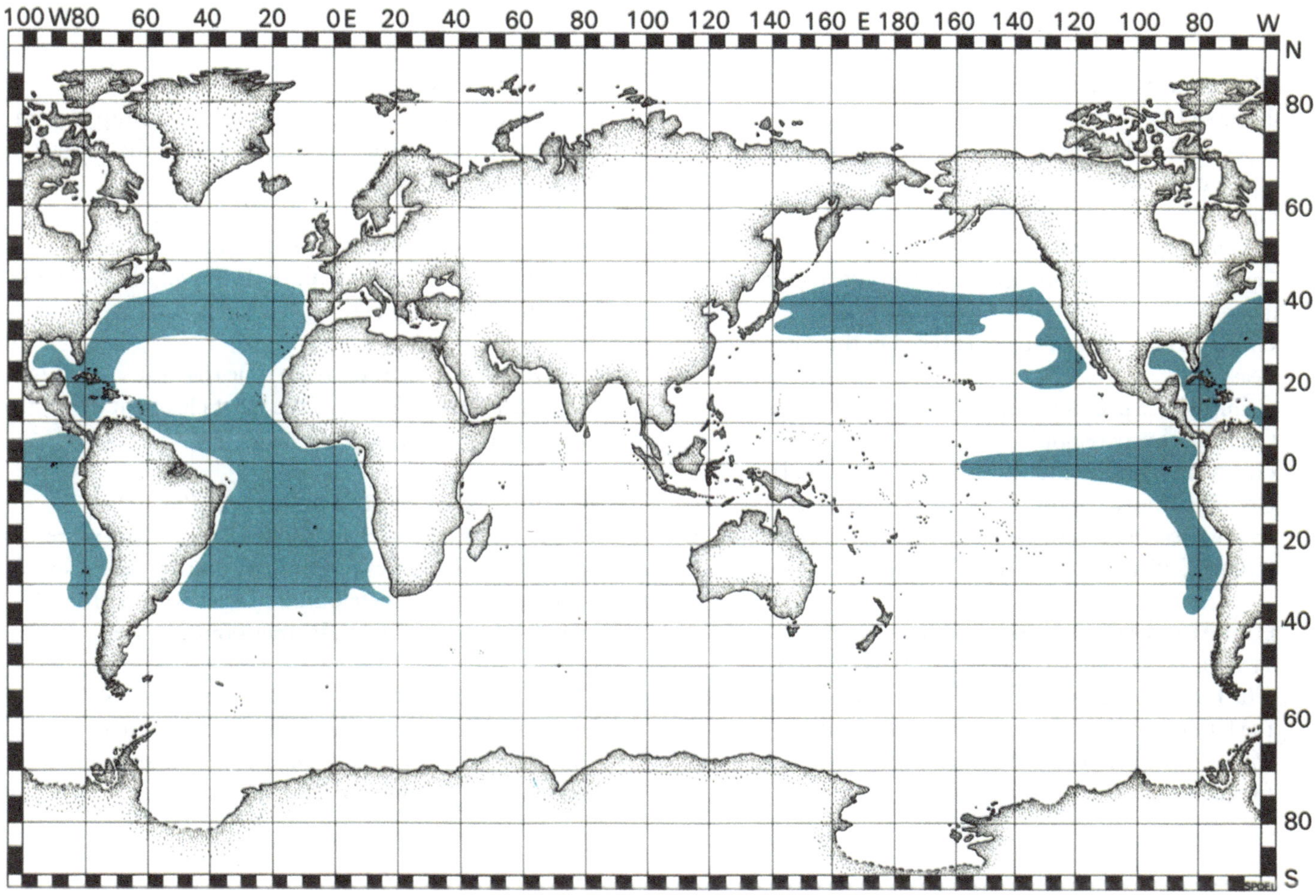
100 W80 60 40 20 0 E 20 40 60 80 100 120 140 160 E 180 160 140 120 100 80 W
N
80
60
40
20
0
20
40
60
80
S

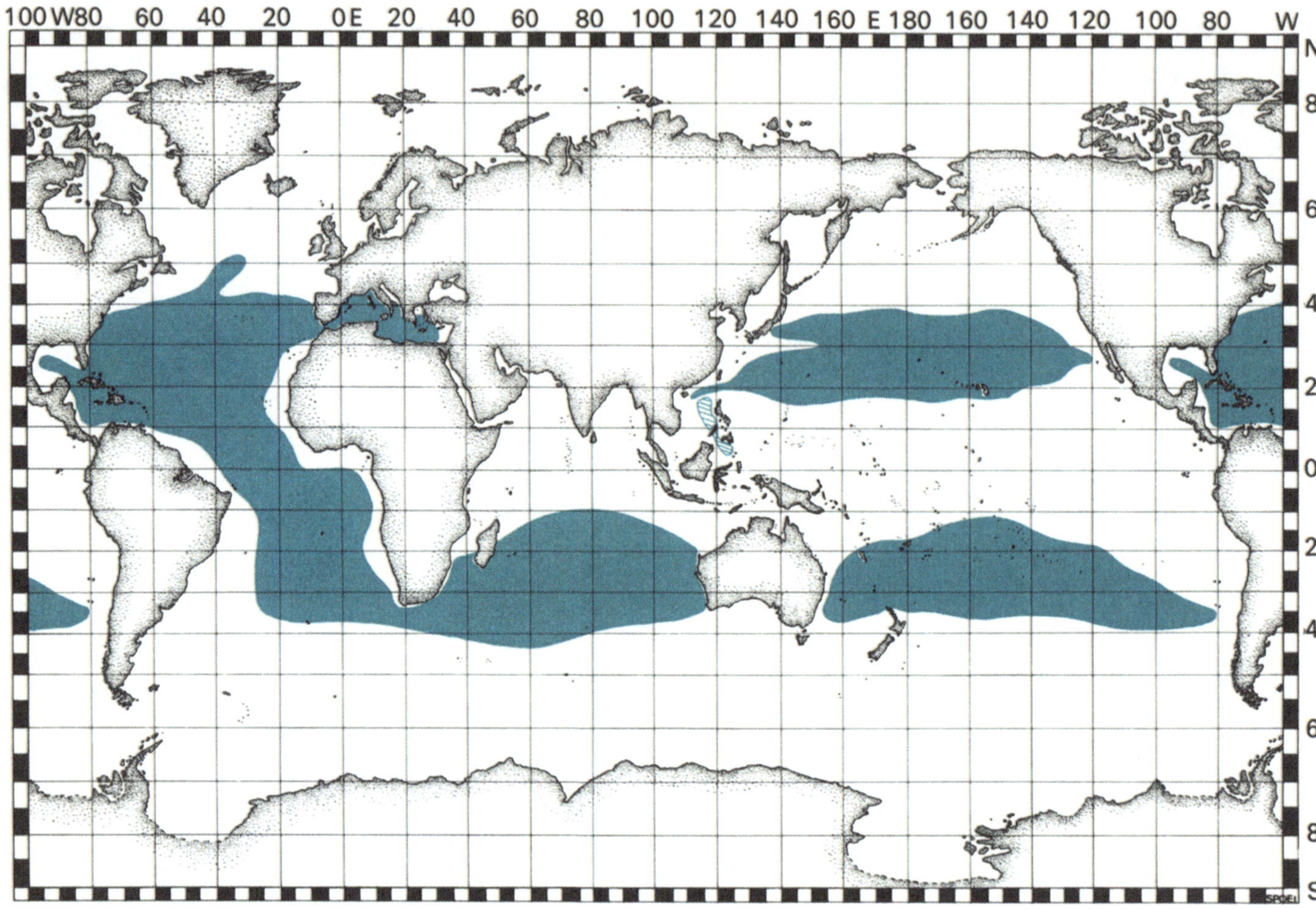

The connection in the Indo-Malayan waters was discussed by Brinton (1975) who distinguished five types of distribution also represented by five groups of taxa, viz.:

Group I: 'Widely ranging species living throughout region; upper layers'

Group II: 'Living throughout region; deep (200 m) layers'

Group III: 'Endemic to or centered in the Inter-Ocean Seas'

Group IV: 'Species largely restricted from the Inter-Ocean pathway region'

Group V: 'Species totally restricted from the Inter-Ocean pathway region'

In the present atlas these groups are represented by the following distributional types.

EUPHAUSIA TENEROIDES type (fig. 75) and STYLOCHEIRON CARINATOIDES type (fig. 63)
The former is more restricted to the warmer waters than the latter, but they show the same low-latitude connections between the basins, mentioned by Brinton for group I.

STYLOCHEIRON MAXIMOIDES type (fig. 72)
This species is a typical representative of group II with small, more or less isolated populations in the Indo-Malayan Archipelago.

THALIA SIBOGAOIDES type (fig. 122)
Like *Stylocheiron insulare* (fig. 63), *Thalia sibogae* shows the pattern of group III.

STYLOCHEIRON SUHMOIDES type (fig. 90)
Stylocheiron suhmi is one of the most clear-cut representatives of group IV.

STYLOCHEIRON ROBUSTOIDES type (fig. 94)
Central water taxa such as *Stylocheiron robustum* are typical representatives of group V. As there are also much more widely distributed species within group V, the following type should be mentioned as well.

CLAUSOCALANUS PAUPULOIDES type (fig. 56)
The widely distributed taxa usually belong to either group I or II, but *Clausocalanus paupulus* shows that they may also belong to group V.

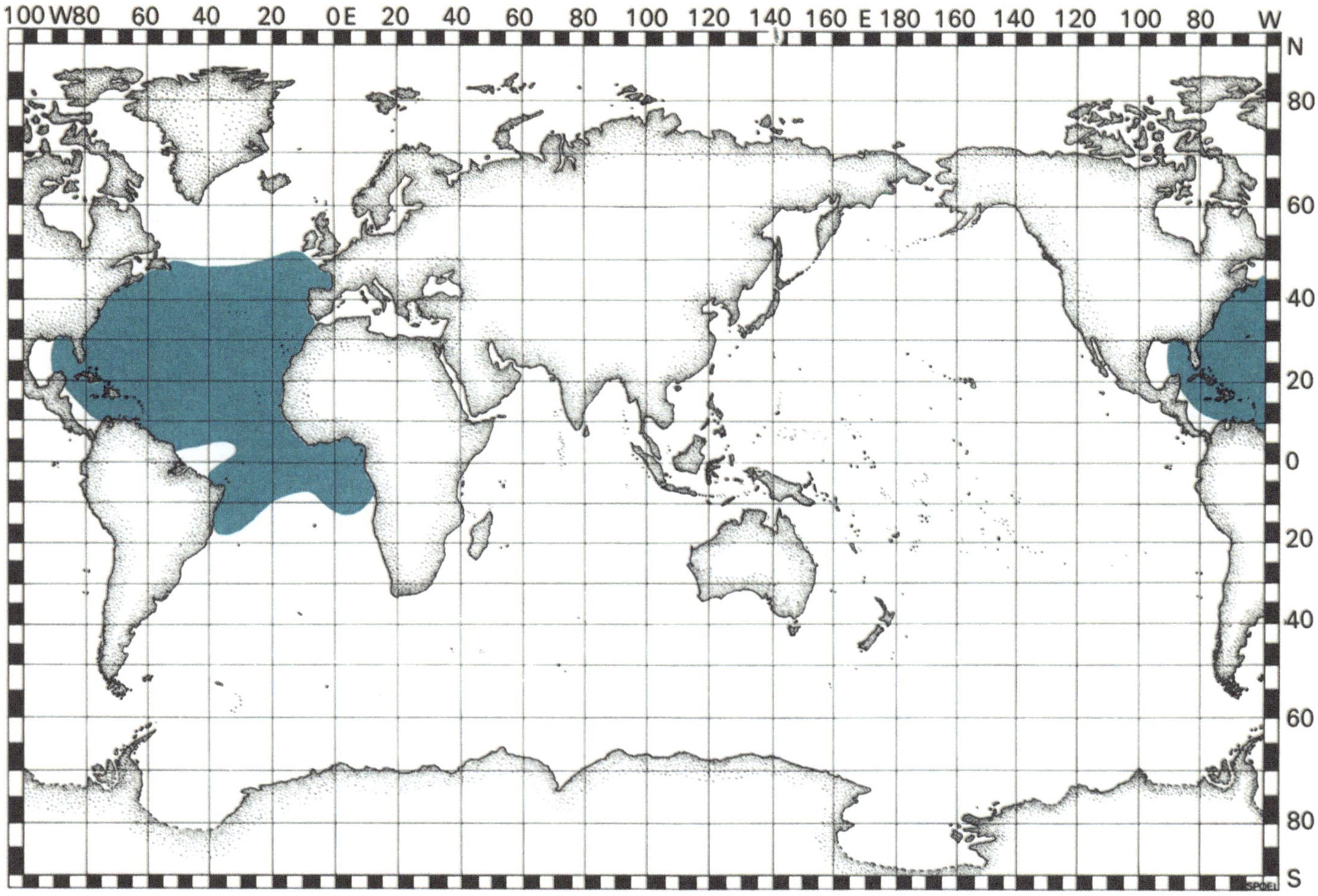

The distributional pattern of the two closely related species *Nematoscelis atlantica* and *N. lobata* is a combination of the THALIA SIBOGAOIDES type (fig. 122) and the STYLOCHEIRON SUHMOIDES type (fig. 90). This shows the great affinity of the patterns of groups III and IV.

Not-beltshaped patterns

Although the neritic taxa are not beltshaped in distribution they are excluded from this paragraph and so are the narrowly distributed endemic taxa. Only the larger oceanic ranges are considered here.

EUPHAUSIA AMERICANOIDES type
Restricted to the warm-water belt of the Atlantic Ocean between 50°N and 20°S, *Euphausia americana* (fig. 109) is distributed differently from the real beltshaped patterns, chiefly as it is restricted to one ocean, and asymmetrically distributed over the northern and southern hemispheres.

FIG. 108 Distribution of *Nematoscelis atlantica* (black) and *N. lobata* (hatched).
FIG. 109 Distribution of *Euphausia americana*.

SAGITTA SERRATODENTATOIDES type
Sagitta serratodentata serratodentata (fig. 118) shows a distribution related to the preceding one, but with a much wider north-south range: from 50°N to 45°S. This range cannot be part of a beltshaped range either, as in that case it would not have remained restricted to the Atlantic Ocean (cf. also the SAGITTA PACIFICOIDES type, fig. 117).

GLOBOROTALIA HIRSUTOIDES type
This type, represented by *Globorotalia hirsuta* (fig. 38) in surface sediment, resembles the next type, but the living plankton populations show only a N. Subtropical and S. Transitional distribution and no special concentration in the E. Atlantic. In addition, there are no indications that this pattern belongs to the group of ranges situated around the Dakar Centre. *Eucalanus hyalinus* (fig. 58) shows the present pattern also in the living plankton perfectly.

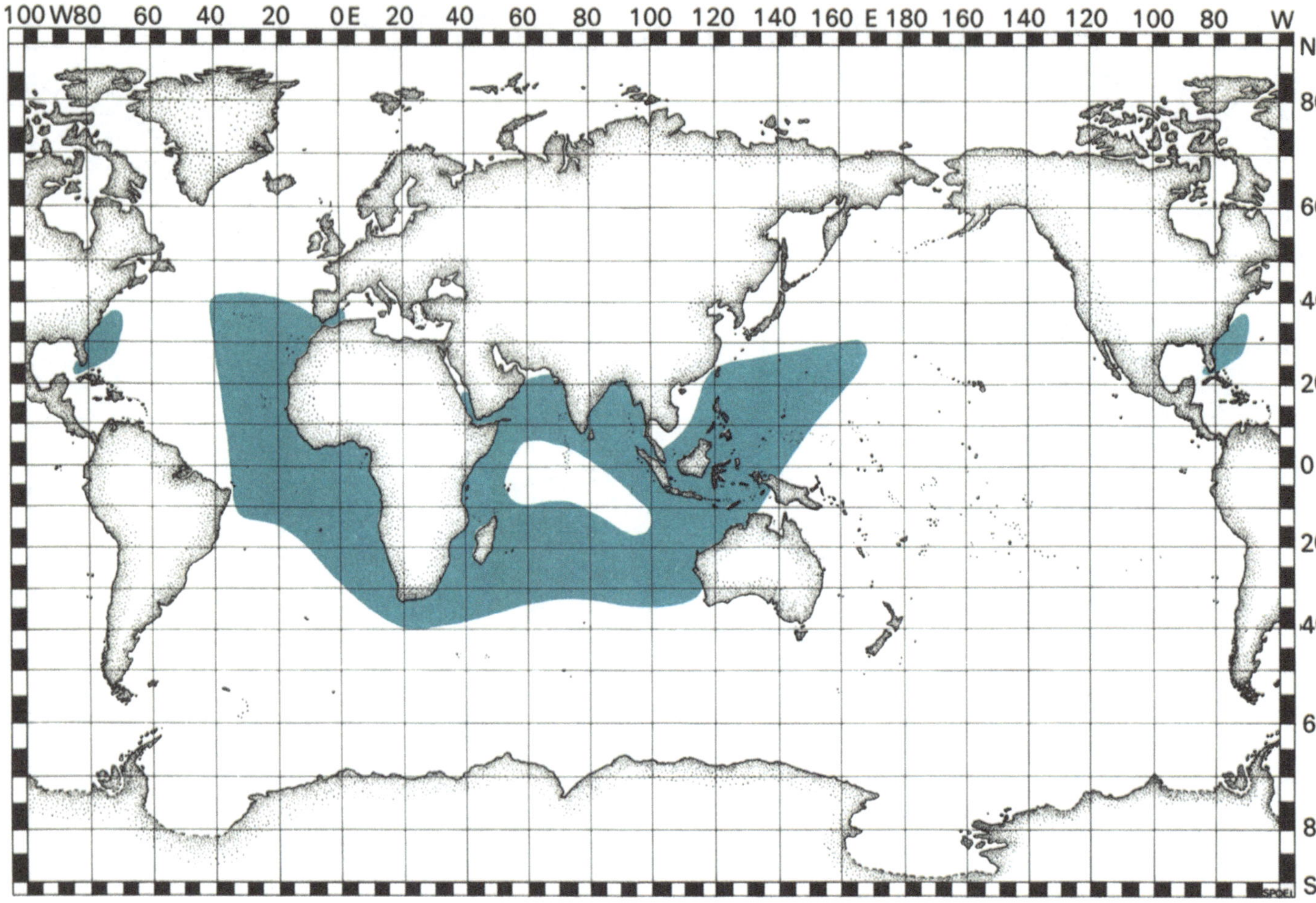

PHRONIMA SOLITAROIDES type
Phronima solitaria (fig. 110) is a representative of a genus distributed around the Dakar Centre (see fig. 16), common in the E. Atlantic and rare in the W. Atlantic, and it penetrates into the Indo-Pacific. Comparable patterns are shown by *Peraclis* (figs 59 & 60, p. 48) and *Clio cuspidata* (fig. 48) though the latter species is clearly circum-global in distribution.

PERACLIS BISPINOSOIDES type
In the genus *Peraclis* different degrees of dispersal out of the Dakar Centre are found. The pattern only occupying Atlantic and Mediterranean waters, also found in epiplankton, is represented by *Peraclis bispinosa* (fig. 60).

SERGESTES EDWARDSOIDES type
Sergestes edwardsi (fig. 101) penetrated the Atlantic through the Panama area (p. 78), so that it is not related to beltshaped patterns or *Phronima*- and *Peraclis*-like patterns (p. 48).

DESMOPTERUS GARDINEROIDES type
Desmopterus gardineri (fig. 111) is restricted to the Indian Ocean north of 30° S, so it is much more widely distributed than the STYLOCHEIRON LONGICORNOIDES type (fig. 66). When also the other species of the genus are considered, *Desmopterus gardineri* may form the central species from which *D. papilio* and *D. pacificus* developed into the Atlantic and Pacific Oceans respectively. *D. papilio*, however, occupies a range of the PHRONIMA SOLITAROIDES type (fig. 110), which is of another historical origin. *D. pacificus* shows a typical endemic distribution (see p. 96).

EUCALANUS ELONGATOIDES type
This type, represented by *Eucalanus elongatus* (fig. 58), covers the Equatorial Indian Ocean, the Indo-Malayan Archipelago and part of the W. Pacific. Therefore it shares some characters with the HYDROMYLES GLOBULOSOIDES type (fig. 112), the DESMOPTERUS GARDINEROIDES type (fig. 111) and the PHRONIMA SOLITAROIDES type (fig. 110).

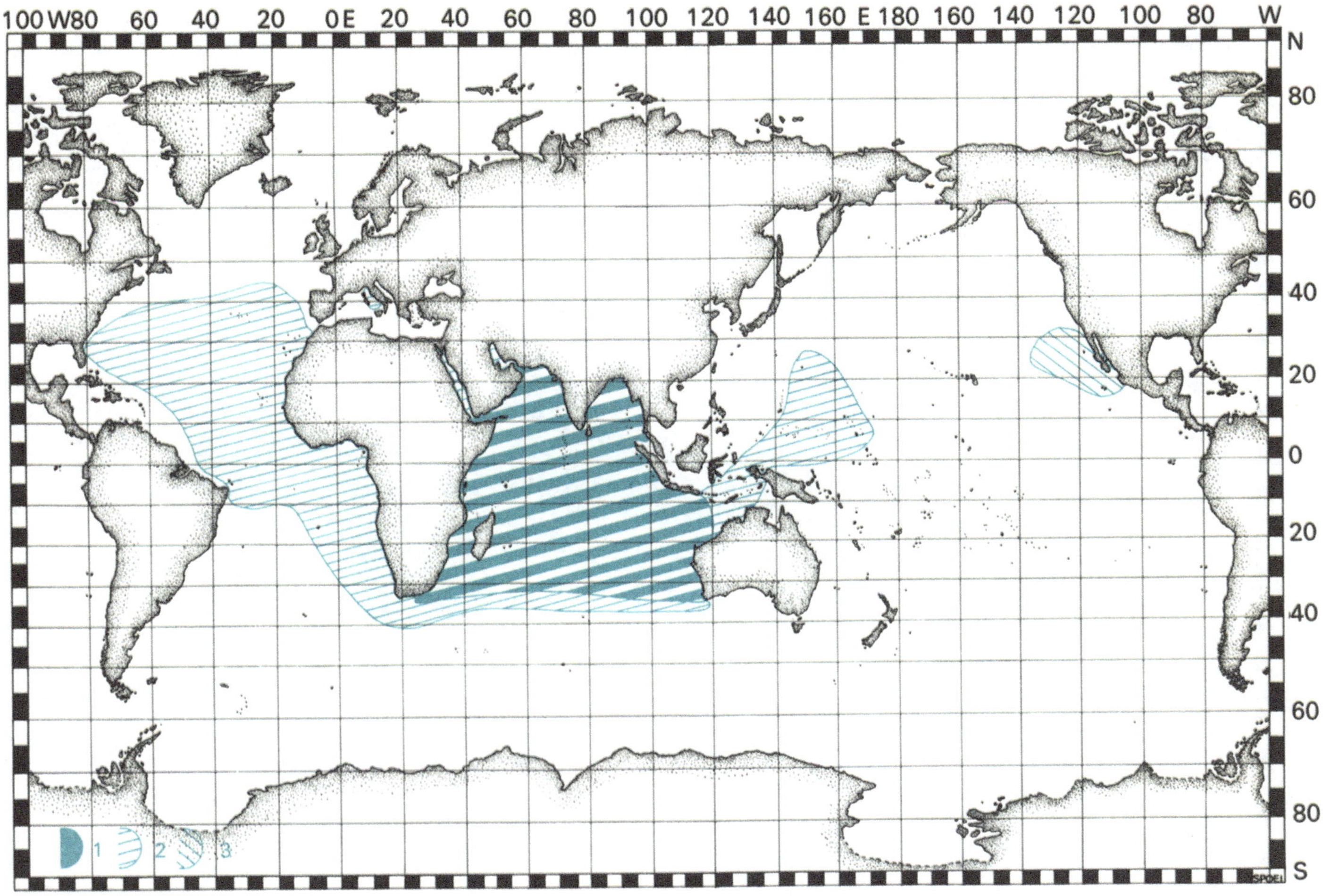

FIG. 110 Distribution of *Phronima solitaria*.
FIG. 111 Distribution of the genus *Desmopterus*.
1 = *D. gardineri*, 2 = *D. papilio*, 3 = *D. pacificus*.

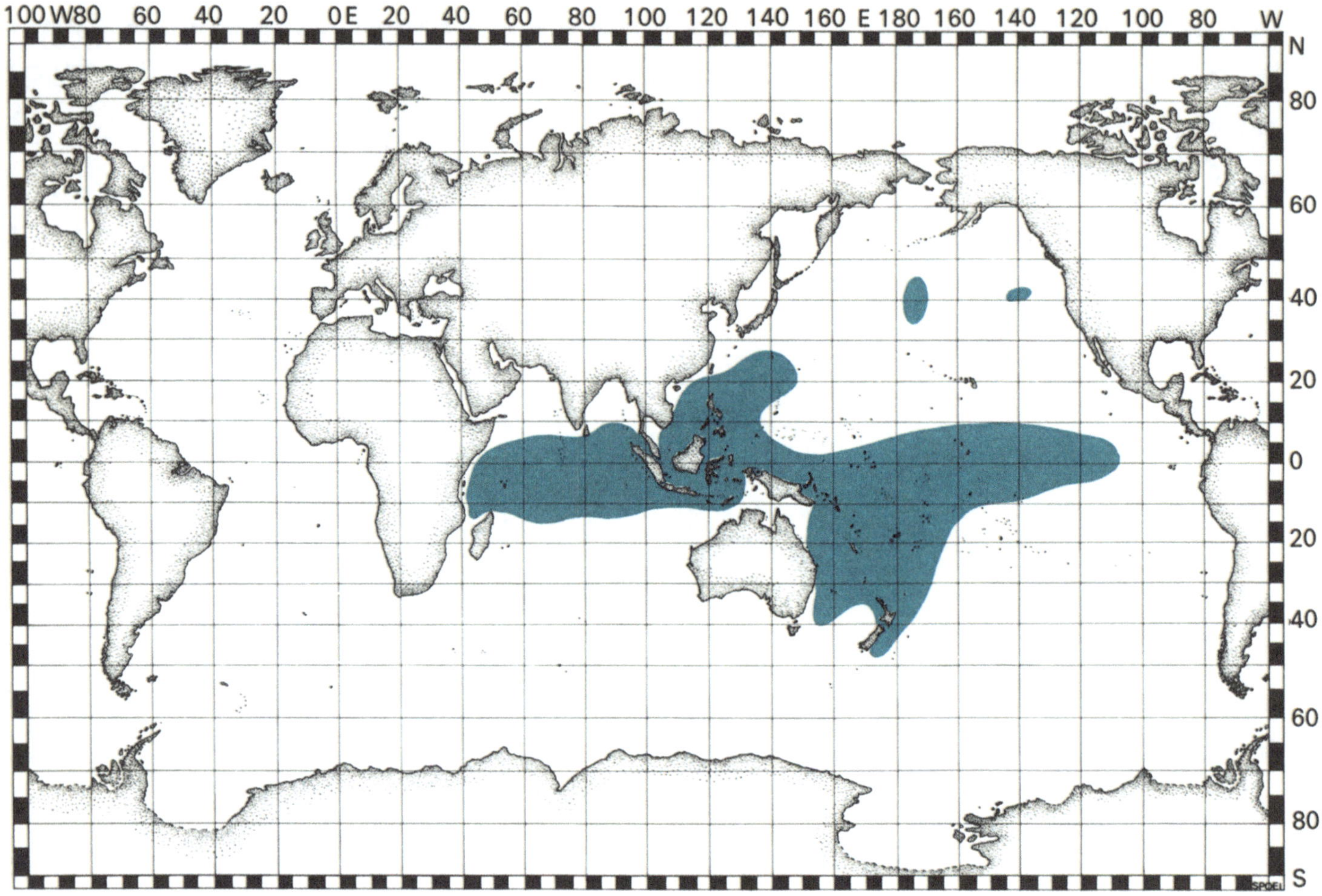

HYDROMYLES GLOBULOSOIDES type

The distribution of *Hydromyles globulosa* (fig. 112) in the Indo-Pacific between 20°N and 20°S is treated separately as it is one of the types which dispersed from a centre in the Indo-Malayan Archipelago, without, however, reaching the E. Pacific or the Agulhas Current and without showing a relation to the beltshaped patterns. In all these characteristics it differs from the CAVOLINIA GLOBULOSOIDES type (fig. 98) and the CLIO CONVEXOIDES type (fig. 57).

CAVOLINIA TESCHOIDES type

Cavolinia tridentata forma *teschi* (fig. 53) is sympatric with the eastern part of the range of *Hydromyles globulosa* (fig. 112); the remaining western part is then identical with the distribution of *Cavolinia tridentata* forma *danae* (DESMOPTERUS GARDINEROIDES type) (fig. 53). As this subdivision of the large range is so typical, the separate CAVOLINIA TESCHOIDES type is proposed.

CARINARIA CRISTATOIDES type

The type of distribution as reflected by *Carinaria cristata* (fig. 113) closely resembles the preceding one, but from the Indo-Malayan waters it penetrates the N. Pacific, through which it finally reaches the E. Tropical Pacific. In the Indian Ocean it is an equatorial species. If this pattern further developes, it may also penetrate the Atlantic Ocean, as shown by the next example.

FIG. 112 Distribution of *Hydromyles globulosa*.
FIG. 113 Distribution of *Carinaria cristata*.
FIG. 114 Distribution of *Phormospyris stabilis* (black) and *Lophospyris pentagona* (hatched).

PHORMOSPYRIS STABILOIDES type

In fig. 114 a variant of the preceding type is given. The two radiolarian species *Phormospyris stabilis* and *Lophospyris pentagona* show, in addition to the area occupied by the type of fig. 113, an extension over the E. Atlantic Boundary waters and the Atlantic Transitional waters.

Some taxa developed distributions which much more closely resemble the beltshaped patterns than the preceding one, though they are, like the present two, essentially not-beltshaped in origin. The next figures are clear examples of such distributions.

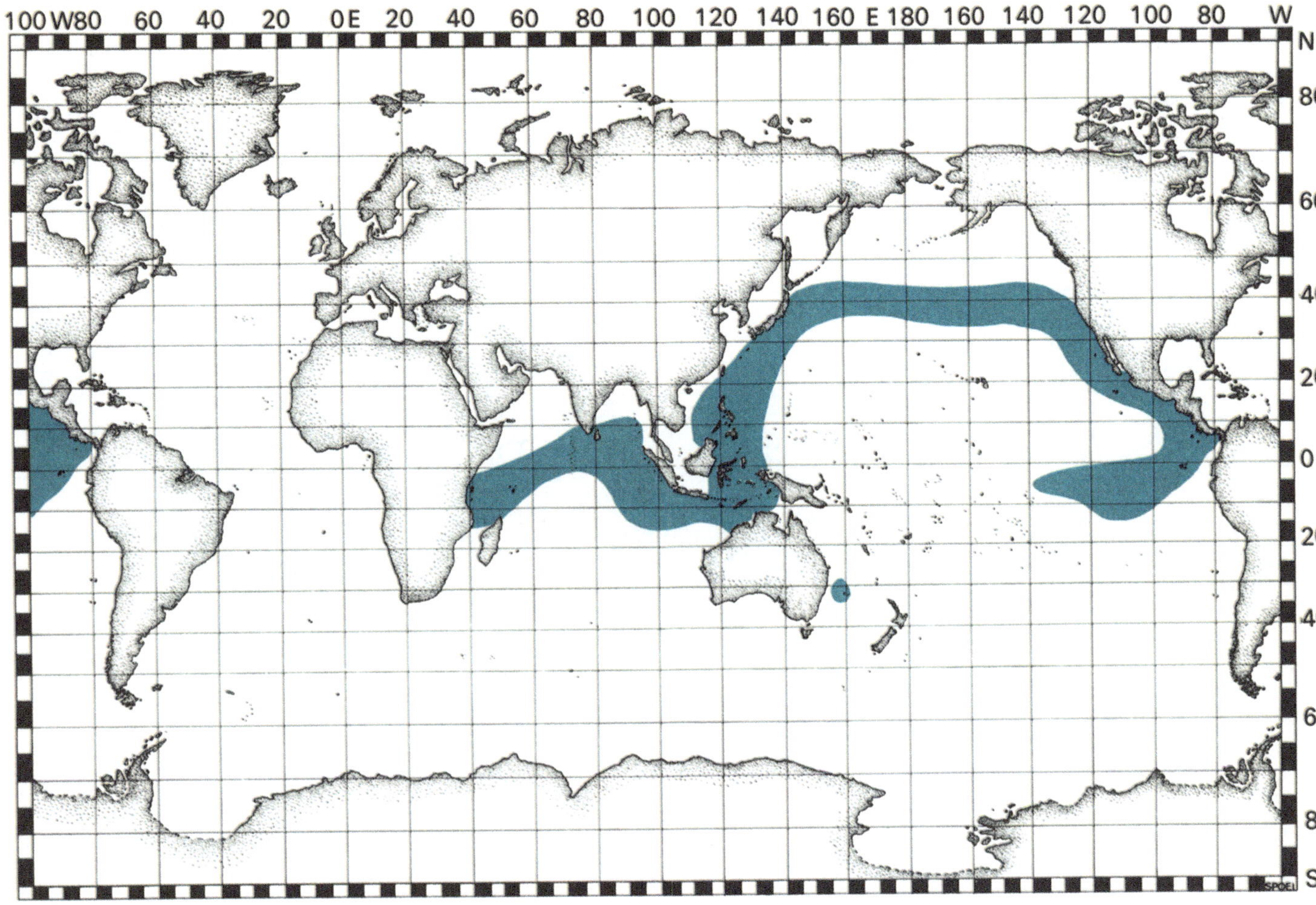
100 W80 60 40 20 0 E 20 40 60 80 100 120 140 160 E 180 160 140 120 100 80 W
N 80 60 40 20 0 20 40 60 80 S
SPOEL

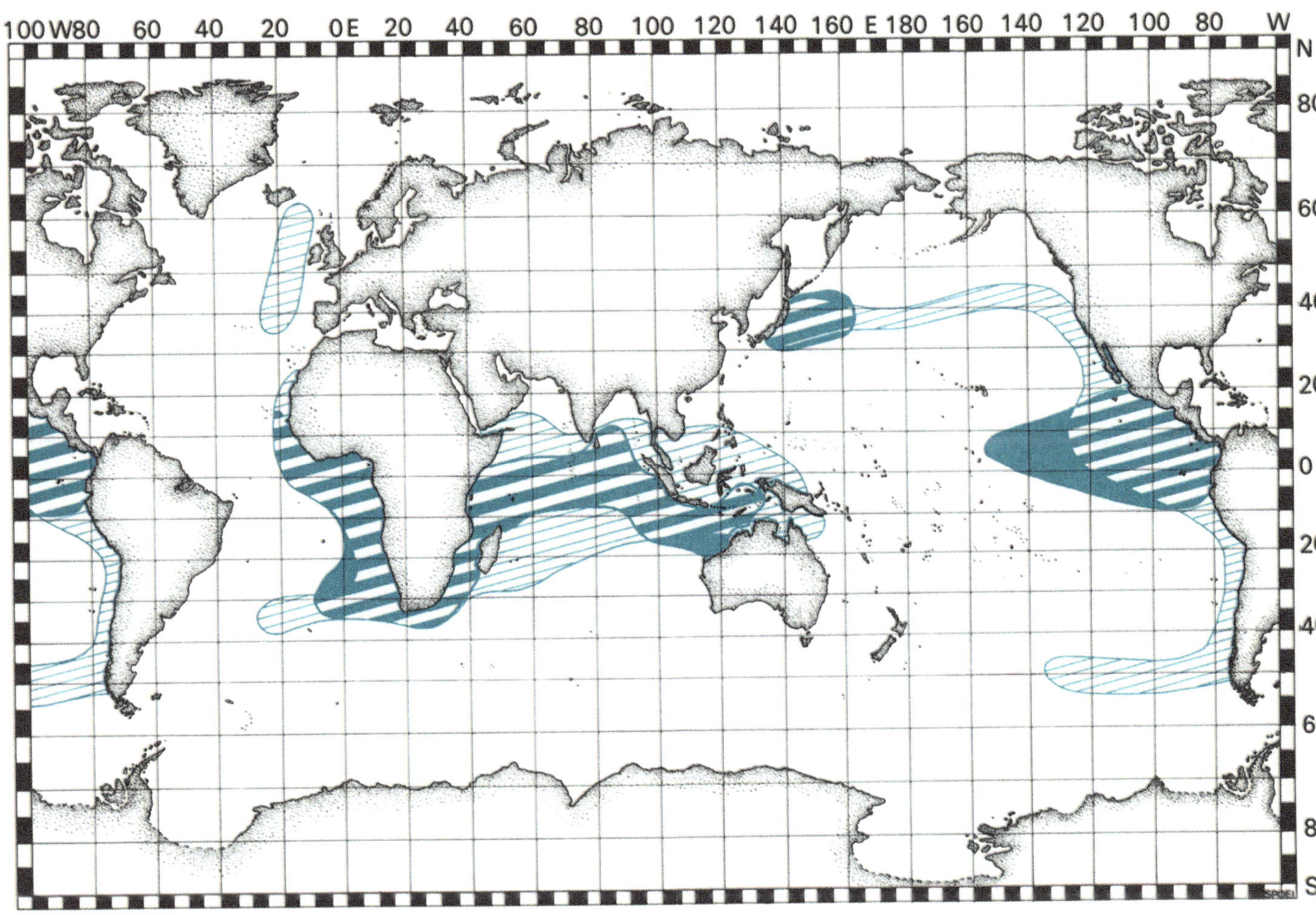
100 W80 60 40 20 0 E 20 40 60 80 100 120 140 160 E 180 160 140 120 100 80 W
N 80 60 40 20 0 20 40 60 80 S
SPOEL

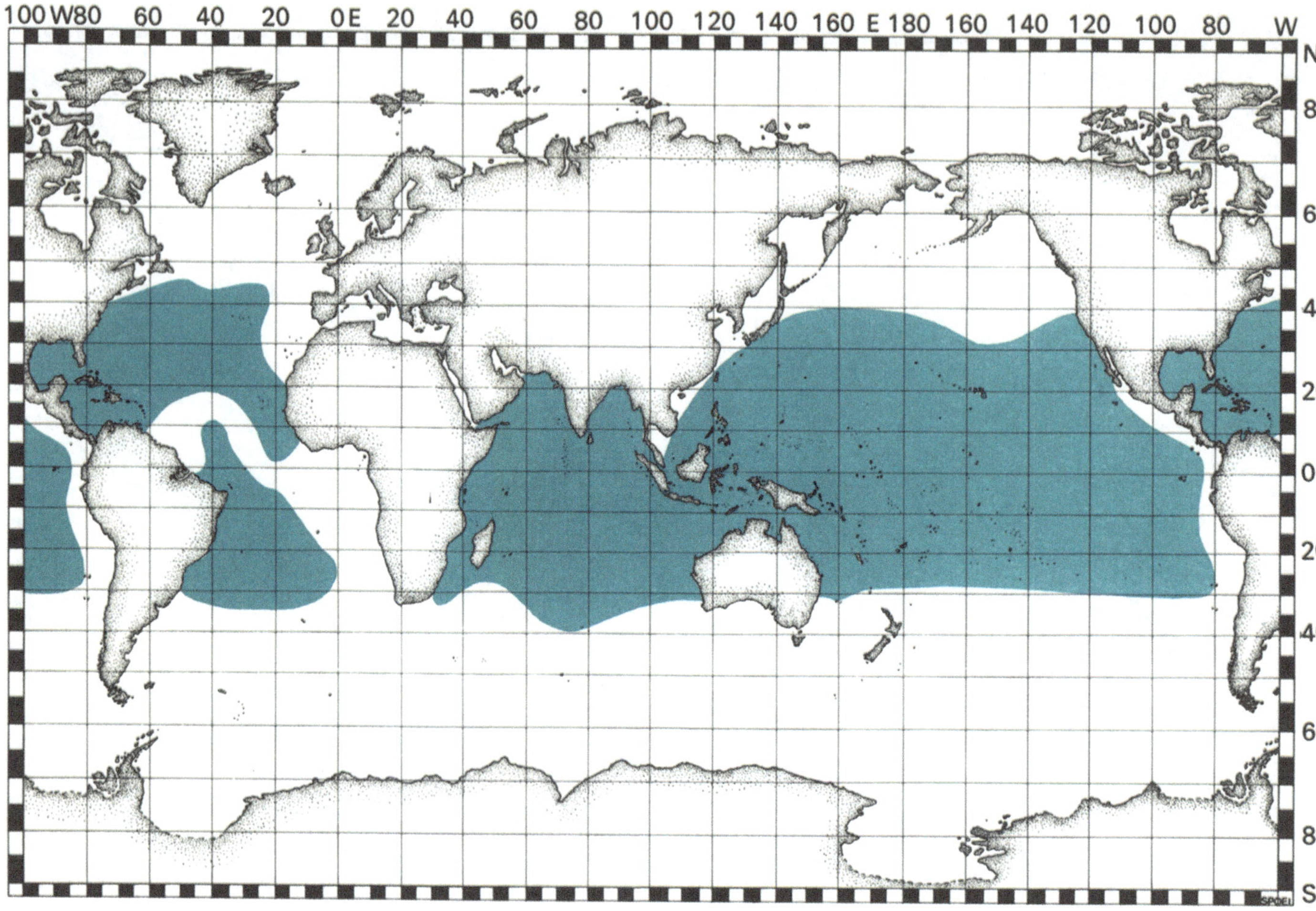

PONTELLINA PLUMATOIDES type

The distribution of *Pontellina plumata* (fig. 115) between 40°N and 40°S resembles the warm-water beltshaped patterns. The strong influence of cold-water currents (west of Africa, west of Peru), however, forms an additional threshold, besides the continental barriers. Moreover, this species belongs to the species-group *P. plumata* (*P. plumata*, *P. platychela*, *P. sobrina*, *P. morii*, figs 98 & 106) whose members populated the oceans separately.

PNEUMODERMA ATLANTICOIDES type

The pattern given in fig. 88 may easily be mistaken for a modified beltshaped pattern. *Pneumoderma atlanticum* is, however, represented in the N. Pacific Ocean by a separate forma *pacificum*. A comparison with the EUPHAUSIA HEMIGIBBOIDES type (fig. 97) cannot be made as the separate taxon is found in the northern instead of southern Central waters. More important is that the species spread over Indian and Atlantic Oceans in Equatorial and Tropical Terminal waters. Hence, a special type of distribution is proposed here.

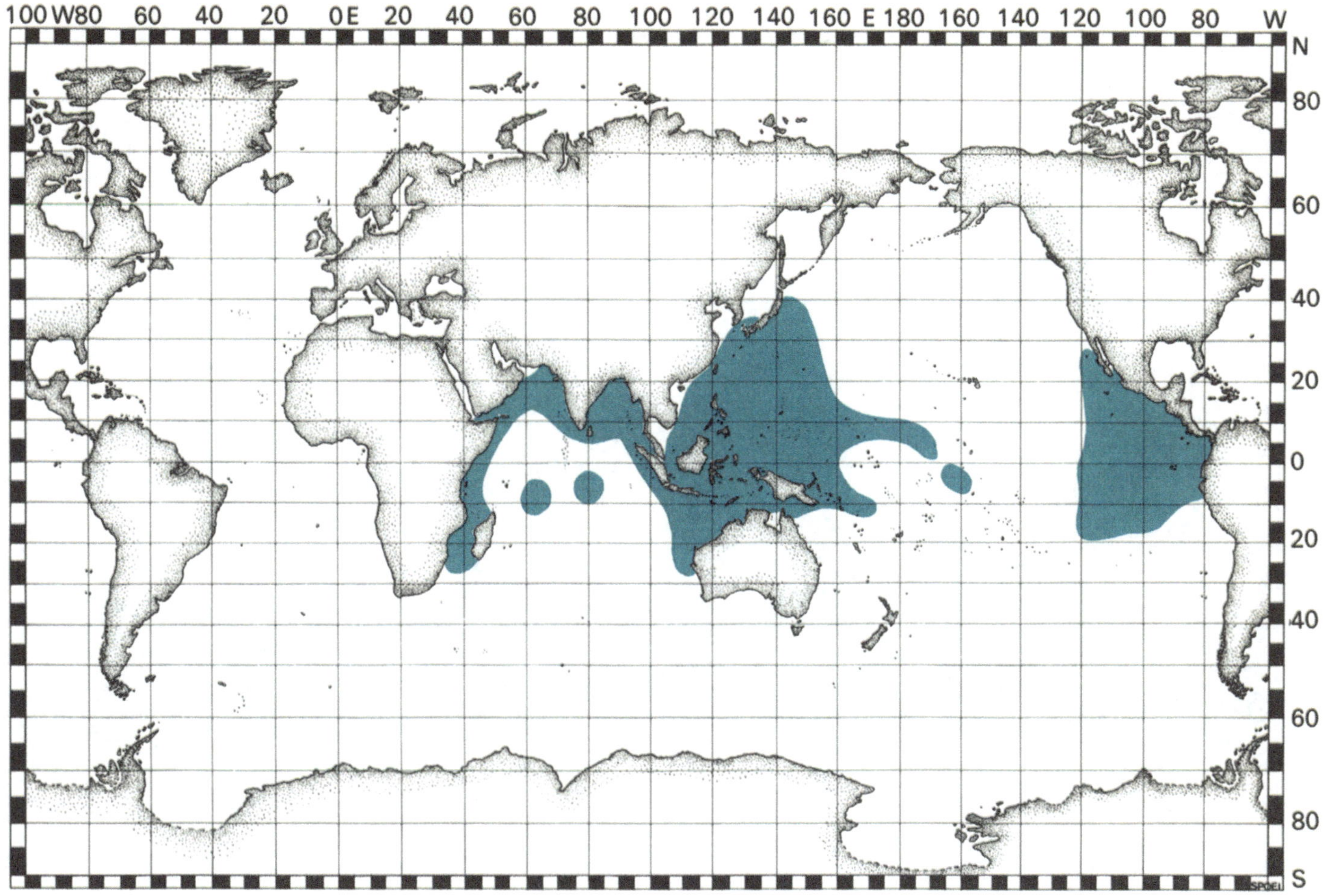

FIG. 115 Distribution of *Pontellina plumata*.
FIG. 116 Distribution of *Melamphaes spinifer*.

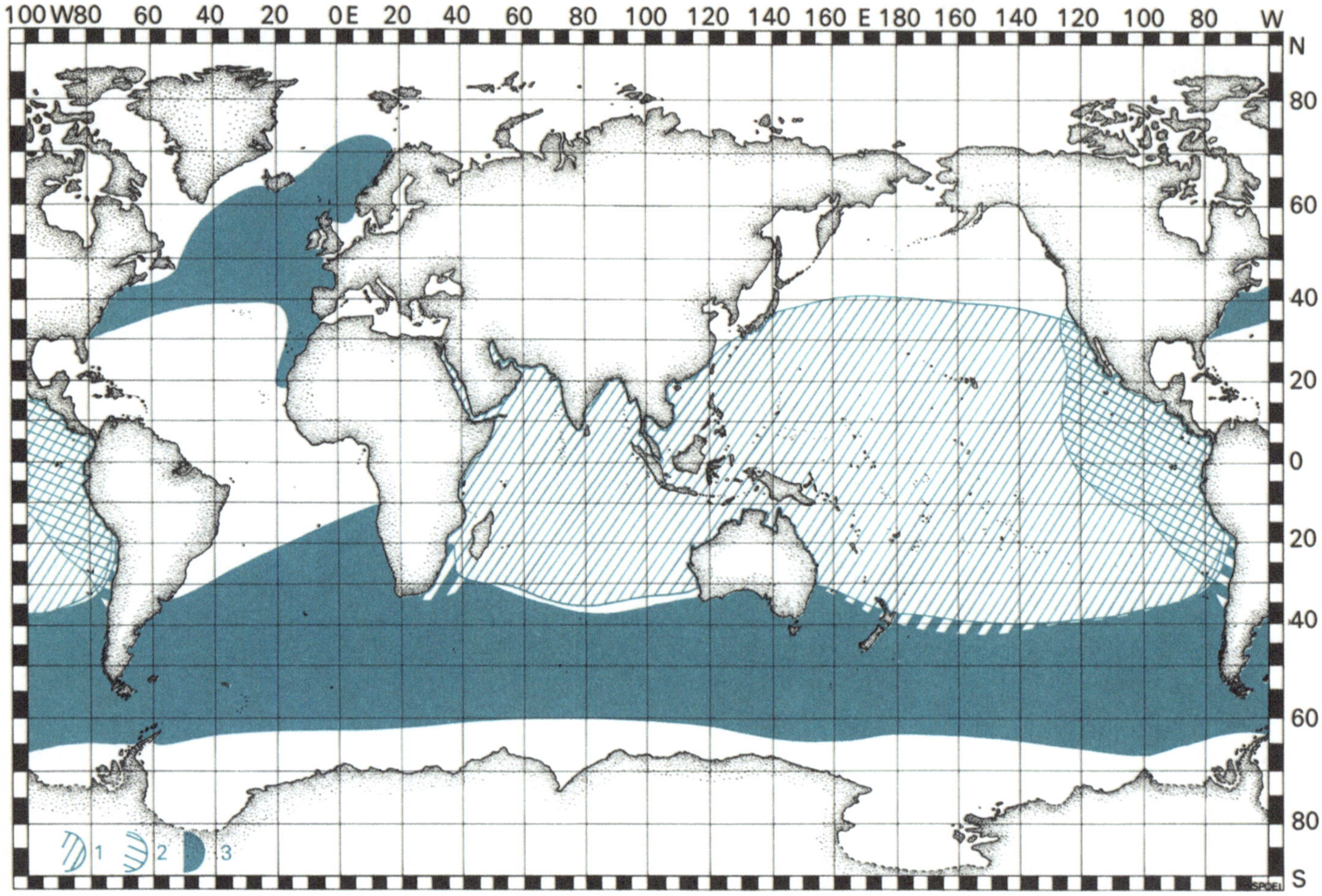

SAGITTA PACIFICOIDES type

In the *Sagitta serratodentata* species-group (figs 117 & 118) a tendency is seen to form beltshaped ranges, but closer examination of the group proves that no common belts, nor real latitudinal variation, is found. The SAGITTA PACIFICOIDES type seems a clear Indo-Pacific part of a circum-global taxon which split off the Atlantic *S.s.serratodentata*. The present type can then be compared with the CLIO CONVEXOIDES type (fig. 57). *S.pacifica* occurs between 40°N and 40°S, so that its range is too wide to remain restricted to the Indo-Pacific area because of mere isolation by the continents.

Melamphaes spinifer (fig. 116) seems to show the SAGITTA PACIFICOIDES type distribution, though in a more discontinuous way. The distribution of *Melamphaes spinifer* seems to represent an intermediate phase between the SAGITTA PACIFICOIDES type and the DIACRIA ERYTHROIDES plus DIACRIA OCCIDENTALOIDES types (cf. figs 116, 117 and 123). This regressive phase was already considered to form the CAVOLINIA GLOBULOSOIDES type (fig. 98).

SAGITTA TASMANICOIDES type

This type also looks like a beltshaped pattern. It differs, however, from the most related pattern of the NITZSCHIA CYLINDRICOIDES type (fig. 81) in occurring in less cold water without living in the Pacific 'Subarctic'. *S.bierii* is endemic and will be discussed later (p. 96).

For the sake of completeness, the remaining taxa of the *Sagitta serratodentata* species-group are given in fig. 118. *S.s.serratodentata* shows the SAGITTA SERRATODENTATOIDES type related to the EUPHAUSIA AMERICANOIDES type (fig. 109). The other two species are given as endemic types, viz. the SAGITTA ATLANTICOIDES type and the SAGITTA PSEUDOSERRATODENTATOIDES type. No beltshaped pattern is found in these taxa.

LIMACINA ACUTOIDES type

The Pacific 'Subarctic' is difficult to explain as a hydrographic area, as it is a beltshaped area essentially different from the Atlantic Subarctic. For this reason, the patterns restricted to this area are given separately. *Limacina helicina acuta* is found in the western part of the N. Pacific 'Subarctic', and is the example of the present type (fig. 82).

LIMACINA PACIFICOIDES type

In the eastern part of the 'Subarctic' N. Pacific, forms separated from the western representatives are frequently found, as exemplified by *Limacina helicina*

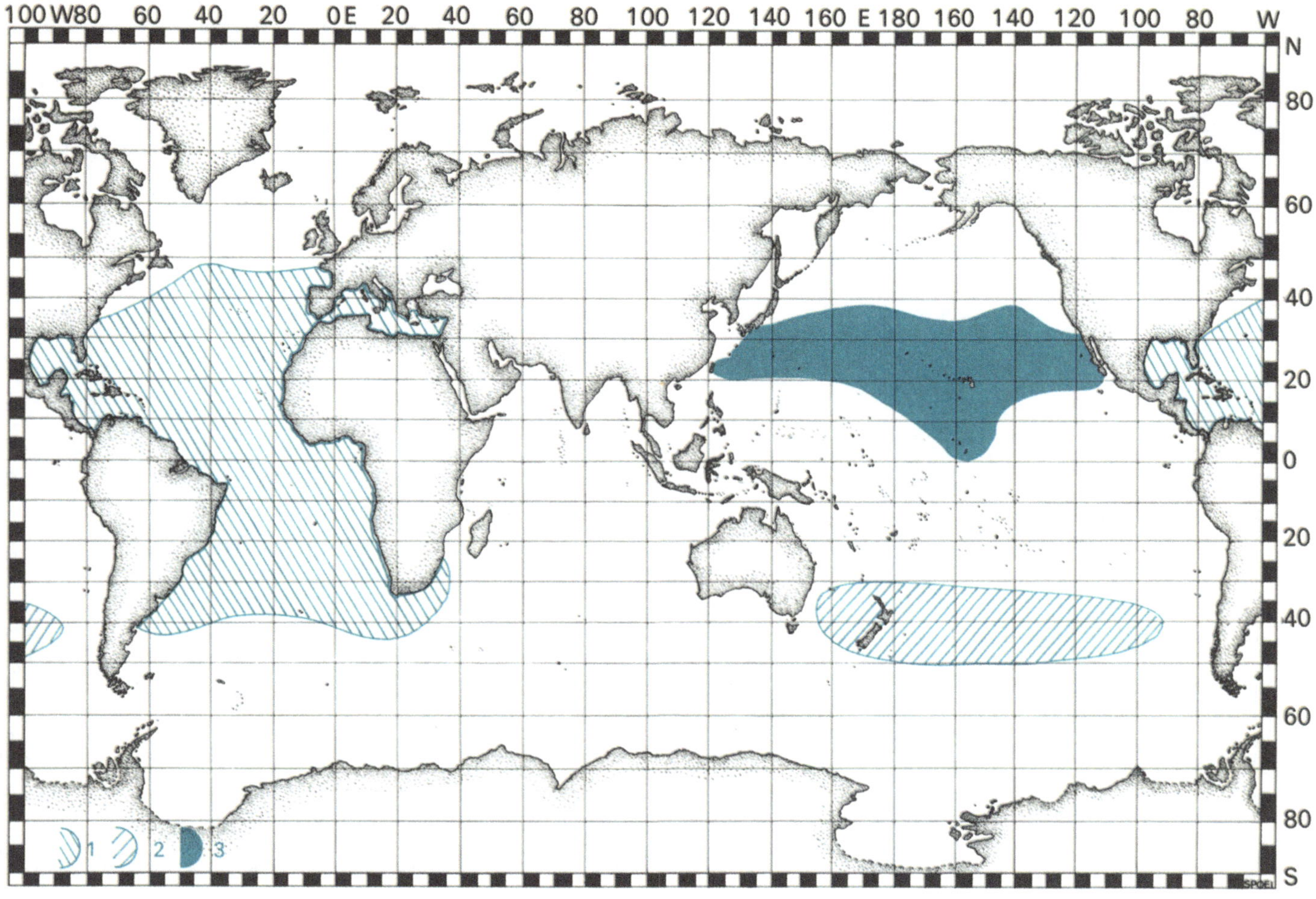

pacifica (fig. 82). Sometimes only a taxon occurs in the eastern part without relatives in the western part of the 'Subarctic' as shown by *Calanus marshallae* (fig. 78). Usually the major part of the ranges of this type are situated within the Transitional waters.

CALANUS HELGOLANDICOIDES type

As a rule, the N. Atlantic shows somewhat more connection with the cold waters of the southern hemisphere than the N. Pacific. Still, the Subarctic N. Atlantic usually docs not fit in with beltshaped patterns and a separate type of distribution is concerned, best represented by *Calanus helgolandicus* (fig. 79).

FIG. 117 Distribution of the *Sagitta serratodentata* species-group (see also fig. 118). 1 = *S. pacifica*, 2 = *S. bierii*, 3 = *S. tasmanica*.

FIG. 118 Distribution of the *Sagitta serratodentata* species-group (see also fig. 117). 1 = *S.s. serratodentata*, 2 = *S.s. atlantica*, 3 = *S. pseudoserratodentata*.

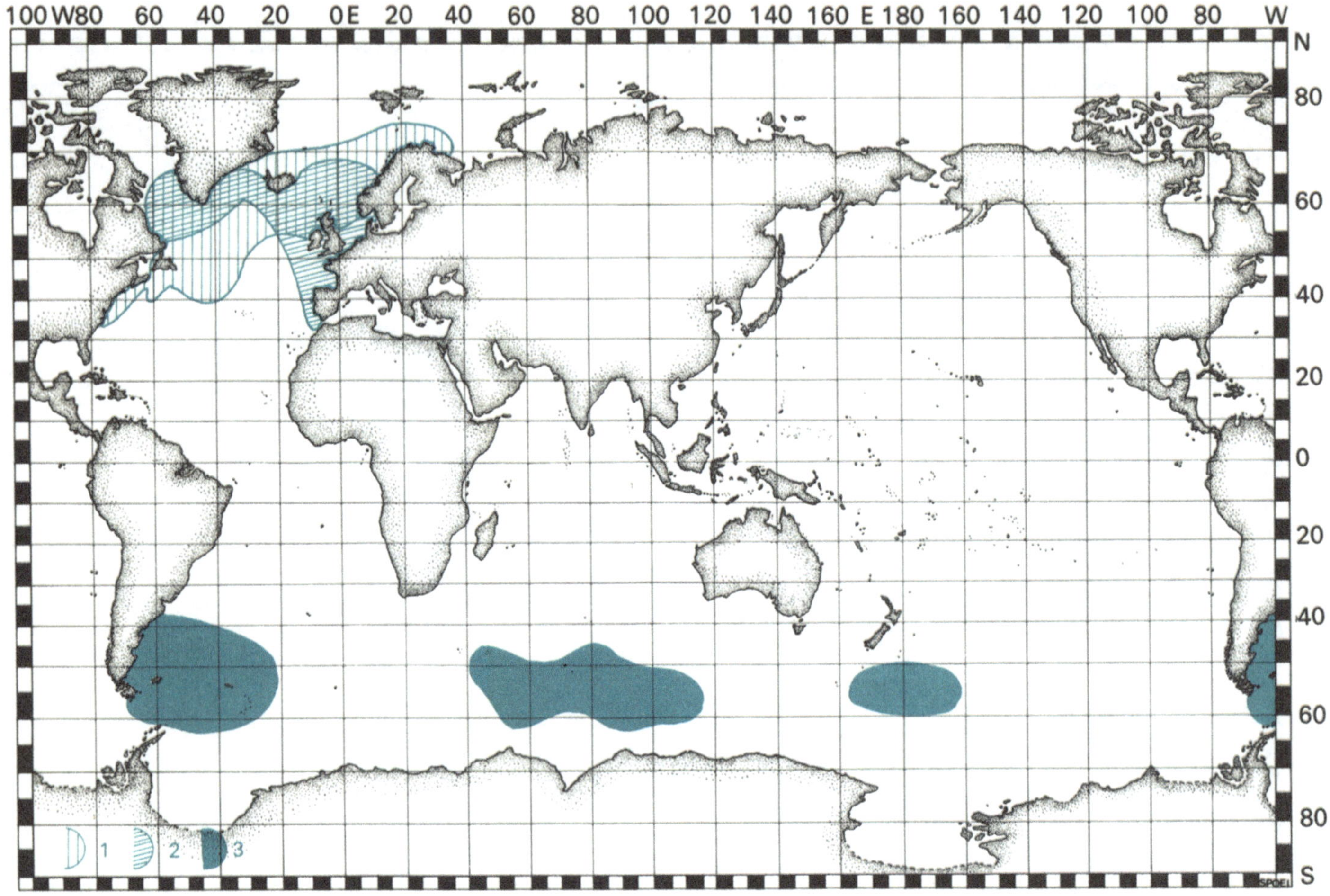

LIMACINA RETROVERSOIDES type

As in the N. Pacific, sometimes in the N. Atlantic a division of the Subarctic can be made. *Limacina r.* forma *retroversa* (fig. 119) is an N.E. Atlantic taxon with little extension to the west. Like the N. Pacific *L. helicina pacifica* (fig. 82), *L.r.* forma *retroversa* penetrates far into the Transitional waters.

LIMACINA BALEOIDES type

The range of *Limacina retroversa* forma *balea* (fig. 119) is more concentrated in the mid and E. Atlantic, and penetrates only slightly to the west. The distribution, though incompletely known, of *L.r.* forma *australis* is also given in fig. 119 for the sake of completeness.

EUCALANUS PARKOIDES type

Like the Subarctic taxa, the isolated northern Transitional taxa in the Pacific cannot be considered to be beltshaped. *Eucalanus parki* (fig. 54) is the typical representative of this range also occupied by e.g. *Pneumoderma atlanticum pacificum* (fig. 88) and *Eucalanus bungi californicus* (fig. 58). The LIMACINA PACIFICOIDES type and LIMACINA ACUTOIDES type (fig. 82) distributions may become more or less sympatric with the present pattern.

EUCALANUS BUNGOIDES type

Ranges that cover really Transitional waters in the Pacific are considered to form the present type as exemplified by *Eucalanus bungi bungi* (fig. 58)

CALANUS PACIFICOIDES type

Resembling the above type is the distribution of *Calanus pacificus* (fig. 79), but this range is distinctly wider and, as many species show this wider distribution, a special type is proposed here.

EUCALANUS LANGOIDES type

In *Eucalanus langae* (fig. 54) a southern Transitional species is found to be restricted to the Indo-Pacific Oceans. This restriction and the fact that this pattern is shown by a member of a group with a not-beltshaped distribution were the reason to consider it a separate type.

CLAUSOCALANUS JOBEOIDES type

A very complicated pattern, not related to any beltshaped distribution is the one represented by *Clausocalanus jobei* (fig. 55). It covers the warm Atlantic, South Subtropical Indian and Pacific Oceans and the

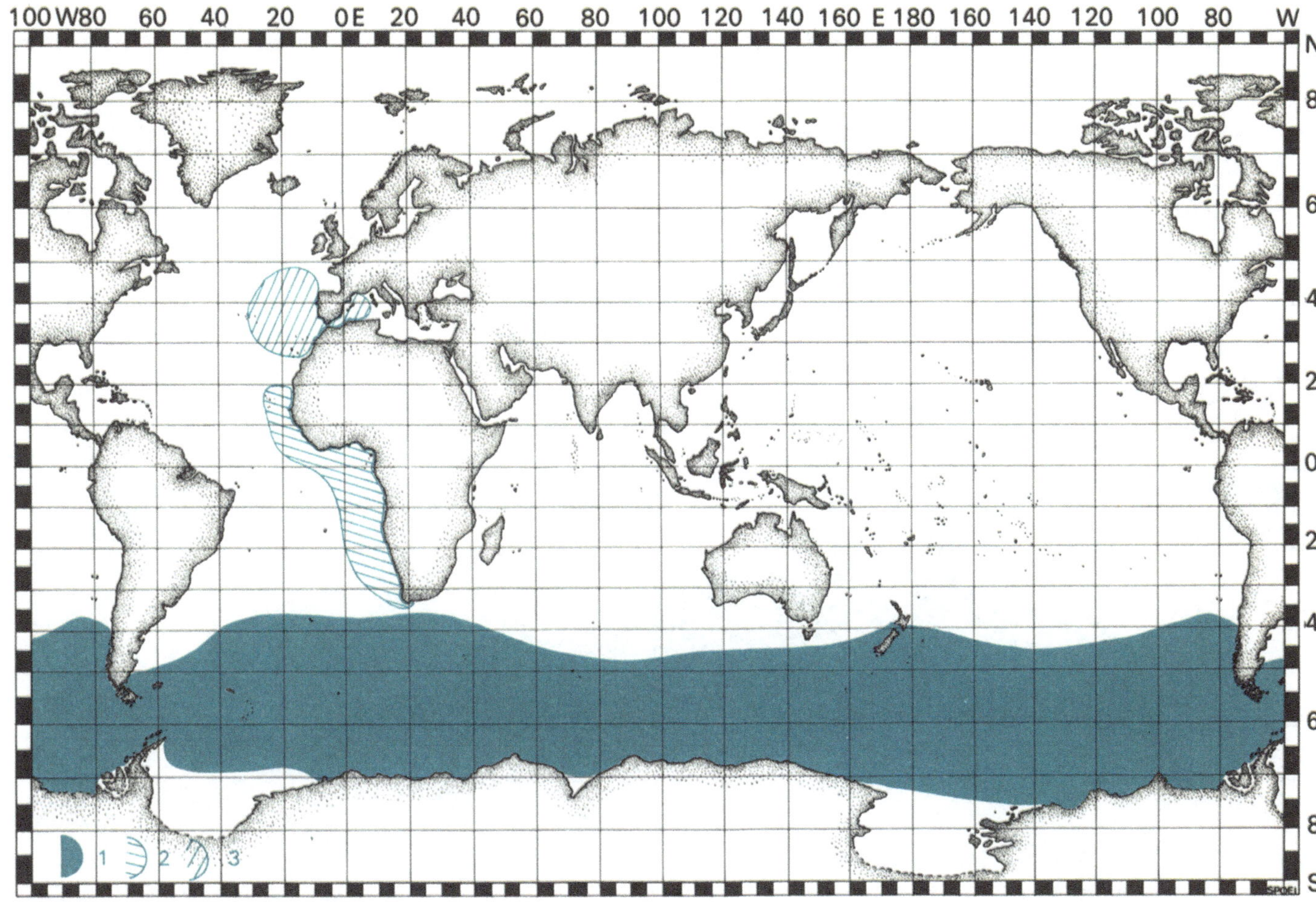

FIG. 119 Distribution of the polytypic species *Limacina retroversa*. 1 = *L.r.* forma *retroversa*, 2 = *L.r.* forma *balea*, 3 = *L.r.* forma *australis*.

FIG. 120 Distribution of *Spongiobranchaea australis* (1), *Euphausia hanseni* (2) and *Fowlerina zetezios* (3).

area also covered by the EUCALANUS INERMOIDES type (fig. 58). It is difficult to explain how this pattern can develop and how it can be covered by one monotypic taxon. The same can be stated with regard to the following type.

EUCALANUS SEWELLOIDES type

This type, represented by *Eucalanus sewelli* (fig. 54), is a combination of the EUPHAUSIA AMERICANOIDES type (fig. 109), the EUCALANUS ELONGATOIDES type (fig. 58), the DESMOPTERUS PACIFICOIDES type (fig. 111) and the CAVOLINIA AFFINOIDES type (fig. 53), with very evident discontinuities. Also with regard to this pattern it is hard to believe, though possible, that it can be occupied by a monotypic species.

Endemics

Planktonic species with very small distributional ranges are rare, as planktonic organisms tend to be dispersed over greater distances by way of currents. Biotopes of limited geographic extent, except for some neritic ones (see next paragraph) are not found in the sea. When a species is found to have a restricted endemic distribution it is always questionable whether sufficient sampling has taken place. In other words, real proof of restricted occurrence is hard to give. As this atlas tries to give a summary of patterns, the most typical restricted ranges are given here.

SPONGIOBRANCHAEA AUSTRALOIDES type

Spongiobranchaea australis (fig. 120) shows a distribution around the Antarctic continent, related to that of the EUPHAUSIA CRYSTALLOROPHIOIDES type and the LIMACINOIDES type (figs 135 & 82, resp.). The present type, however, concerns a species without close relatives elsewhere. It cannot be considered a cold-water form of a taxon also occurring in other areas, such as *Clio pyramidata* forma *sulcata* (fig. 57) or *Limacina helicina antarctica* (fig. 82). Moreover, it is not a taxon restricted to ice-covered areas, so that it is believed to form a really separate type.

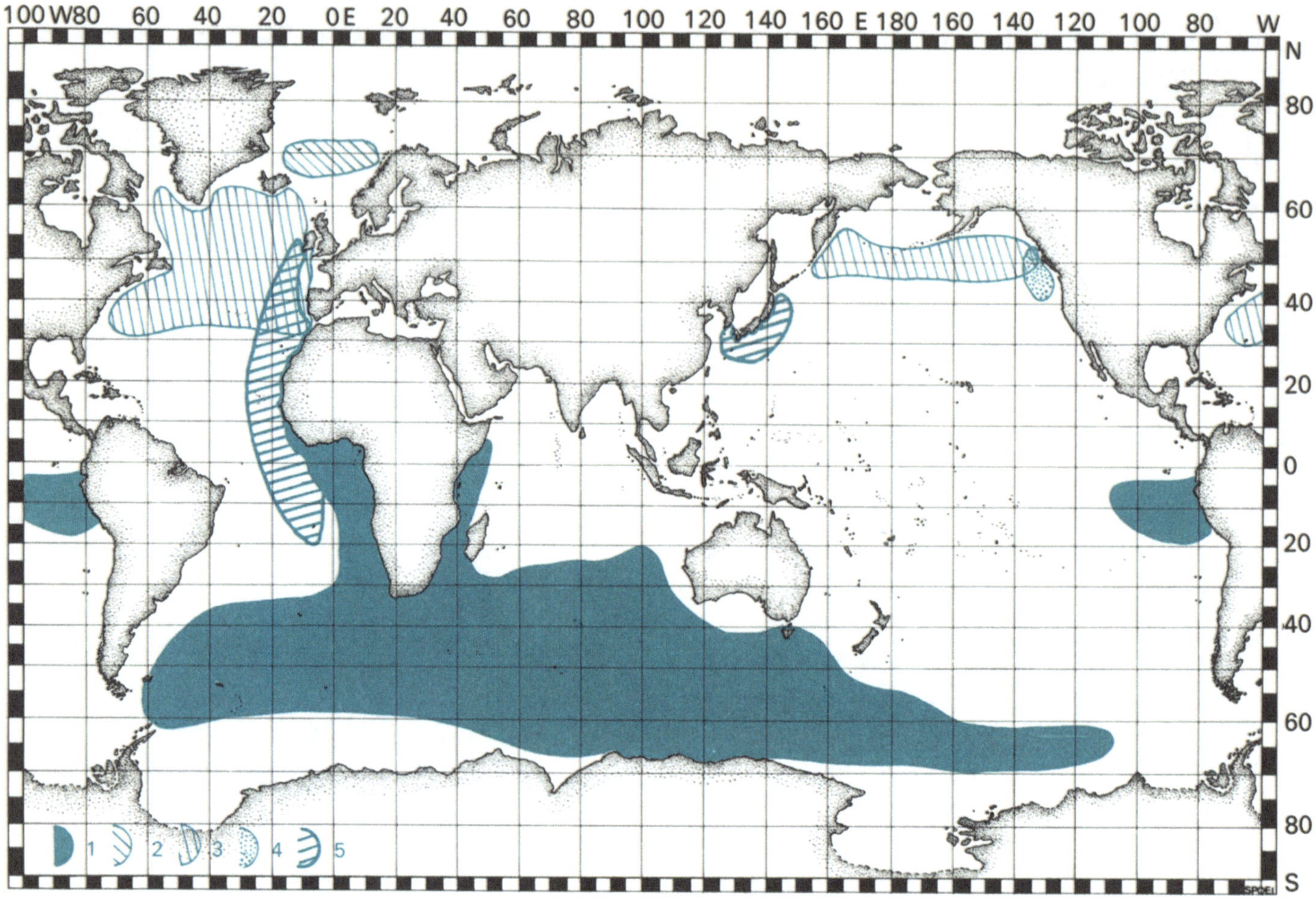

FOWLERINA ZETEZIOIDES type
Until now, *Fowlerina zetezios* (fig. 120) has only been found in a small area around the Atlanto-Mediterranean Centre and it is considered to represent a unique type.

EUPHAUSIA HANSENOIDES type
Euphausia hanseni (fig. 120) is not a typical neritic species (cf. fig. 129 for *Sagitta tenuis*) though it shows a comparable restricted distribution around the East Equatorial Atlantic Centre.

CROSSOTA ALBOIDES type
This distributional type over the Atlantic and Indo-Malayan areas (fig. 121) reminds one of a Tethys Sea dispersal. In plankton, however, it is hard to prove that patterns as shown by *Crossota alba* did not develop from a migration around South Africa. However, the clear-cut appearance of this type led to its separate treatment.

SAGITTA BIERIOIDES type
In fig. 117 *Sagitta bierii* is shown as it is distributed over the E. Tropical Pacific, a range comparable with the preceding one in the Atlantic. This area forms a very distinct province in the Pacific, which can be subdivided into three smaller areas as shown below.

DESMOPTERUS PACIFICOIDES type
The waters of the California Current form a separate part of the E. Pacific and they are occupied by some endemic taxa, of which *Desmopterus pacificus* has been selected as an example (fig. 111). In some distributional patterns, e.g. that of the COROLLA SPECTABILOIDES type, this area constitutes a distinctly separate part (cf. *C. spectabilis*, fig. 106, and *Sergestes consobrinus*, fig. 102).

SERGESTES BREVISPINATOIDES type
The southern part of the SAGITTA BIERIOIDES type (fig. 117) also forms a separate range, as shown by *Sergestes brevispinatus* (fig. 101).

EUCALANUS INERMOIDES type
In between the two preceding types a small range can be distinguised of exclusively Equatorial E. Pacific taxa, such as *Eucalanus inermis* (fig. 58).

SERGESTES TANTILLOIDES type
The three preceding types are linked to the Boundary Current area and their pendants are three open ocean ranges. *Sergestes tantillus* is the Equatorial open ocean counterpart of the EUCALANUS INERMOIDES type (fig. 58) and is the example of the present range (fig. 101).

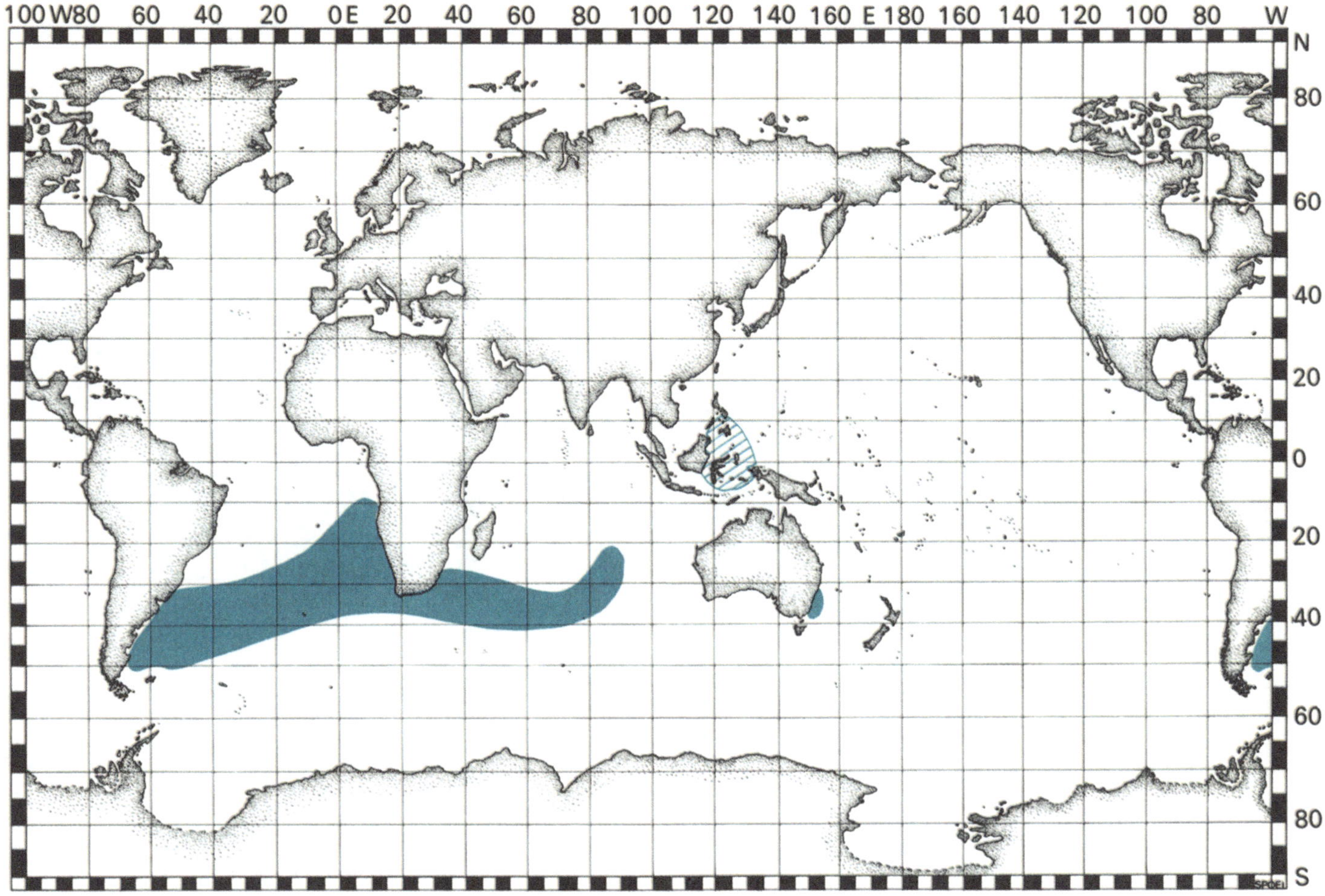

SERGESTES CONSOBRINOIDES type

This type, represented by *Sergestes consobrinus* 'Central water form' (fig. 102) is the oceanic counterpart of the more neritic DESMOPTERUS PACIFICOIDES type (fig. 111). This is underlined by the fact that *Sergestes consobrinus* split off a 'California Current form', sympatric with *Desmopterus pacificus*.

SERGESTES GIBBILOBATOIDES type

Sergestes gibbilobatus (fig. 102) is the representative of the oceanic range which is the counterpart of the SERGESTES BREVISPINATOIDES type (fig. 101).

THALIA SIBOGAOIDES type

A great number of species are endemic to the area around the Indo-Malayan Centre, e.g.: *Thalia sibogae*, *Stylocheiron insulare* and numerous neritic taxa. They all show the THALIA SIBOGAOIDES type (fig. 122).

THALIA LONGICAUDOIDES type

Thalia longicauda lives in the South Transitional belt, without occurring throughout the belt. It is restricted to the Atlantic and Indian Oceans and the Tasman Sea (fig. 122). As this pattern is not at all unique for this species (cf. *Electrona paucirastra*, fig. 148), and as it is different from the beltshaped pattern, though it may be strongly related with it, a separate type has been created.

FIG. 121 Distribution of *Crossota* species. 1 = *C. brunnea*, 2 = *C. norvegica*, 3 = *C. rufobrunnea*, 4 = *C. pedunculata*, 5 = *C. alba*.

FIG. 122 Distribution of *Thalia longicauda* (black) and *Th. sibogae* (hatched).

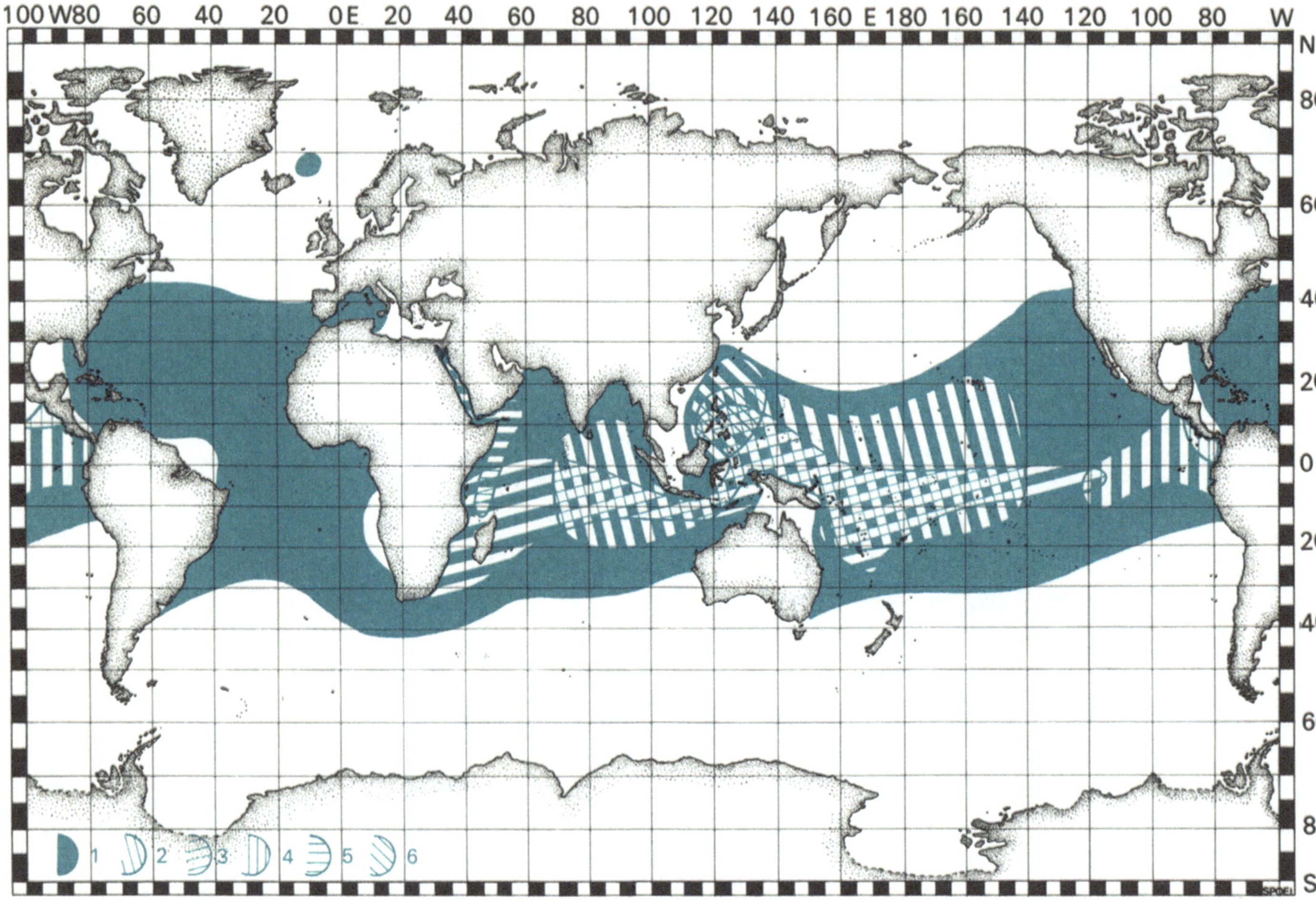

STYLOCHEIRON LONGICORNOIDES type
The distribution in the N. Indian Ocean north of 5°N is represented by *Stylocheiron longicorne* 'N. Indian Ocean form' (fig. 66) but also by *Sergestes semissis* (fig. 102) and by many neritic forms. A great number of species cover both the N. Indian Ocean and the Indo-Malayan waters, e.g. *Stylocheiron indicus* (fig. 66) and *S. affine* 'Indo-Australian form' (fig. 67); these species, however, can better be grouped under the neritic patterns (cf. p. 100).

SALPA GERLACHOIDES type
Restricted ranges for epiplankton in the Circum-Antarctic waters are hard to explain. Sometimes they are in all probability due to insufficient sampling (cf. *Limacina retroversa* forma *australis*, fig. 119). In the S.E. Pacific Ocean, however, an isolated area is found with a few endemic forms. *Salpa gerlachei* (fig. 51) is the representative of this type.

SAGITTA ATLANTICOIDES type
Sagitta serratodentata atlantica is found in the Transitional S. Pacific waters as an endemic taxon most closely related to the Atlantic populations of the species-group. However, as it does not show any traces of a real Transitional beltshaped pattern (cf. fig. 118), it is considered an endemic pattern.

SAGITTA PSEUDOSERRATODENTATOIDES type
Sagitta pseudoserratodentata (fig. 118) is an isolated N. Pacific Central water taxon. As there is no relation to other Central water forms, it is considered an endemic range.

The most common endemics are taxa restricted to one ocean or enclosed sea. The following types are distinguished:
CAVOLINIA TRIDENTATOIDES type from the Mediterranean Sea (p. 99, fig. 53)
SAGITTA EUXIMOIDES type from the Black Sea (p. 99, fig. 124)
SAGITTA BALTHICOIDES type from the Baltic Sea (p. 100, fig. 124)
DIACRIA ERYTHROIDES type from the Red Sea (p. 99, fig. 123)
EUPHAUSIA AMERICANOIDES type from the Atlantic Ocean (p. 85, fig. 109)
SAGITTA SERRATODENTATOIDES type from the Atlantic Ocean (p. 92, fig. 118)
DESMOPTERUS GARDINEROIDES type from the Indian Ocean (p. 86, fig. 111).

No taxa are known which could be considered to be endemic to the Pacific Ocean, occupying larger parts of this ocean.

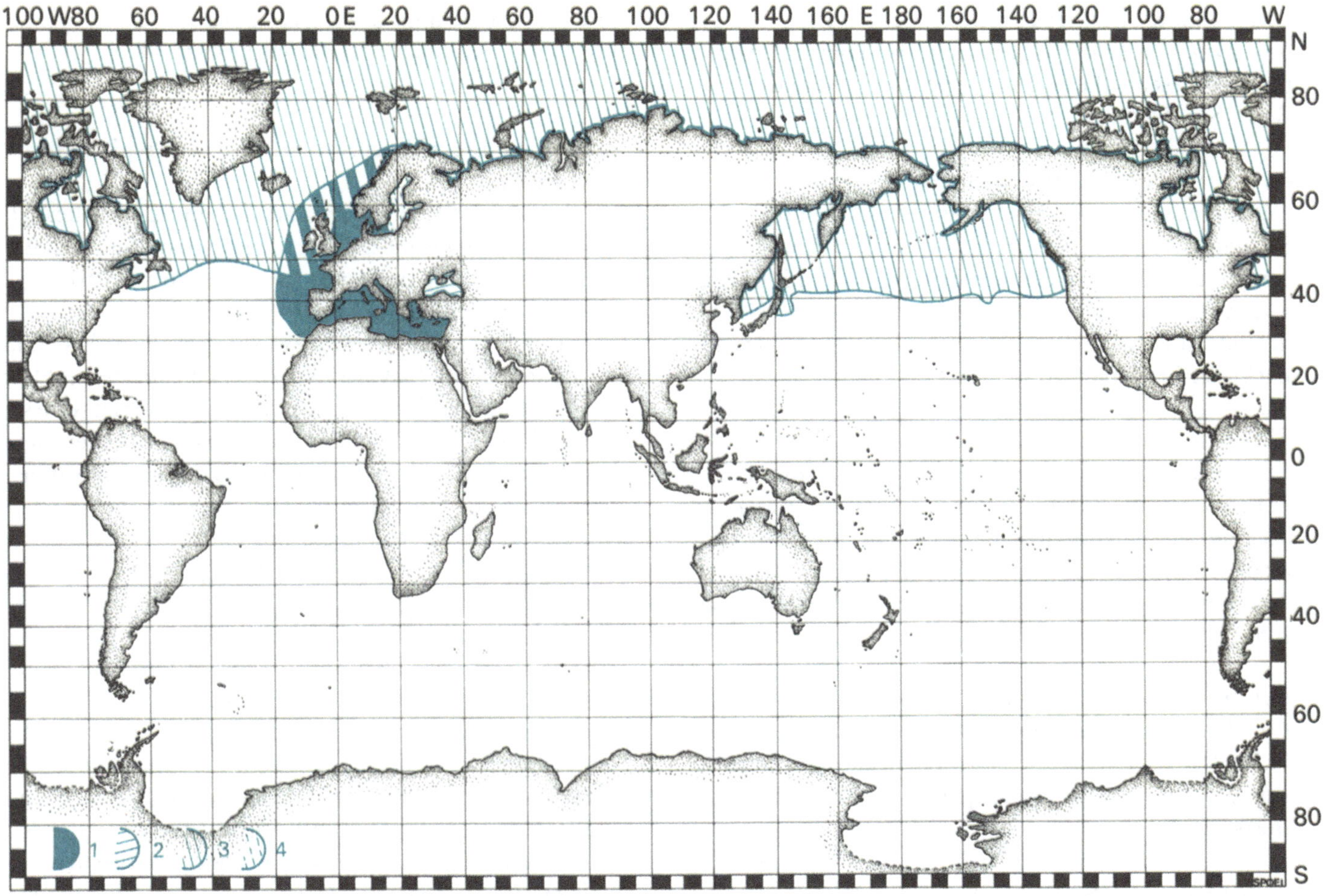

The *Diacria quadridentata* species-group (fig. 123) shows some phenomena not met with in the other species-groups. The species-group is clearly grouped around the Indo-Malayan Centre (fig. 16, centre 33). Yet, *Diacria danae* shows the STYLOCHEIRON CARINATOIDES type (fig. 63) of distribution, which is usually considered to be beltshaped and not to radiate from one centre. *Diacria costata* and *D. quadridentata* are of the THALIA SIBOGAOIDES type (fig. 122) with an inclination to the HYDROMYLES GLOBULOSOIDES type (fig. 112), which are both distributions spreading out of the Indo-Malayan Centre.

DIACRIA ERYTHROIDES type
Diacria erythra (fig. 123) is endemic to the Red Sea and adjacent Indian Ocean, with a strong neritic tendency.

DIACRIA OCCIDENTALOIDES type
Diacria schmidti (fig. 123), described from the E. Pacific, shows the EUCALANUS INERMOIDES type (fig. 58) but this only holds for the recent population. A subfossil population, taxonomically slightly different (forma *occidentalis*), is found in the northern Indo-Malayan waters. As it is a subfossil range, found in a rather high latitudinal position, compared with the THALIA SIBOGAOIDES type (fig. 122), it is considered a separate type (fig. 123). The ranges of the taxa *D. erythra* and *D. schmidti* forma *schmidti* both developed in all probability from the centre 33 (fig. 16), but lost contact as the populations of these taxa disappeared from the E. Indian and W. Pacific Oceans. This is shown by the ancestral subfossil *D. schmidti* forma *occidentalis* occurring in the area from which the species has now disappeared.

FIG. 123 Distribution of the *Diacria quadridentata* species-group. 1 = *D. danae*, 2 = *D. costata*, 3 = *D. quadridentata*, 4 = *D. s. schmidti*, 5 = *D. erythra*, 6 = *D. s. occidentalis*.
FIG. 124 Distribution of *Sagitta setosa* (1), *S. euxima* (2), *S. e. elegans* (3) and *S. e. balthica* (4).

CAVOLINIA TRIDENTATOIDES type
The endemic range occupying only the Mediterranean is exemplified by *Cavolinia tridentata* forma *tridentata* (fig. 53). Only few taxa of the species level are restricted to the Mediterranean, but a great number of infraspecific taxa are.

SAGITTA EUXIMOIDES type
Sagitta euxima is endemic to the Black Sea (fig. 124), and though this taxon has many neritic characters, it is presented here as a special type.

SAGITTA BALTHICOIDES type
Sagitta elegans balthica is certainly also a neritic taxon, but it represents the endemic type restricted to the Baltic Sea (fig. 124).

CAVOLINIA ATLANTICOIDES type
Cavolinia tridentata forma *atlantica* (fig. 53) of the S. Central Atlantic represents a type which is not so commonly found as an endemic range. That this range and the two following ones are still worth to be distinguished, is justified, for example, in the paper of Backus & Craddock (1977).

CAVOLINIA DAKARENSOIDES type
Cavolinia tridentata forma *dakarensis* (fig. 53) of the N.E. Atlantic Tropical, Central and Terminal waters, occupies the endemic area of the present type. This range is a sibling of the next one, and probably determined by the 18 °C Mode Water (cf. Worthington, 1976).

CAVOLINIA BERMUDENSOIDES type
Cavolinia tridentata forma *bermudensis* (fig. 53) of the N.W. Atlantic lives in the same waters at the western side as the former forma occupies at the eastern side of the ocean.

CAVOLINIA AFFINOIDES type
Cavolinia tridentata forma *affinis* (fig. 53) of the eastern part of the S. Central Pacific is an example of a rare endemic pattern. Most taxa showing this pattern are closely related to S. Transitional water taxa, and are also found more closely to the S. American coasts.

POEOBIUS MESEROIDES type
A typical endemic taxon is *Poeobius meseres* (fig. 125), a polychaetous species without known relatives in the ocean. It is found in the N. Pacific 'Subarctic', Transitional and Boundary Current waters. It is usually considered an oceanic species, though its distribution slightly resembles the neritic distribution of *Carinaria cristata* (cf. fig. 113).

Neritic distribution patterns

The neritic patterns can be divided roughly into the warm-water neritic patterns between 50 °N and 40 °S and the cold-water neritic patterns at latitudes higher than 30 °N or 30 °S.

In principle, each oceanic coast has its own neritic fauna, which shows a latitudinal succession of Polar, Subpolar, Transitional, Subtropical and Tropical elements. The ecomorphology of the coasts makes it possible to distinguish, within these belts, smaller faunal areas, especially for neritic-inlet species and near-coastal

FIG. 125 Distribution of *Poeobius meseres*.
FIG. 126 Distribution of the genus *Pterotrachea*.

taxa. The more wide-spread taxa are dealt with in this paragraph.

The W. Pacific neritic area is much more complicated, from a zoogeographical point of view, than the E. Pacific. First there are the two types: THALIA SIBOGAOIDES (fig. 122) and LUCIFER HANSENOIDES (fig. 130) (cf. p. 104), furthermore there are the ranges off the E. China coasts and off the E. Australian coasts. From the first two mentioned types, however, a number of different types can be derived, of which it is questionable whether they really constitute distinct patterns. As this cannot yet be proven the two most clear patterns are given below as ACETES JAPONICOIDES type and ACETES SERRULATOIDES type (figs 103 & 104, resp.).

WARM-WATER PATTERNS

The narrowly restricted warm-water patterns are numerous and only a selection of the most representative ones is given. The ranges based on local populations, and the ranges of no importance to the general zoogeography discussed in this atlas, have been omitted. Some neritic patterns were already mentioned as endemic ranges (cf. the SAGITTA BALTHICOIDES fig. 124, the SAGITTA EUXIMOIDES fig. 124, the DIACRIA ERYTHROIDES fig. 123 types) and the patterns restricted to the E. Tropical Pacific and those around the Indo-Malayan Archipelago (cf. the SAGITTA BIERIOIDES fig. 117, the DESMOPTERUS PACIFICOIDES fig. 111, the SERGESTES-BREVISPINATOIDES fig. 101, the EUCALANUS INERMOIDES fig. 58 and the THALIA SIBOGAOIDES fig. 122 types) can very well be grouped with the distant-neritic taxa as well. For most of the other more coast-bound neritic taxa with a small degree of dispersal, it would not have been necessary to avoid geographic names, and to use the naming introduced for the oceanic taxa, but for the sake of uniformity, the same system of naming is used throughout.

PTEROTRACHOIDES type
The total area of all distant-neritic warm-water species is nicely represented by the range of the genus *Pterotrachea* (fig. 126) (cf. fig. 7). There are a large number of genera and species-groups (cf. *Acetes*, figs 103, 104 & 105) with this pattern. This range and those of most other distant-neritic warm-water taxa show three discontinuities also found for the separate species of *Pterotrachea* viz.: the centre of the Pacific Ocean, the centre of the Atlantic Ocean and the Panama area. In many species also discontinuities are found west of S.W. Africa and in the N. Indian Ocean, which may be so extensive that, as in *Euphausia sanzoi* (cf. Brinton, 1975), the impression is given that there are two separate ranges in the W. Indian and W. Pacific Oceans.

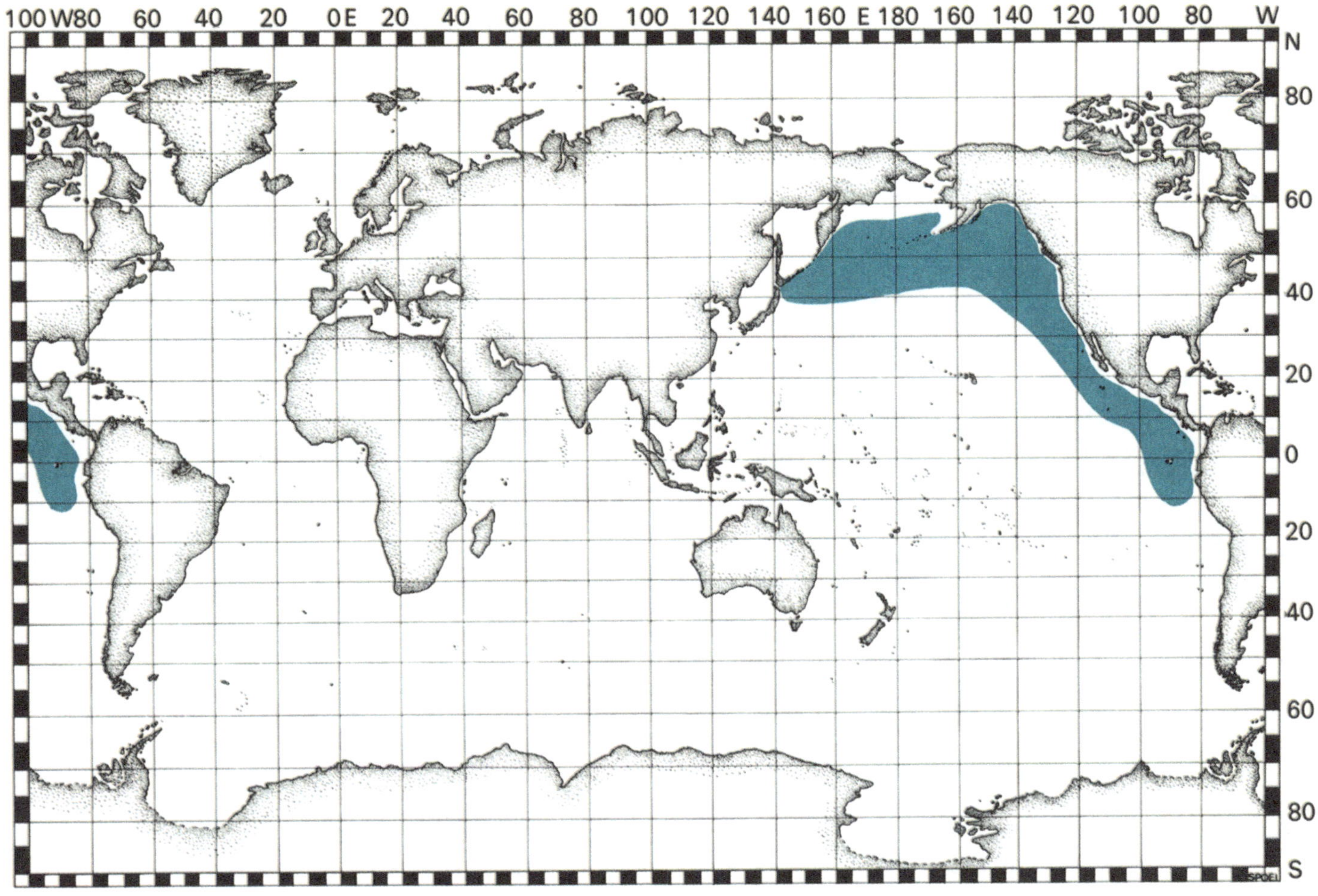
100 W80 60 40 20 0E 20 40 60 80 100 120 140 160 E 180 160 140 120 100 80 W
N 80 60 40 20 0 20 40 60 80 S
SPOEL

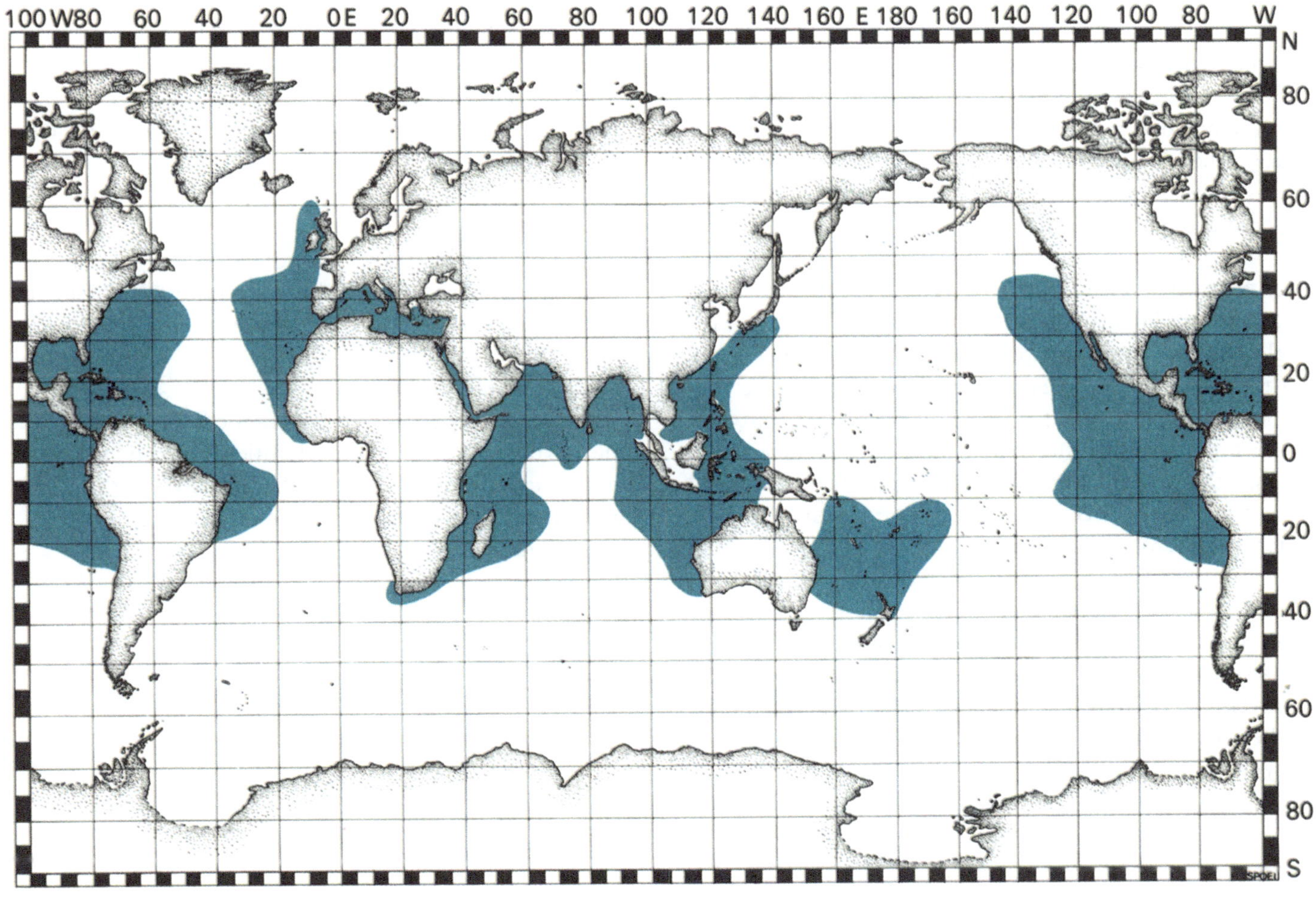
100 W80 60 40 20 0E 20 40 60 80 100 120 140 160 E 180 160 140 120 100 80 W
N 80 60 40 20 0 20 40 60 80 S
SPOEL

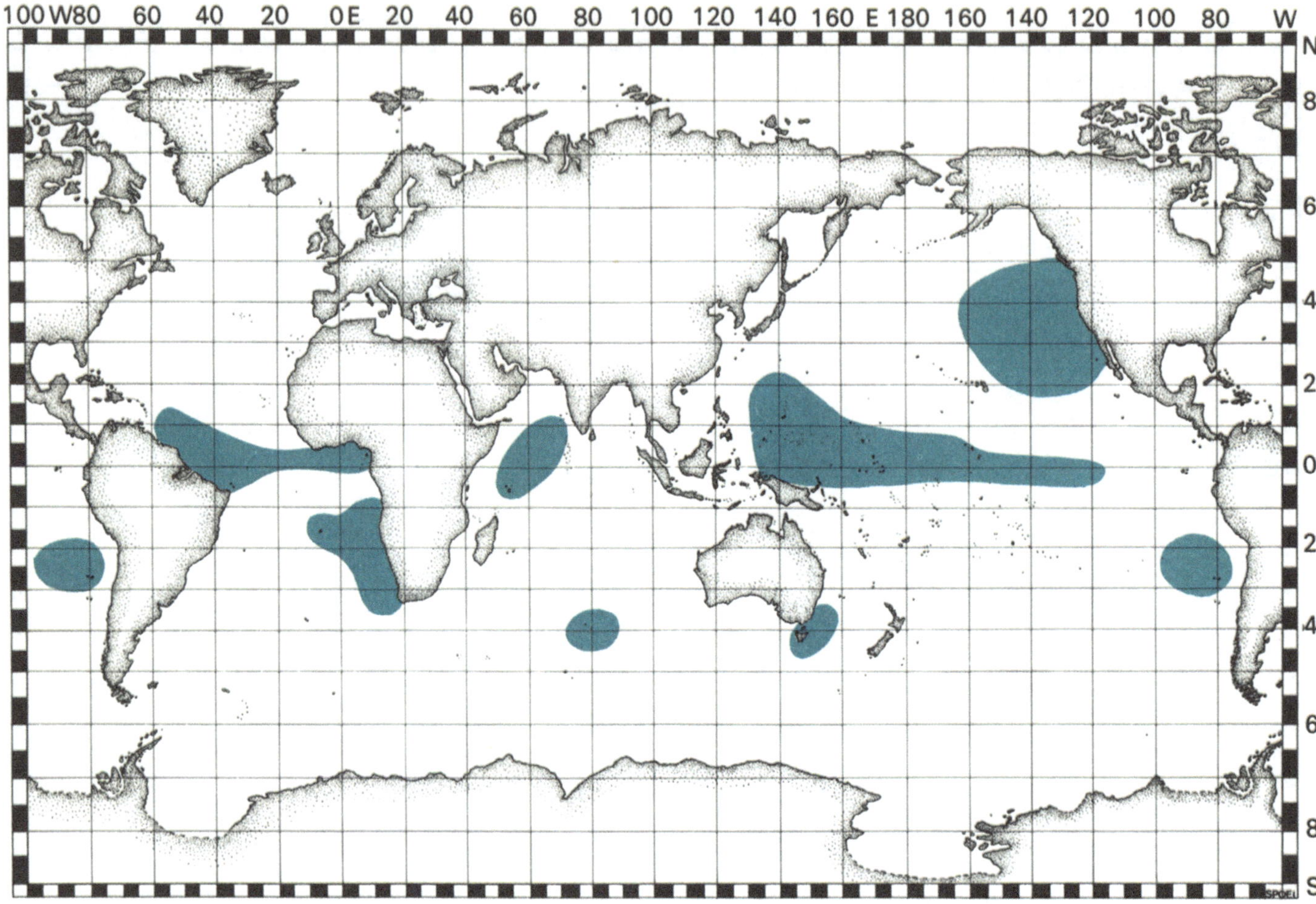

ELECTRONA RISSOOIDES type

Some distant-neritic species have, in contrast with the preceding one, very wide but discontinuous and scattered ranges. *Electrona rissoi* (fig. 127) has been chosen as an example, though it is difficult to indicate a very fixed pattern for this 'type'.

CARDIAPODA PLACENTOIDES type

The tropical distant-neritic distribution of the circum-global type is best exemplified by *Cardiapoda placenta* (fig. 128) which lives between 40° N and 40° S. The distribution is disjunct, as in many neritic taxa, but in this type especially due to influences of cold currents along the west sides of the southern continents. This type is more restricted than the PTEROTRACHOIDES type (fig. 126).

SAGITTA BEDOTOIDES type

In fig. 129 three discontinuous warm-water neritic patterns are shown. The SAGITTA BEDOTOIDES type covers the Indo-W. Pacific and E. Pacific and resembles the oceanic pattern of the CAVOLINIA GLOBULOSOIDES type (fig. 98).

FIG. 127 Distribution of *Electrona rissoi*.
FIG. 128 Distribution of *Cardiapoda placenta*.
FIG. 129 Distribution of *Sagitta tenuis* (1), *S. bedoti* (2) and *S. friderici* (3).

SAGITTA FRIDERICOIDES type

This type (fig. 129) covers the E. Atlantic and E. Pacific and is closely related to the next one.

SAGITTA TENUOIDES type

The present type (fig. 129) covers the E. Pacific and occurs on both sides of the Atlantic. Probably its geological history is linked with the EUPHAUSIA GIBBOIDOIDES type (fig. 107) in the oceanic environment.

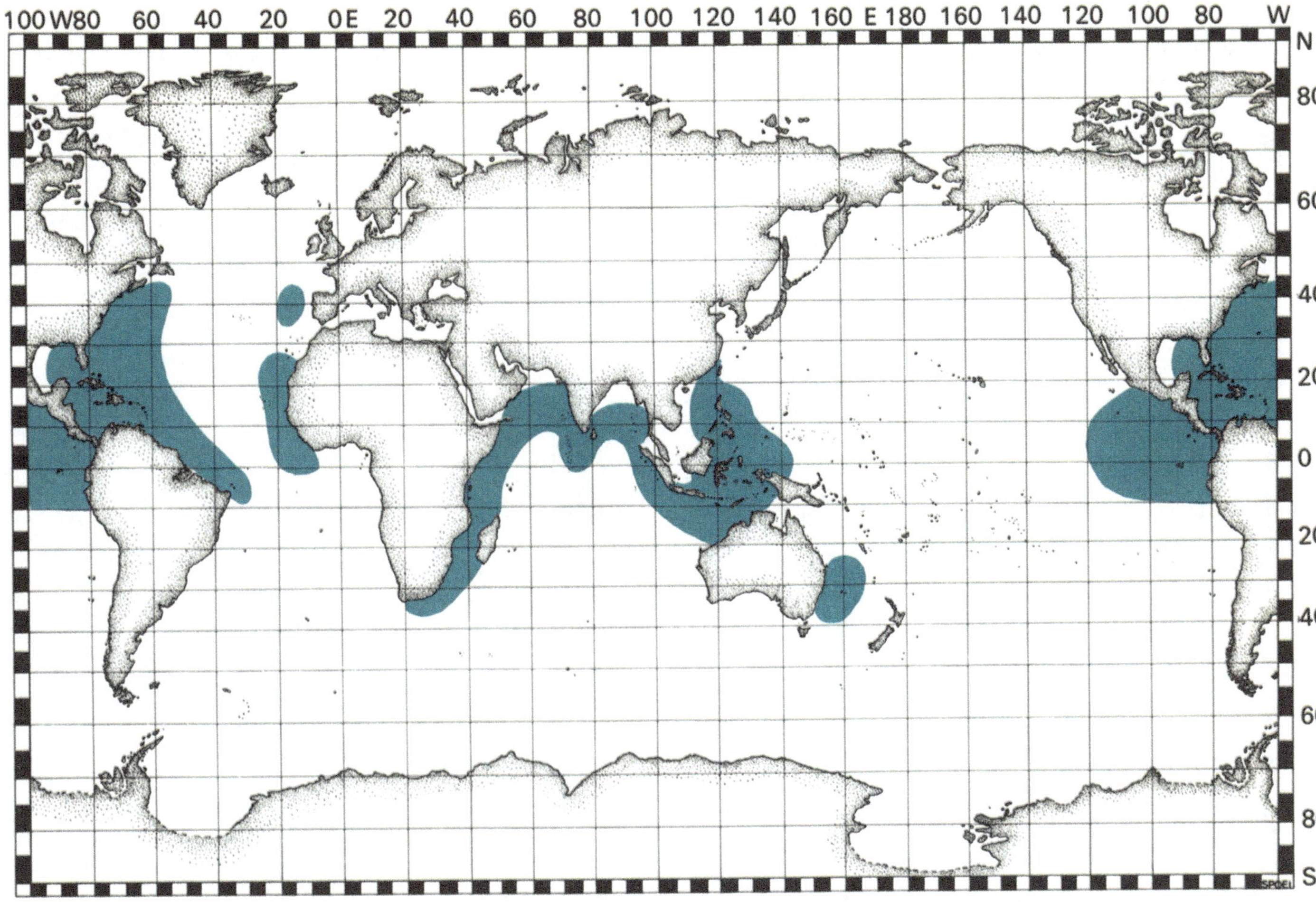
100 W80 60 40 20 0E 20 40 60 80 100 120 140 160 E 180 160 140 120 100 80 W
N 80 60 40 20 0 20 40 60 80 S

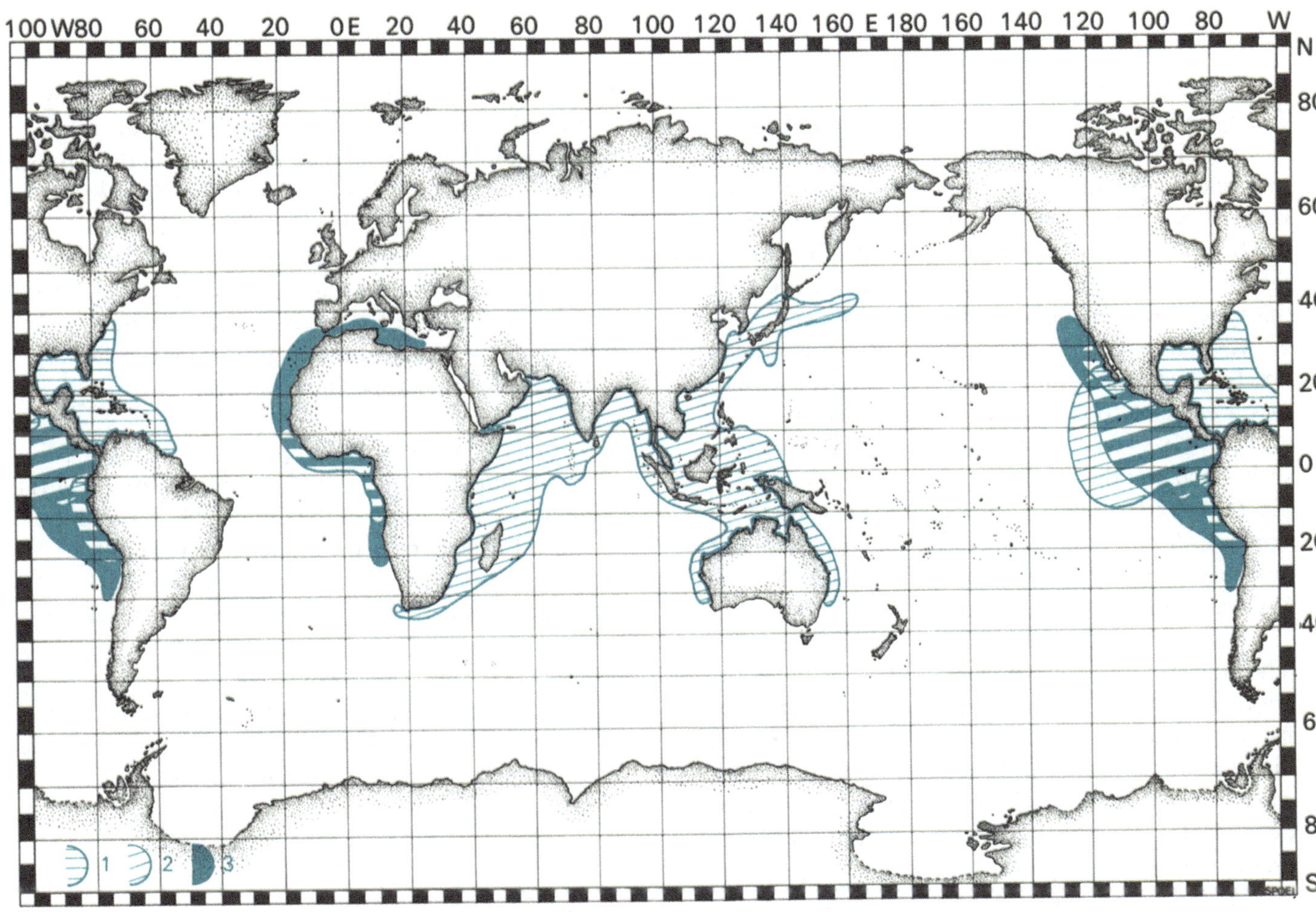
100 W80 60 40 20 0E 20 40 60 80 100 120 140 160 E 180 160 140 120 100 80 W
N 80 60 40 20 0 20 40 60 80 S
1
2
3

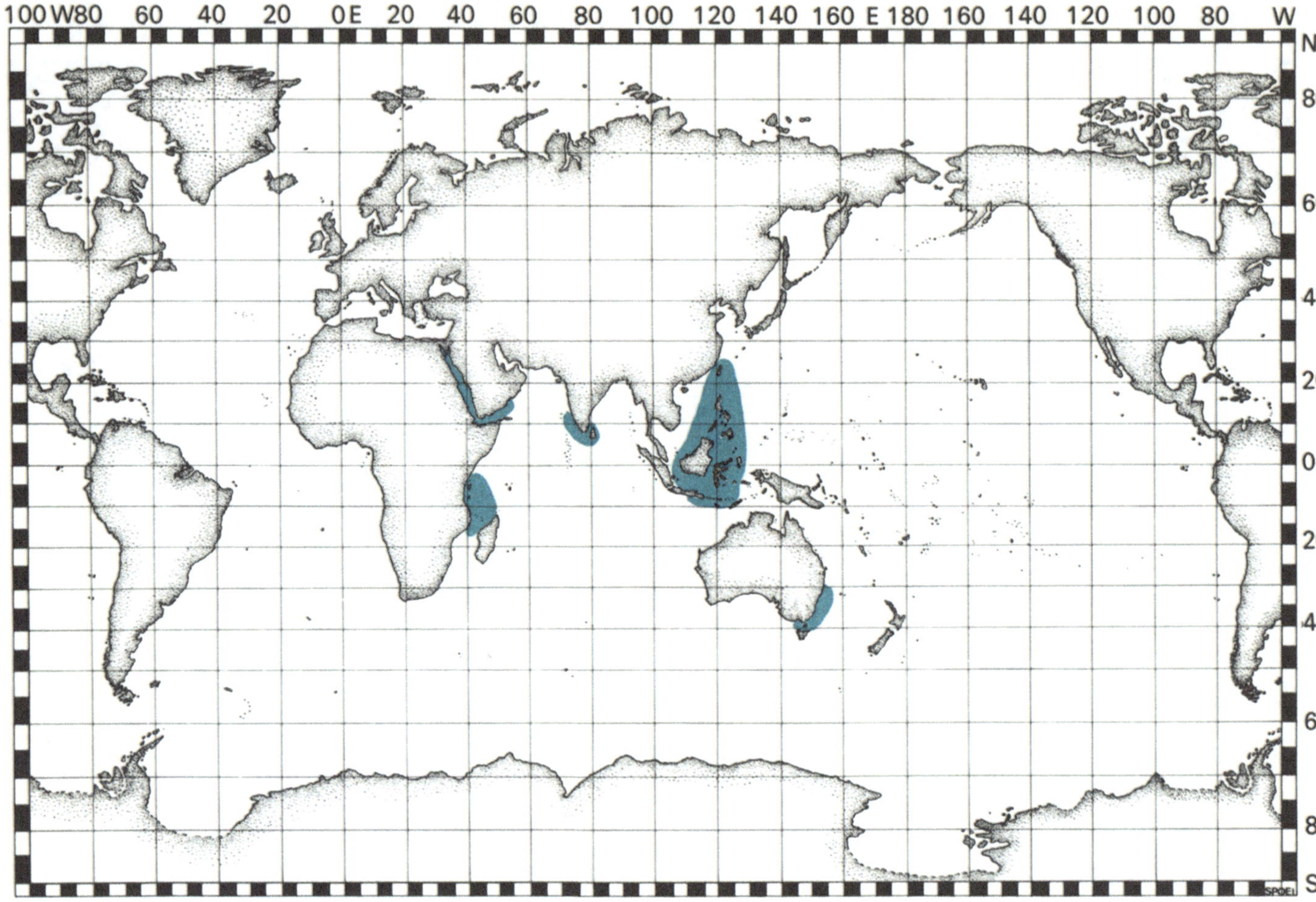

LUCIFER HANSENOIDES type

This type concerns disjunct patterns of the Indian Ocean and Indo-Malayan waters. As e.g. in *Lucifer hanseni* (fig. 130), this range is always composed of an Indo-Malayan population, a population near S. India, an E. African and a W. Australian population. Sometimes the Red Sea and the waters around Japan are enclosed in this type, and the populations near E. Africa and W. Australia may be found at other latitudes in other taxa (cf. *Lucifer penicillifer*, fig. 131). Comparison of the SAGITTA BEDOTOIDES (fig. 129) and LUCIFER HANSENOIDES types with the ranges of oceanic taxa such as those of the *Diacria quadridentata* group (fig. 123) and *Hydromyles globulosa* (fig. 112) provides a firm basis for the assumption that either continuous or discontinuous ranges may develop from the Indo-Malayan Centre (fig. 16, centre 33).

The ranges of warm-neritic species depicted in fig. 131, are all of moderate size: they are smaller than those in fig. 128 and larger than the types in fig. 132. Probably the distributions here are chiefly determined by the climatic zones, while the distributions given in fig. 132 are more dependent on the local hydrological conditions. For example, *Litopenaeus* is to be considered a tropical species while *Acetes americana* and *A. marinus* (figs 103 & 105) are primarily determined in their dispersal by hydrological conditions. The same comparison can be made for the Tropical *Temora discaudata* in the Tropical belt and *Labidocera jollae*, *L. diandra* and *L. lubbocki* (fig. 132), found in the influence of the California Current, of the Equatorial waters and of the Panama Gulf water and the Peru Current offshoots, resp. The hydrological data are too scarce to make a rigid subdivision between these two groups so that they are given here without a specific sequence.

MELICERTUS AZTECOIDES type

This Subtropical pattern of *Melicertus aztecus* (fig. 131) is confined to the north of the Gulf of Mexico and the coasts of Florida under the influence of the Gulf Stream; its northern border coïncides with the transition between Subtropical and Subarctic waters.

LITOPENAEUS SETIFEROIDES type

Litopenaeus setiferus (fig. 131) represents the Tropical W. Atlantic neritic distribution occupying an area under direct influence of the Equatorial Current.

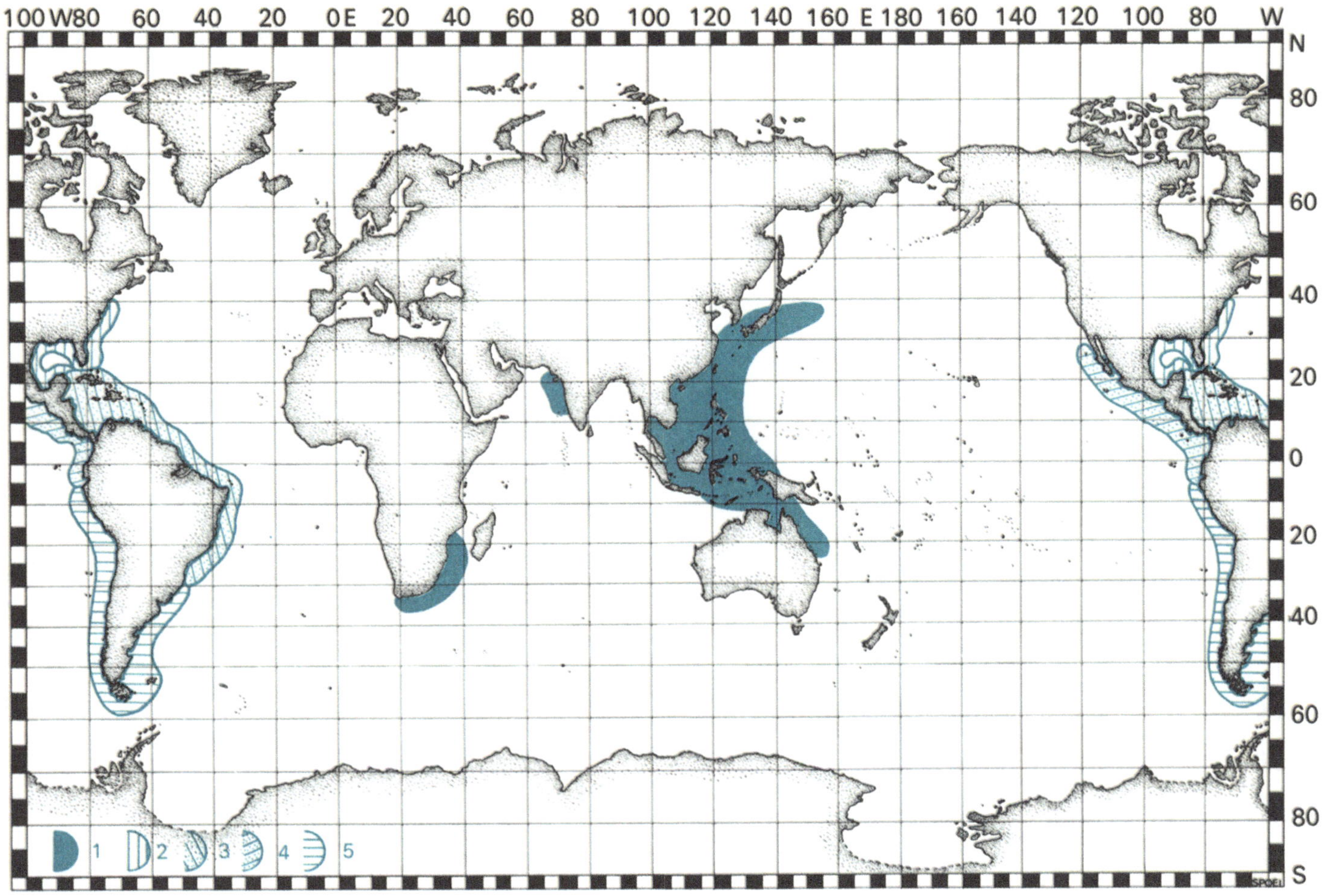

FIG. 130 Distribution of *Lucifer hanseni*.
FIG. 131 Small ranges of neritic warm-water species.
1 = *Lucifer penicillifer*, 2 = *Melicertus aztecus*,
3 = *Litopenaeus setiferus*, 4 = *Temora discaudata*,
5 = *Centropagus brachiatus*.

CENTROPAGUS BRANCHIATOIDES type
This range around S. America covers the S. Transitional and Subantarctic waters, as illustrated by *Centropagus branchiatus* (fig. 131). A subdivision of this range is still possible, as shown by the next pattern and the CALANUS AUSTRALOIDES type (fig. 79).

CALANUS CHILENSOIDES type
Calanus chilensis (fig. 79) proves that the western coast of S. America, in the Transitional belt, can form a separate area with taxa not occurring in the Atlantic so that a specific type is proposed for these taxa.

ACETES MARINOIDES type
Part of the Tropical W. Atlantic is strongly influenced by the Amazonas outflow; salinity in this area is low and nutrient levels are high. *Acetes marinus* (fig. 105) is a typical representative of this area (cf. p. 80).

TEMORA DISCAUDATOIDES type
Temora discaudata (fig. 131) represents the Tropical E. Pacific neritic area. Most closely linked with the present distribution pattern is the EUCALANUS INERMOIDES type (fig. 58), which represents the distant-neritic taxa of the Tropical E. Pacific.

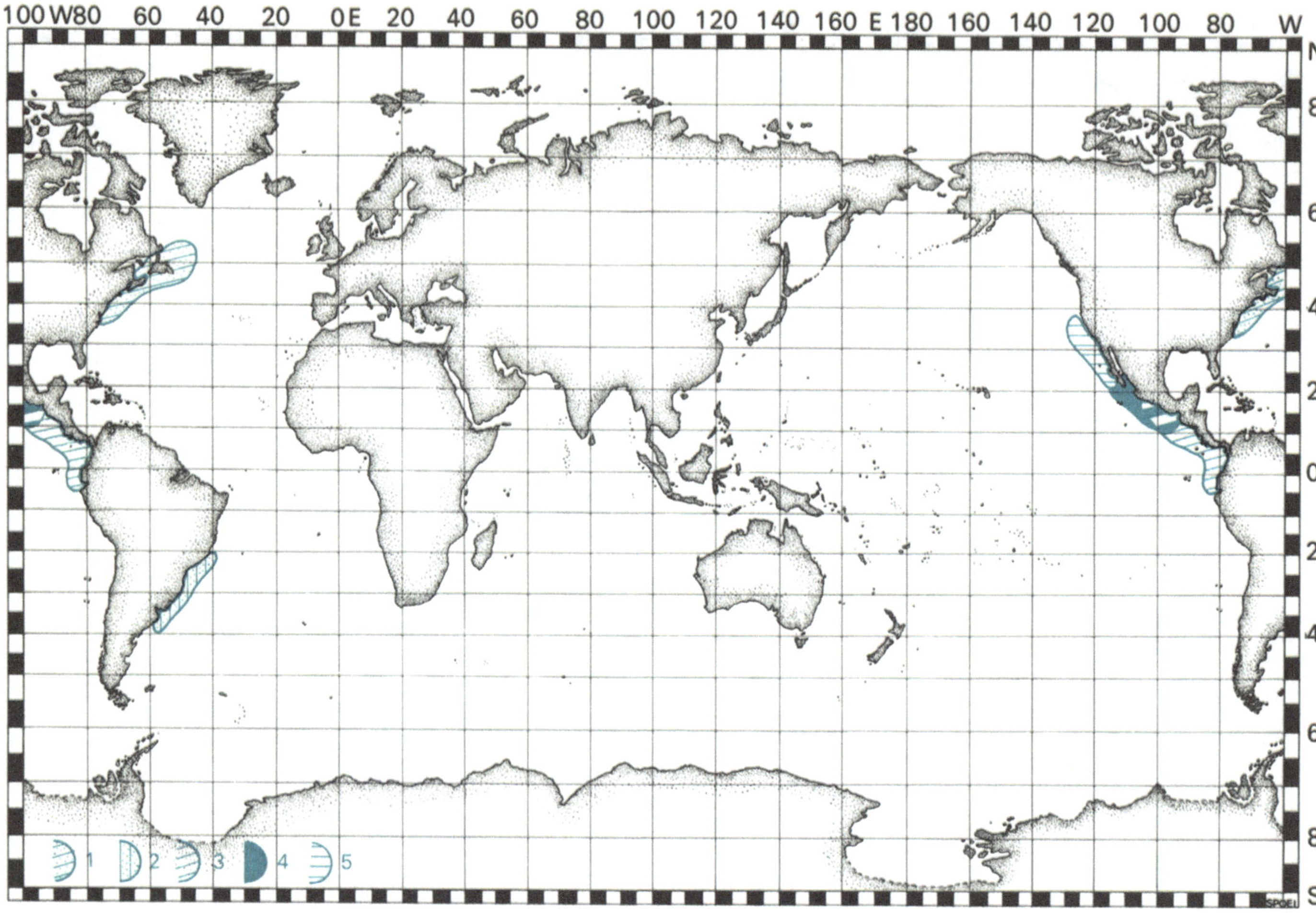

In fig. 132 the small ranges are shown into which the TEMORA DISCAUDATOIDES type can be split up. This map also gives two small-sized W. Atlantic ranges.

FIG. 132 Small ranges of neritic warm-water species. 1 = *Temora longicornis*, 2 = *Melicertus paulensis*, 3 = *Labidocera lubbocki*, 4 = *L. diandra*, 5 = *L. jollae*.

LABIDOCERA LUBBOCKOIDES type

Labidocera lubbocki (fig. 132) from the Panama Gulf forms the southern section of the neritic Tropical E. Pacific. Probably this range should be located some ten degrees more southwards, but there was no good example of such a distribution.

LABIDOCERA DIANDROIDES type

The area where the Equatorial Current originates, is marked by the distribution of *Labidocera diandra* (fig. 132) and it forms the central part of the Tropical E. Pacific.

LABIDOCERA JOLLAOIDES type

The northern section of this E. Pacific neritic distribution is occupied by *Labidocera jollae* (fig. 132).

For the E. Pacific, a pattern of distributions can be distinguished between 30°N and 30°S, comprising three series of related ranges composed of oceanic, distant-neritic and neritic ranges, as shown in diagram III. In the other oceans such regular sequences of types are not found.

LABIDOCERA KOLPOIDES type

For the sake of completeness, *Labidocera kolpos* has been selected as an example of the distributions restricted to the Gulf of California. This distribution is not depicted, as the range exactly covers the Gulf.

ACETES JAPONICOIDES type

The tropical neritic species *Acetes japonicus* (fig. 103) is found in the Indo-Malayan waters, W. Pacific and N. Indian Oceans. Though this species has been selected as an example for the present pattern, not all representatives show the same extension into the two oceans. *A. indicus*, for example, does not penetrate the Pacific at all. If there is no penetration into either ocean, the distribution is considered to be of the following type.

ACETES SERRULATOIDES type

Typical in the pattern of *Acetes serrulatus* (fig. 104) is that it occurs entirely west of Borneo in the S. China Sea. In this characteristic it differs from the THALIA SIBOGAOIDES type (fig. 122) which is found more east of Borneo.

Diagram III
The three pairs of Pacific Ocean warm-water patterns.

Area:	Oceanic types:	Distant-neritic types:	Neritic types:
North Tropical	SERGESTES CONSOBRINOIDES	DESMOPTERUS PACIFICOIDES	LABIDOCERA JOLLAOIDES
Central Tropical	SERGESTES TANTILLOIDES	EUCALANUS INERMOIDES	LABIDOCERA DIANDROIDES
South Tropical	SERGESTES GIBBILOBATOIDES	SERGESTES BREVISPINATOIDES	LABIDOCERA LUBBOCKOIDES

ACETES CHINENSOIDES type
This neritic warm-water range of *Acetes chinensis* (fig. 103) is the northern Subtropical W. Pacific distribution comparable with the LABIDOCERA JOLLOIDES type (fig. 132) in the east.

ACETES AUSTRALOIDES type
The range of *Acetes sibogae australis* (fig. 104) has been selected as an example of the southern warm-water range of the neritic W. Pacific. For the comparable distribution on the east side of the ocean no good example can be indicated, due to the cold currents brushing the S. American west coast.

ACETES JOHNOIDES type
The range of *Acetes johni* (fig. 105), identical with that of *A. sibogae sibogalis* (fig. 104), forms the present type. This particular area frequently forms part of the wider range of the LUCIFER HANSENOIDES type (fig. 130).

ACETES NATALENSOIDES type
Like the preceding type, the present one, represented by *Acetes natalensis* (fig. 105), in some areas forms part of the LUCIFER HANSENOIDES type (fig. 130). It is a kind of warm-water distribution, though it is partly sympatric with the CALANUS AUSTRALOIDES type (fig. 79).

DIACRIA ERYTHROIDES type
This type has already been mentioned (p. 99) but can be given here too, as a neritic type. Characteristic of this pattern is that representatives usually slightly penetrate into the Indian Ocean, sometimes up to the vicinity of Madagaskar and sometimes only as subfossils, preserved in the surface sediments (fig. 123).

TEMORA LONGICORNOIDES type
The Transitional species *Temora longicornis* (fig. 132) lives partly in the Slope water but is not restricted to this water mass. Its range partly coïncides with Subarctic and with Subtropical ranges.

MELICERTUS PAULENSOIDES type
Melicertus paulensis (fig. 132) forms in its distribution the southern counterpart of the preceding range of the Transitional zone.

The general validity of these warm-water neritic ranges is proved by the barracudas of the American west coast which have a distribution perfectly fitting in with the neritic types given here (Graves & Somero, 1982).

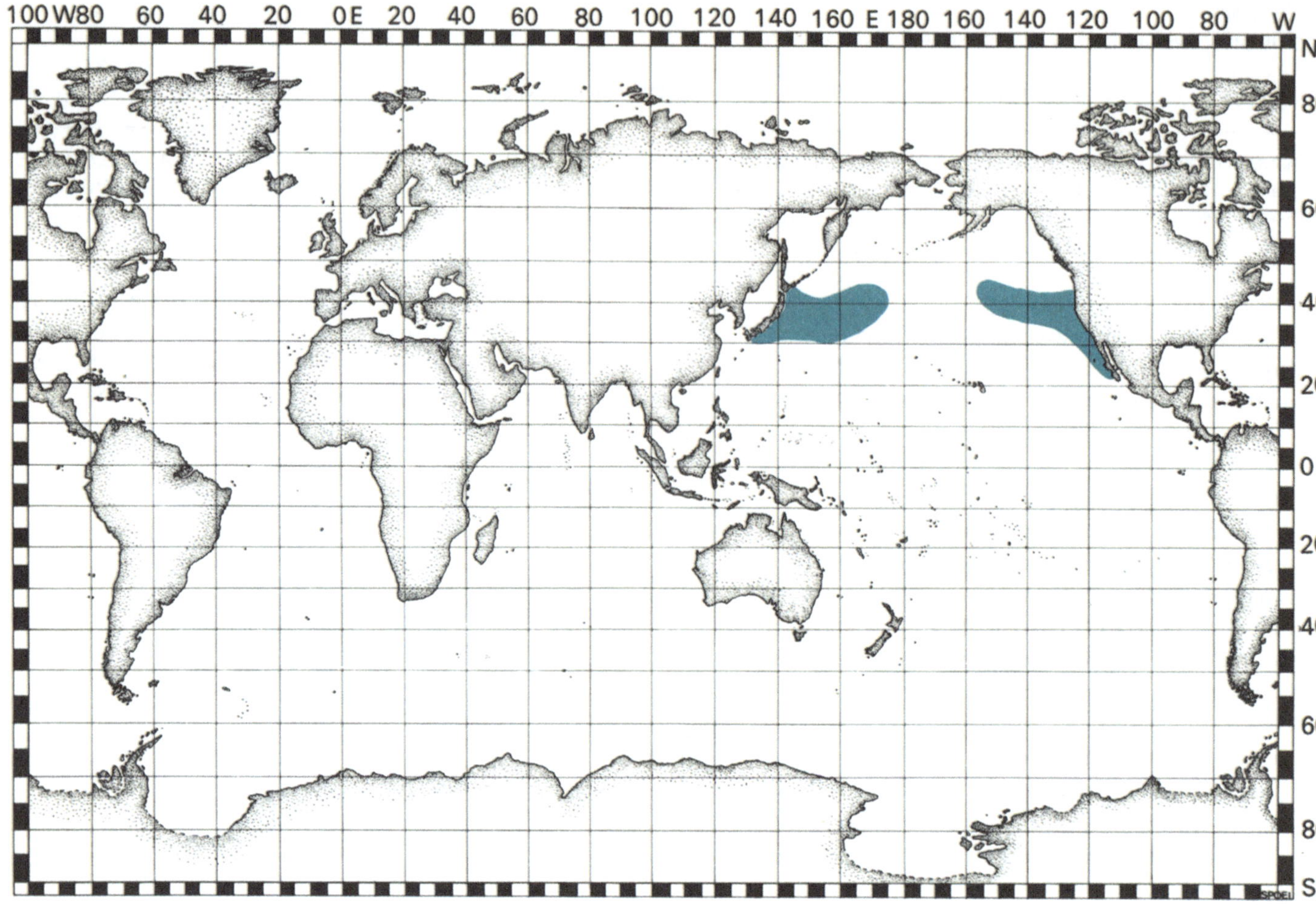

COLD-WATER NERITIC RANGES

CALANUS AUSTRALOIDES type

The neritic and distant-neritic areas in the southern Transitional waters are best represented by *Calanus australis* (fig. 79). This disjunct range is sometimes occupied by one species, sometimes by more species with restricted (endemic) ranges.

IDIACANTHUS ANTROSTOMOIDES type

Idiacanthus antrostomia (fig. 133) is the example of a N. Pacific Transitional water distant-neritic distribution. A comparable pattern is shown by some of the forms of *Calanus plumchrus* (fig. 136).

SAGITTA SETOSOIDES type

Sagitta setosa s.l. (the forma *S. setosa batava* is a coastal form) (fig. 124) is a distant-neritic species of the Transitional zone of the N.E. Atlantic. The Transitional zone at the west side of the Atlantic is so narrow (cf. figs 1-13) that a distant-neritic taxon could not be distinguished.

FIG. 133 Distribution of *Idiacanthus antrostomia.*
FIG. 134 Distribution of *Thysanoessa raschii* (black) and *Clione limacina* forma *antarctica* (hatched).
FIG. 135 Distribution of *Euphausia crystallorophias.*

THYSANOESSA RASCHIOIDES type

The neritic taxa in the Arctic-Subarctic belt of both the Atlantic and Pacific Oceans are best represented by *Thysanoessa raschii* (fig. 134). Characteristic is the interocean connection in this species north of Canada. Real Arctic neritic taxa are difficult to distinguish from oceanic ice-bound taxa. *Clione limacina* forma *antarctica* (fig. 134) in the south, is an Antarctic species of which it is likewise difficult to state whether it is ice-bound or neritic. The forma presented here is part of a much more widely distributed species, *Clione limacina*. The forma *antarctica* shows the EUPHAUSIA TRIACANTHOIDES type of distribution (fig. 86).

EUPHAUSIA CRYSTALLOROPHIOIDES type

Euphausia crystallorophias (fig. 135) is considered a species of ice-covered neritic areas, which is also shown by the fact that it spreads less far northwards from the continent than *Clione limacina* (fig. 134) or *Euphausia triacantha* (fig. 86).

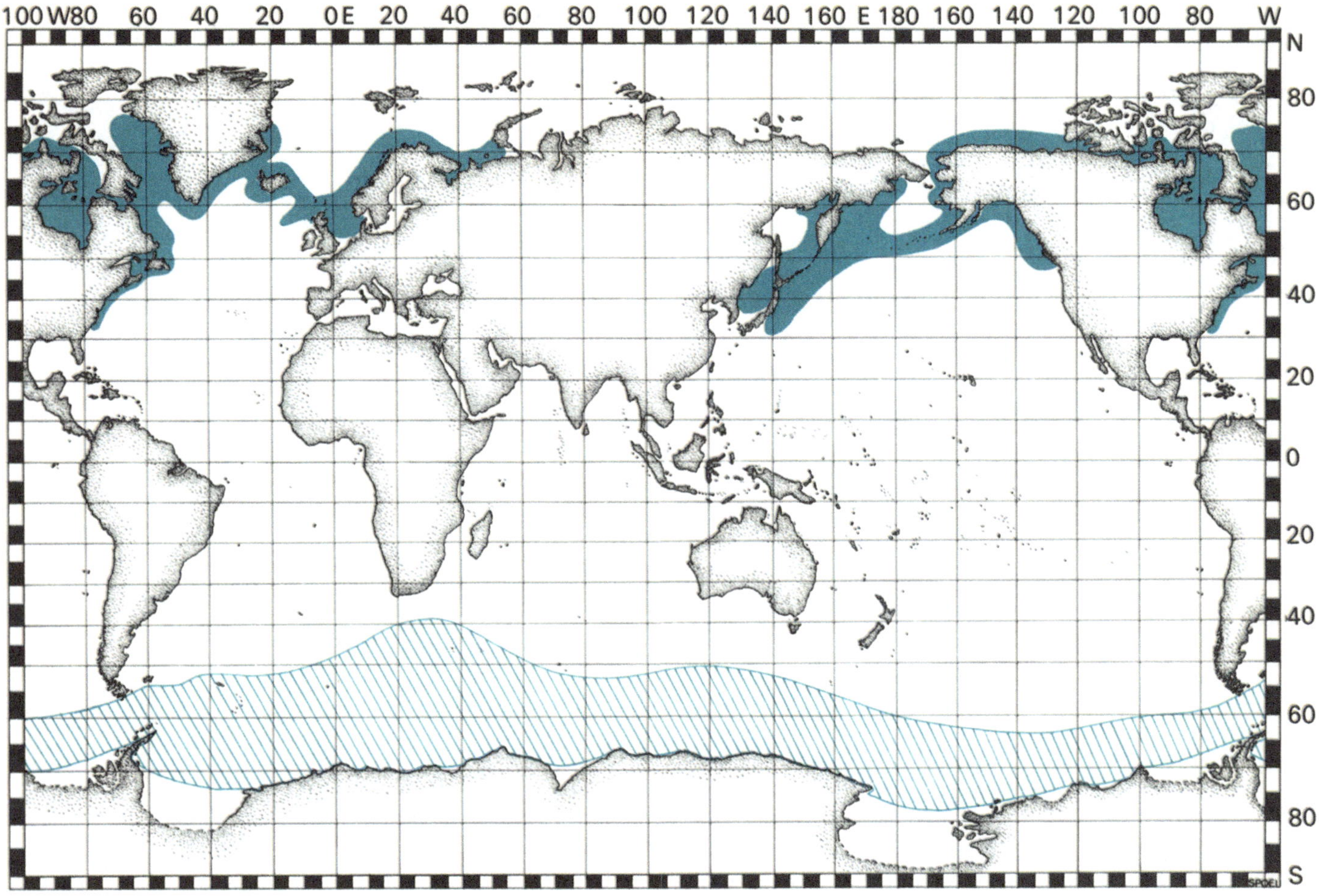
100 W 80 60 40 20 0 E 20 40 60 80 100 120 140 160 E 180 160 140 120 100 80 W
N 80 60 40 20 0 20 40 60 80 S
SPOEL

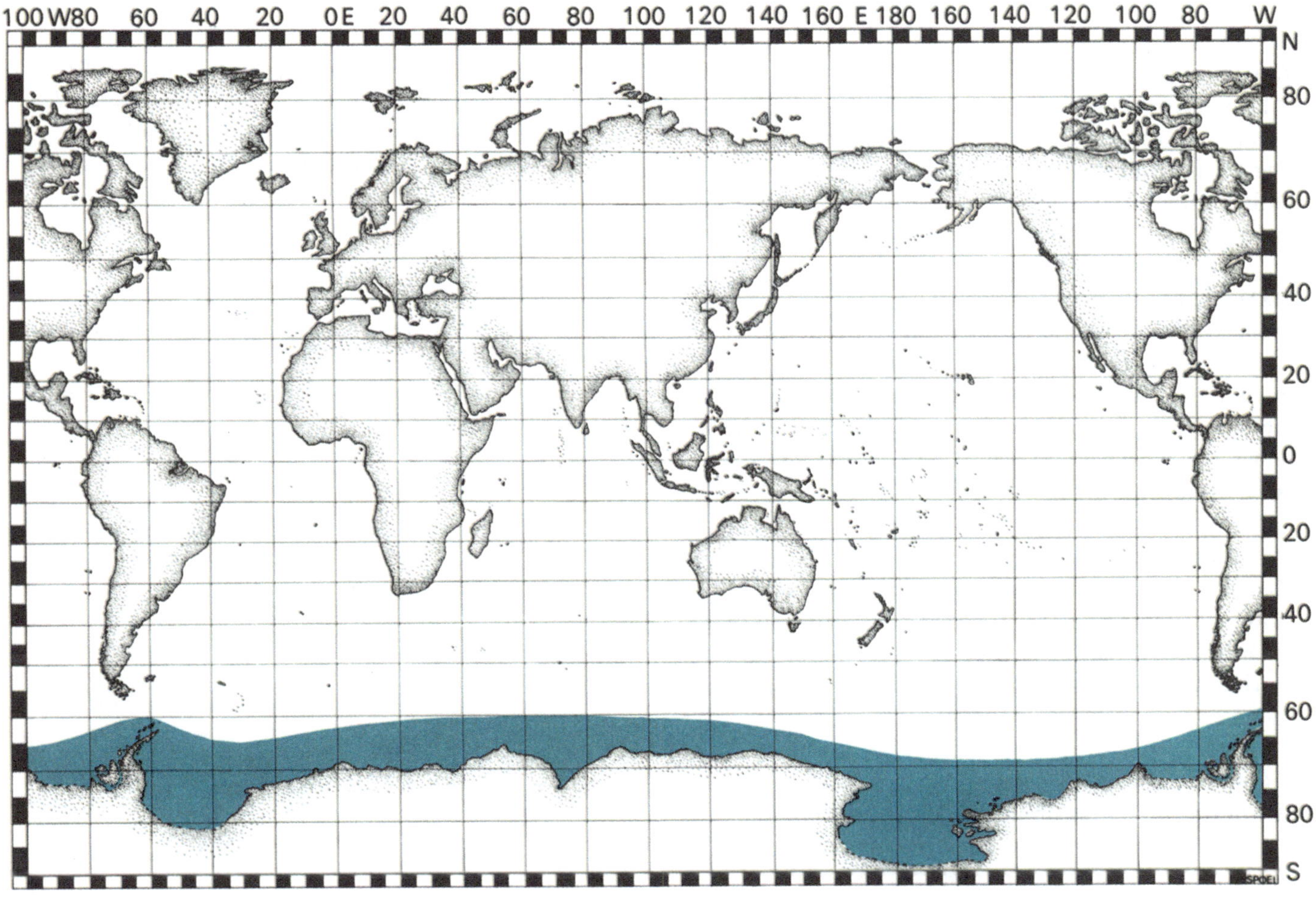
100 W 80 60 40 20 0 E 20 40 60 80 100 120 140 160 E 180 160 140 120 100 80 W
N 80 60 40 20 0 20 40 60 80 S
SPOEL

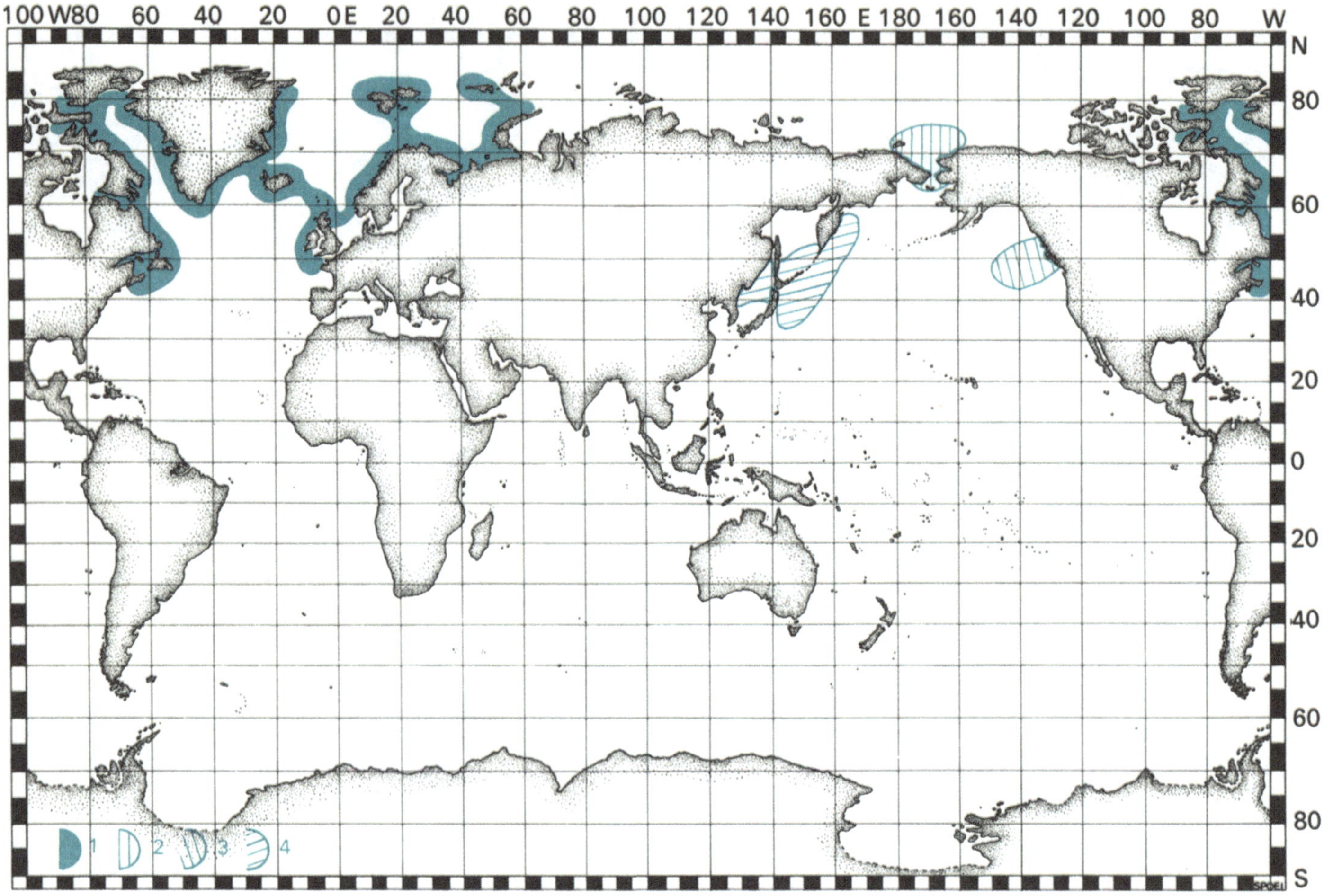

CALANUS MARSHALLOIDES type

The present range is comparable with that of *Sagitta setosa* (fig. 124) in the N. Atlantic. *Calanus marshallae* (fig. 78) is a Subarctic neritic species restricted to the N.E. Pacific without penetrating either westwards or eastwards to the Atlantic (cf. fig. 137).

GAMMARELLUS HOMAROIDES type

Some Subarctic taxa of neritic nature are found on both sides of one ocean basin. In the Atlantic such taxa are monotypic; in the N. Pacific such taxa frequently become polytypic as shown in the two types given here in fig. 136. *Gammarellus homari* is a Subarctic-Arctic species endemic to the Atlantic.

CALANUS PLUMCHROIDES type

Calanus plumchrus (fig. 136) is a polytypic taxon subdivided into a northern, a western and an eastern form. These three forms together represent the Pacific neritic Arctic-Transitional polytypic distribution of this type.

PONTOGENIA INERMOIDES type

Dispersal of N. Transitional Pacific species into the Atlantic Ocean is possible. *Pontogenia inermis* (fig. 137) clearly shows a pattern reflecting this dispersal as the species has not yet reached the E. Atlantic coasts.

An areal subdivision of the neritic seas (cf. p. 8), based on the previously discussed distributions, is given in fig. 138.

FIG. 136 Distribution of *Gammarellus homaris* and *Calanus plumchrus*. 1 = *Gammarellus homaris*, 2 = *Calanus plumchrus* 'Northern form', 3 = *C. plumchrus* 'Eastern form', 4 = *C. plumchrus* 'Western form'.
FIG. 137 Distribution of *Pontegenia inermis*.
FIG. 138 Areal subdivision of the neritic seas.

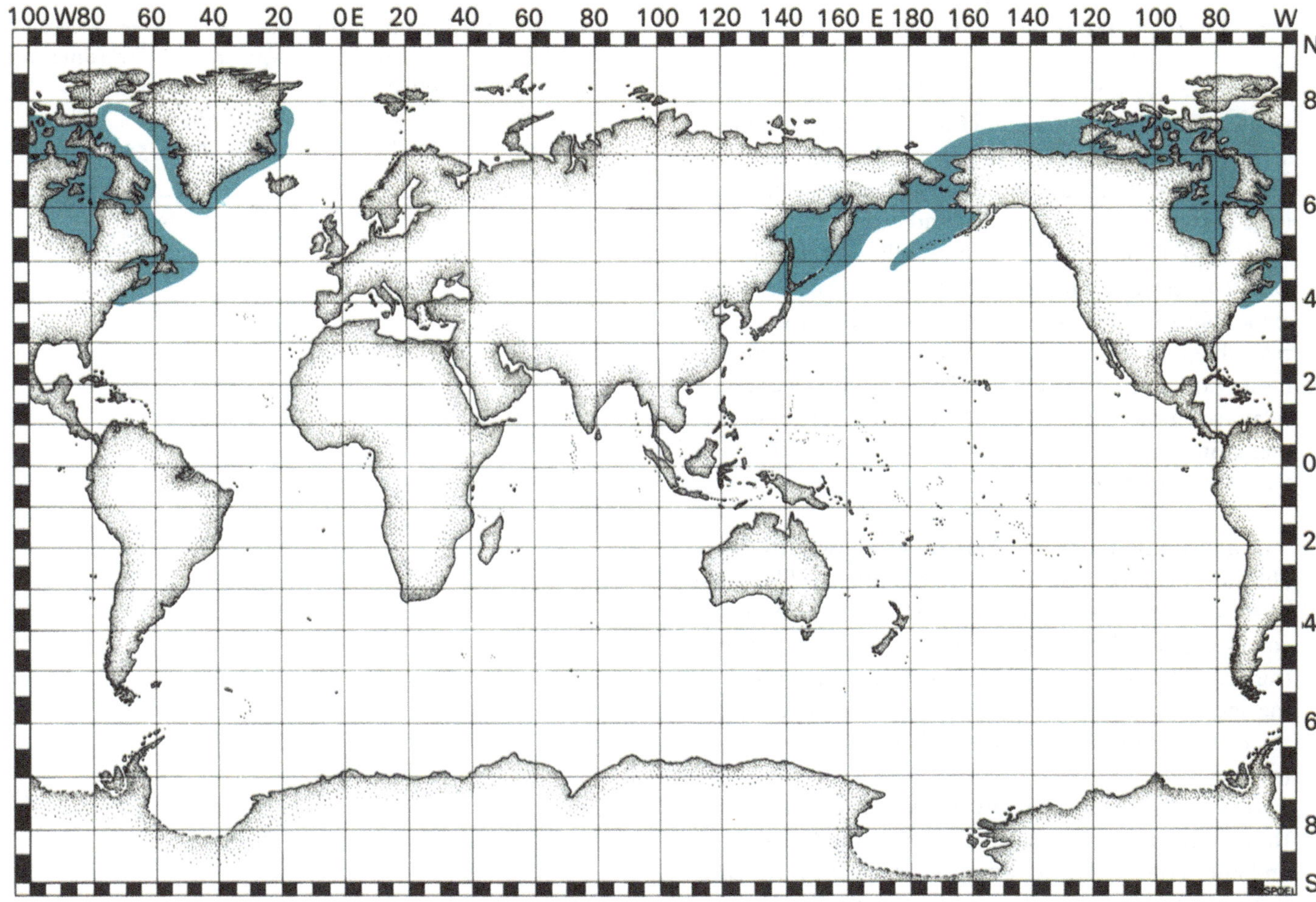
100 W 80 60 40 20 0 E 20 40 60 80 100 120 140 160 E 180 160 140 120 100 80 W
N 80 60 40 20 0 20 40 60 80 S

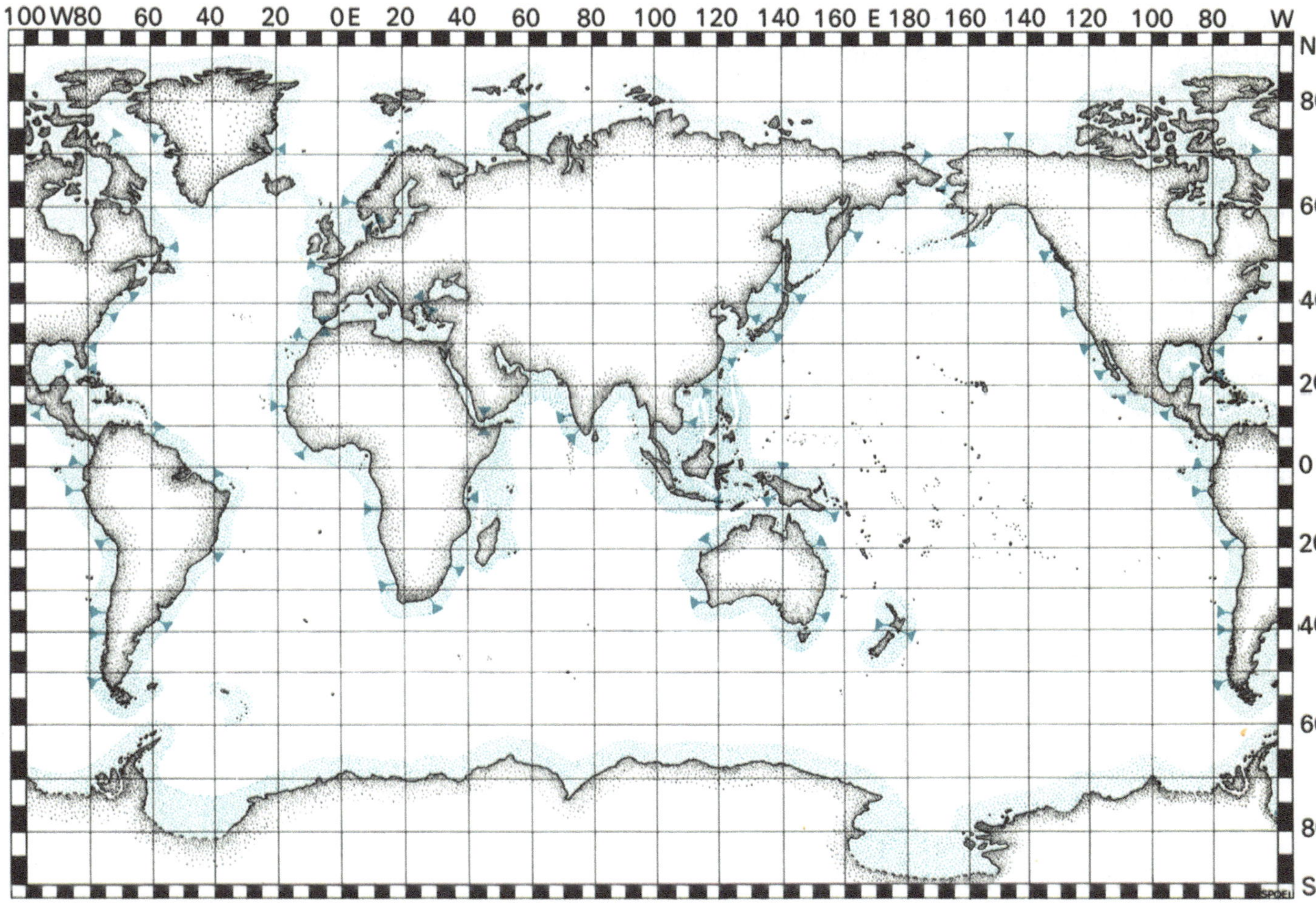
100 W 80 60 40 20 0 E 20 40 60 80 100 120 140 160 E 180 160 140 120 100 80 W
N 80 60 40 20 0 20 40 60 80 S

Vertebrate patterns (fishes excluded)

The marine vertebrates, fishes excluded, comprise the whales, dolphins, reptiles and birds. Reptiles and birds always depend on land for their reproduction. Their reproductive ranges are as a consequence restricted to little islands, rocks and beaches and cannot be considered to be actually oceanic. Still, seven patterns of Albatros distributions are given below, indicating the areas comprising all migratory movements, to show the resemblances with oceanic ranges. Of the mammals, five figures are given, showing the same resemblances and demonstrating that even smaller endemic ranges of plankton can also be found among the completely nektonic mammals.

The distributions of some Albatrosses show patterns absolutely comparable with plankton ranges. The range of *Phoebertia palpebrata* (fig. 139) in Subantarctic and corresponds with the EUPHAUSIA LUCENSOIDES type (fig. 84). This migratory range is very common for marine birds. The Pinguins are found more in a range of the EUPHAUSIA CRYSTALLOROPHIOIDES type (fig. 135).

It is remarkable, that many planktonic animals of these southern regions have relatives in the N. Atlantic and not in the N. Pacific, whereas for marine birds the reverse is found. *Diomedea nigripes* (fig. 139) and *D. albatrus* (fig. 139) in the N. Pacific are examples of such species with relatives in the southern belts. These two species cover the N. Transitional waters in a pattern of the CALANUS PACIFICOIDES type (fig. 79).

Diomedea irrorata (fig. 140) shows the SERGESTES BREVISPINOSOIDES type (fig. 101), *D. immutabilis* (fig. 140) the SAGITTA PSEUDOSERRATODENTATOIDES type (fig. 118), *D. bulleri* (fig. 140) the SAGITTA ATLANTICOIDES type (fig. 118) and *Phoebertia fusca* (fig. 140) resembles in its dispersal the THALIA LONGICAUDOIDES type (fig. 122). It is clear that the sympatry of bird and plankton species points to the existence of larger biotopes in which different groups, influenced by different environmental conditions, and interrelated by special interactions, are found together.

FIG. 139 Distribution of Albatrosses. 1 = *Diomedea nigripes*, 2 = *D. albatrus*, 3 = *Phoebertia palpebrata*, 4 = breeding sites.
FIG. 140 Distribution of Albatrosses (see also fig. 139). 1 = *Phoebertia fusca*, 2 = *Diomedea immutabilis*, 3 = *D. irrorata*, 4 = *D. bulleri*, 5 = breeding sites.

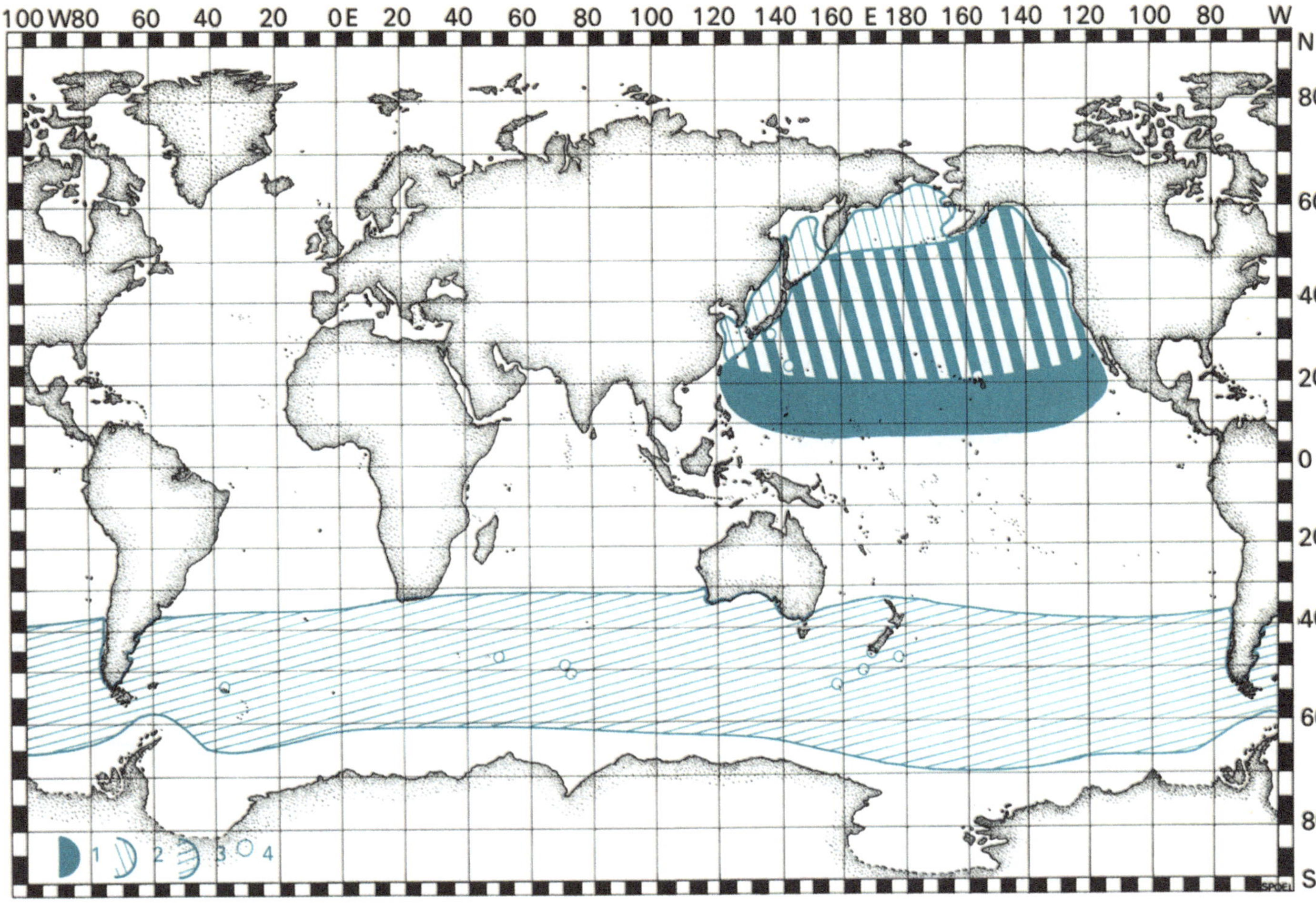
100 W80 60 40 20 0 E 20 40 60 80 100 120 140 160 E 180 160 140 120 100 80 W
N 80 60 40 20 0 20 40 60 80 S
1 2 3 4

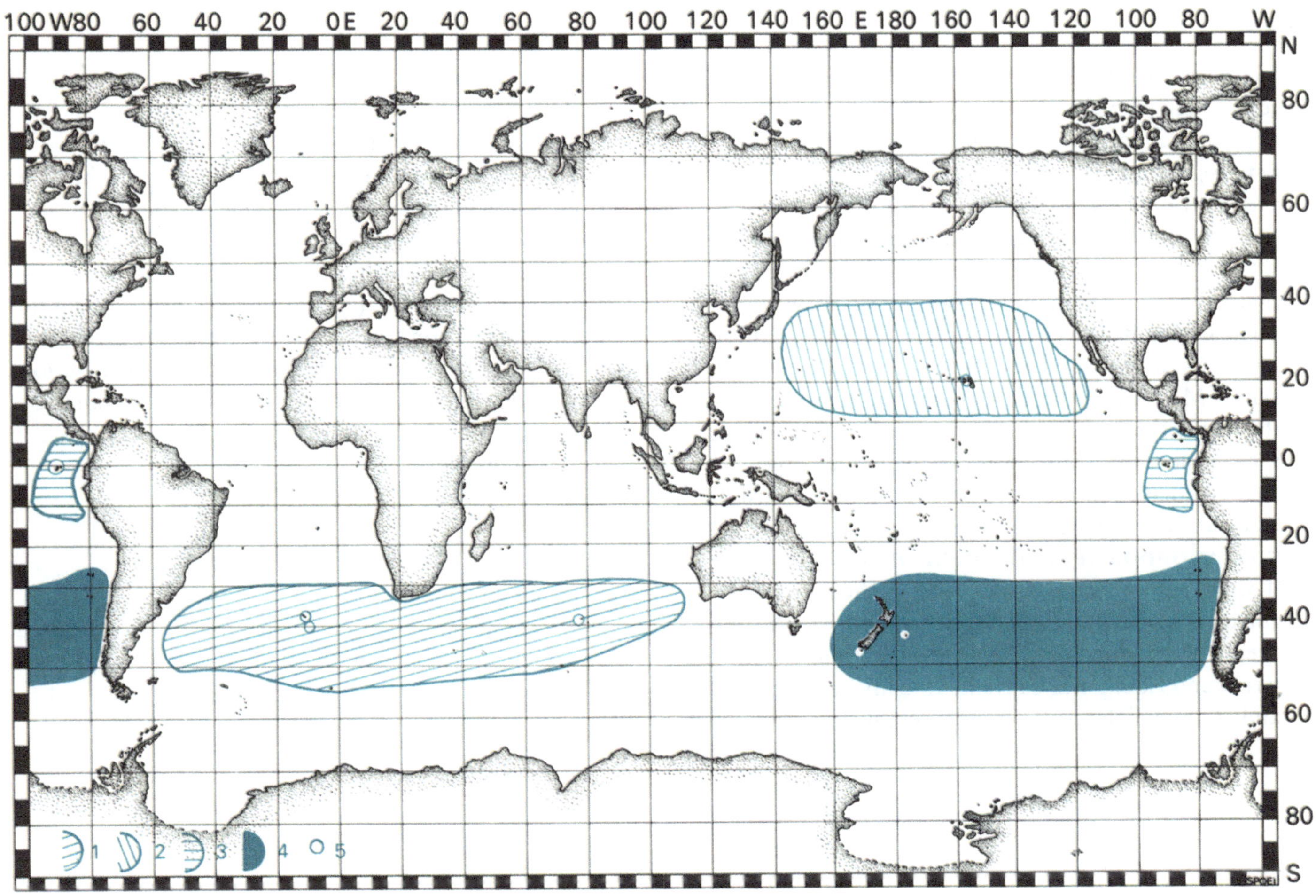
100 W80 60 40 20 0 E 20 40 60 80 100 120 140 160 E 180 160 140 120 100 80 W
N 80 60 40 20 0 20 40 60 80 S
1 2 3 4 5

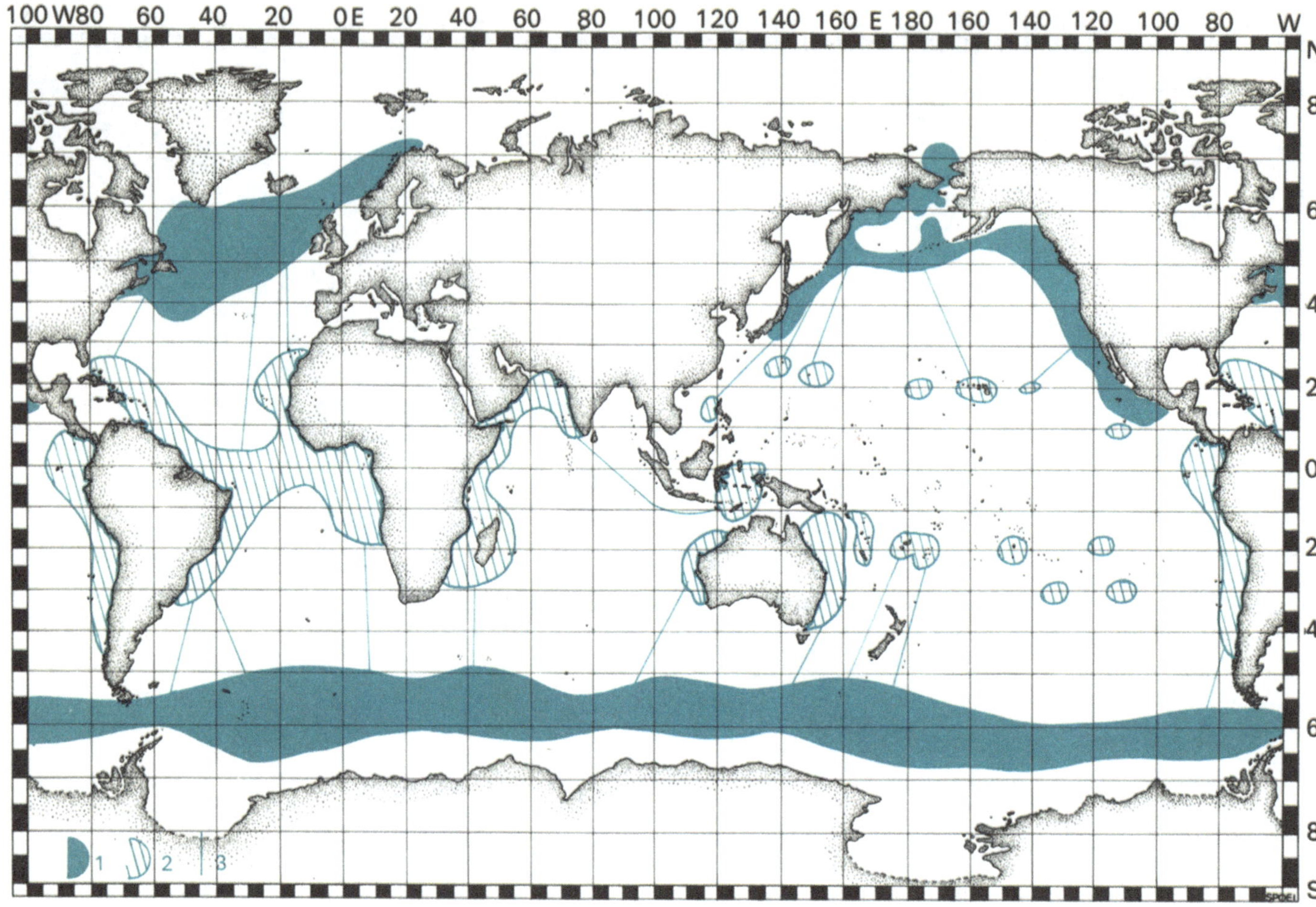

In marine mammals also some peculiar patterns resembling plankton distributions are found. The breeding range of the humback-whale *Megaptera noveanglia* (fig. 141) strikingly resembles the range of the pteropod *Limacina helicina* (fig. 82). In this mollusc, however, the N. Atlantic, N. Pacific and Southern Ocean waters are populated by different infraspecific taxa. Some studies of the humback-whale also showed population differences in the different oceans, but it is not yet possible to give these populations a taxonomic status. Migration is probably responsible for mixing of the populations and consequent geneflow. The winter range is not to be compared with plankton patterns, though it has some characteristics in common with the EUPHAUSIA PSEUDOGIBBOIDES type (fig. 77) and the THYSANOPODA OBTUSIFRONSOIDES type (fig. 93). This does not mean real similarity, as the humback-whales migrate into the centres of upwelling areas for their high productivity. These areas are found along the coasts and Equatorial Currents roughly between 30°N and 30°S, which results in the resemblances mentioned.

The dolphin *Stenella euphrosyne* (fig. 142) shows the 30°N to 30°S distribution of the EUCALANUS SUBTENUOIDES type (fig. 52) with the east-west variation. The distributions of the three forms, however, do not show the characteristics of beltshaped patterns. The 'Atlantic form' shows the PERACLIS BISPINOSOIDES type (fig. 60), the 'Tropical form' the HYDROMYLES GLOBULOSOIDES type (fig. 112) and the 'Pacific form' shows the SERGESTES TANTILLOIDES type (fig. 101). The difference between the distribution of these dolphins and plankton is that in plankton the ranges here given are never occupied by infraspecific groups of one species.

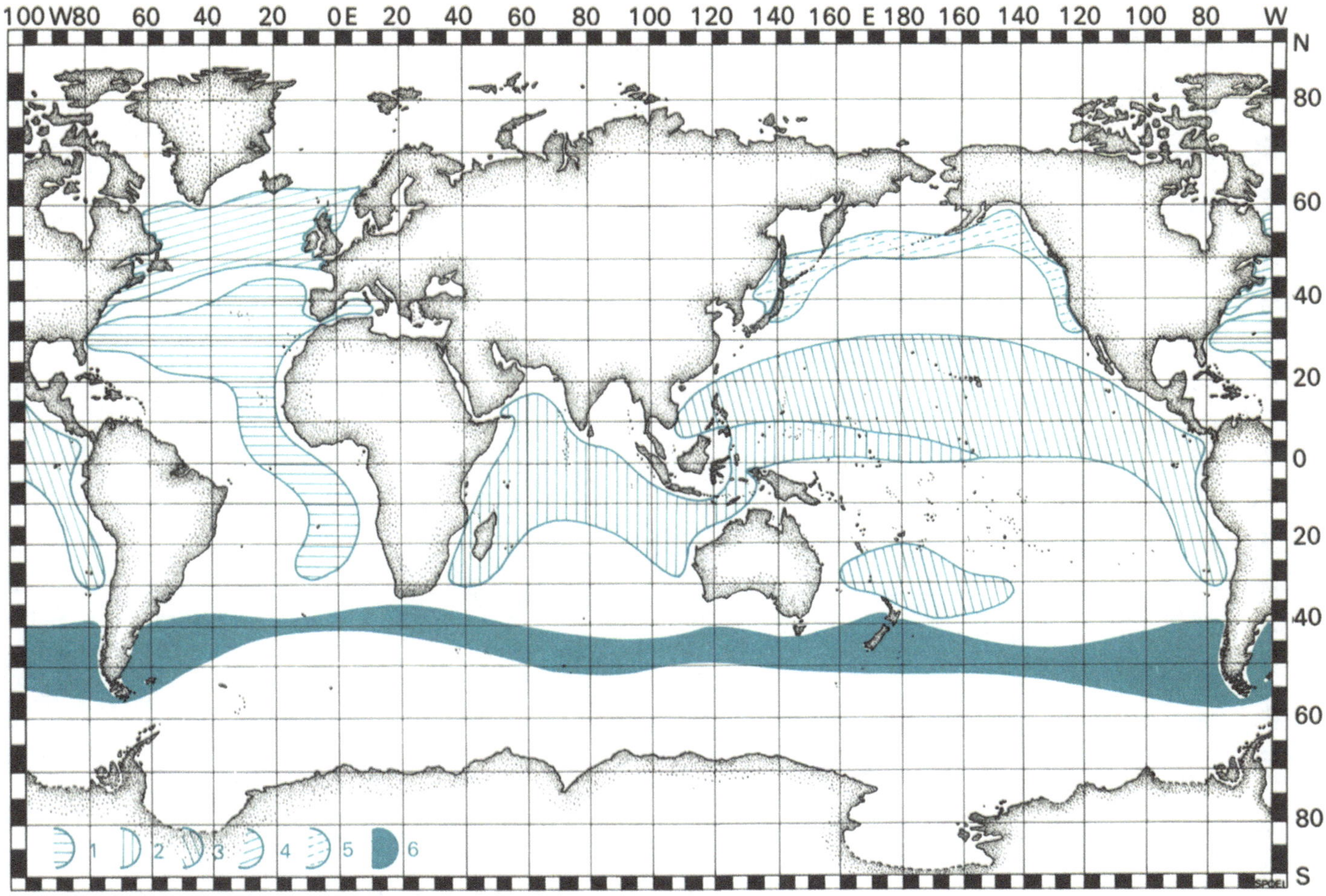

The forms of *Stenella euphrosyne* have a pattern comparable with: a distribution around the Dakar Centre (fig. 16, centre 8), one around the Indo-Malayan Centre (fig. 16, centre 33) and a distribution of an Indo-Pacific origin, respectively. This proves that in these mammals a combination of patterns is found, never met with in a plankton taxon. Therefore, the same patterns may have developed in different groups (whales and plankton), but it is hard to accept that this development has run along the same historical lines.

Laginorhynchus acutus (fig. 142) shows a pattern of the DIACRIA ATLANTICOIDES type (fig. 95) and *L. obliquidens* (fig. 142) a pattern of the CALANUS PACIFICOIDES type (fig. 79). So, both species show patterns which, in plankton, are usually occupied by infraspecific groups. *Lissodelphis peroni* (fig. 142) is comparable with the EUPHAUSIA LUCENSOIDES type (fig. 84) in distribution.

FIG. 141 Distribution of *Megaptera noveangliae*. 1 = summer (is breeding) range, 2 = winter range, 3 = diagrammatic migration routes.

FIG. 142 Distribution of Dolphins. 1 = *Stenella euphrosyne* 'Atlantic form', 2 = *S. euphrosyne* 'Pacific form', 3 = *S. euphrosyne* 'Tropical form', 4 = *Laginorhynchus acutus*, 5 = *L. obliquidens*, 6 = *Lissodelphis peroni*.

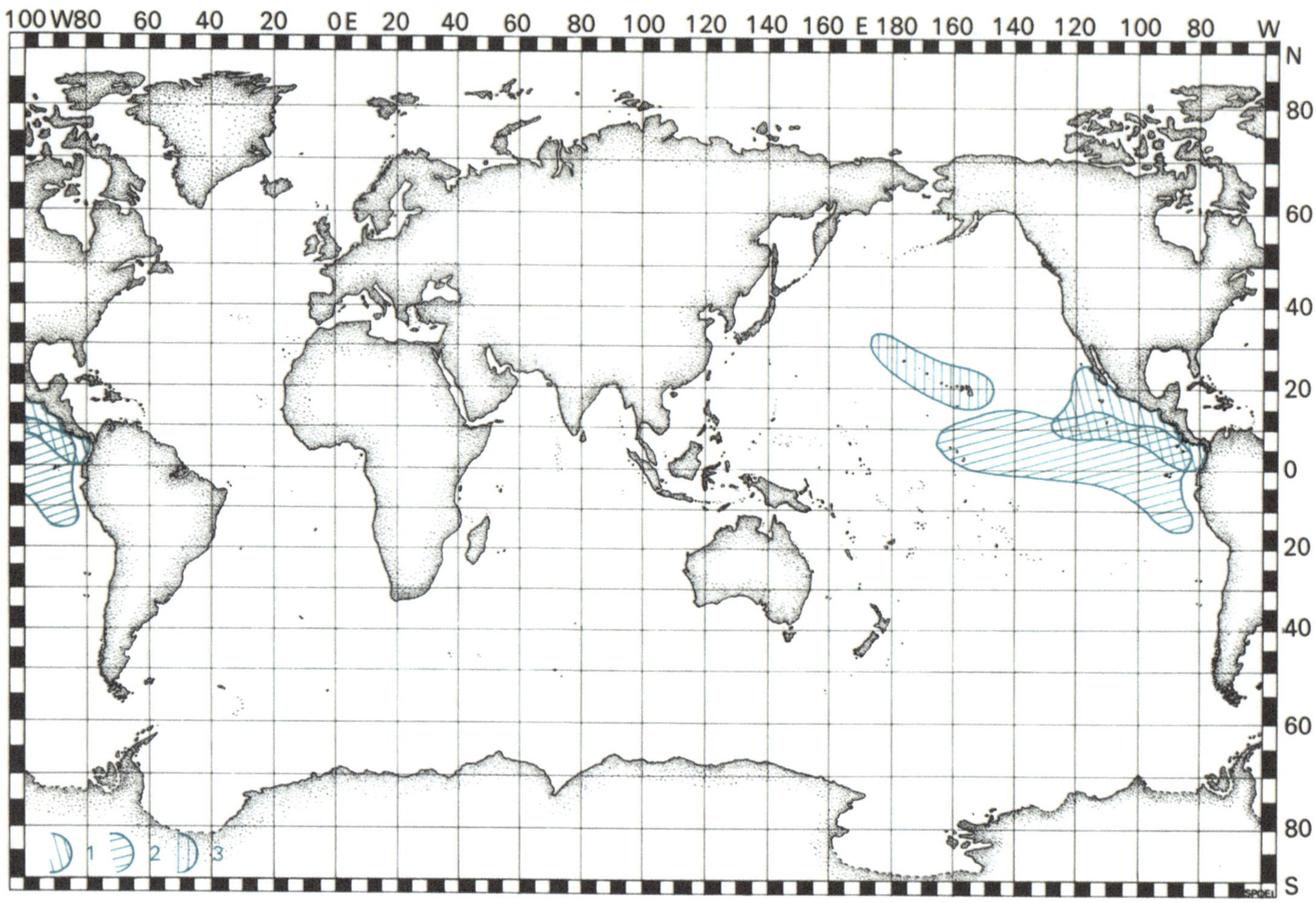

FIG. 143 Distribution of *Stenella microps*. 1 = *S. microps* 'Eastern form', 2 = *S. microps* 'White Belly form', 3 = *S. microps* 'Hawaiian form'.
FIG. 144 Distribution of the polytypic species *Phocoena phocoena*. 1 = *P.p. phocoena*, 2 = *P.p. relicta*, 3 = *P.p. vomerina*.
FIG. 145 Distribution of Sirenia. 1 = *Trichechus m. manatus*, 2 = *T.m. latirostris*, 3 = *T. inunguis*, 4 = *T. senegalensis*, 5 = *Dugong dugong*, 6 = *Rhytina gigas*.

STENELLA MICROPSOIDES type

The distribution of the dolphin *Stenella microps* (fig. 143) resembles Tropical E. Pacific patterns. The separate form around the Hawaiian Islands, however, is typical as in most plankton species the Hawaiian populations are related to West instead of East Pacific taxa (cf. *Sergestes orientalis*, fig. 101, and the Central water taxa). The 'Whitebelly form' shows the SERGESTES BREVISPINATOIDES type (fig. 101) and the 'Eastern form' shows the LABIDOCERA LUBBOCKOIDES type (fig. 132) of distribution, representing a distant-neritic and a neritic pattern, respectively. To consider the 'Eastern form' as being of the EUCALANUS INERMOIDES type (fig. 58) seems incorrect as it is too much coastbound.

PHOCOENA VOMERINOIDES type

The polytypic species *Phocoena phocoena* (fig. 144) is neritic and also shows resemblance with the neritic *Sagitta* species in its distribution. *P. phocoena relicta* shows the SAGITTA EUXIMOIDES type of distribution (fig. 124) and *P.p. phocoena* has a distribution similar to the SAGITTA FRIDERICOIDES type (fig. 129). The populations in the W. Atlantic, but more specifically those in the Tasman Sea, are not found in the type last-mentioned, but they represent common neritic areas as seen from the CARDIAPODA PLACENTOIDES type (fig. 128). There is so much variation in the warm-water neritic ranges of greater extension that it seems practical to distinguish only the seven types given in the figures 126 to 132.

The distribution of *P.p. vomerina* is different from the preceding ones in having a very special pattern. Also in some plankton species a tendency is found to occupy this range, but the best example is found in the present form, so that the PHOCOENA VOMERINOIDES type is proposed here.

A very typical neritic-freshwater pattern is shown by the *Serinia* group (fig. 145). It is composed of: a MELICERTUS AZTECOIDES type (fig. 131) (*Trichechus*

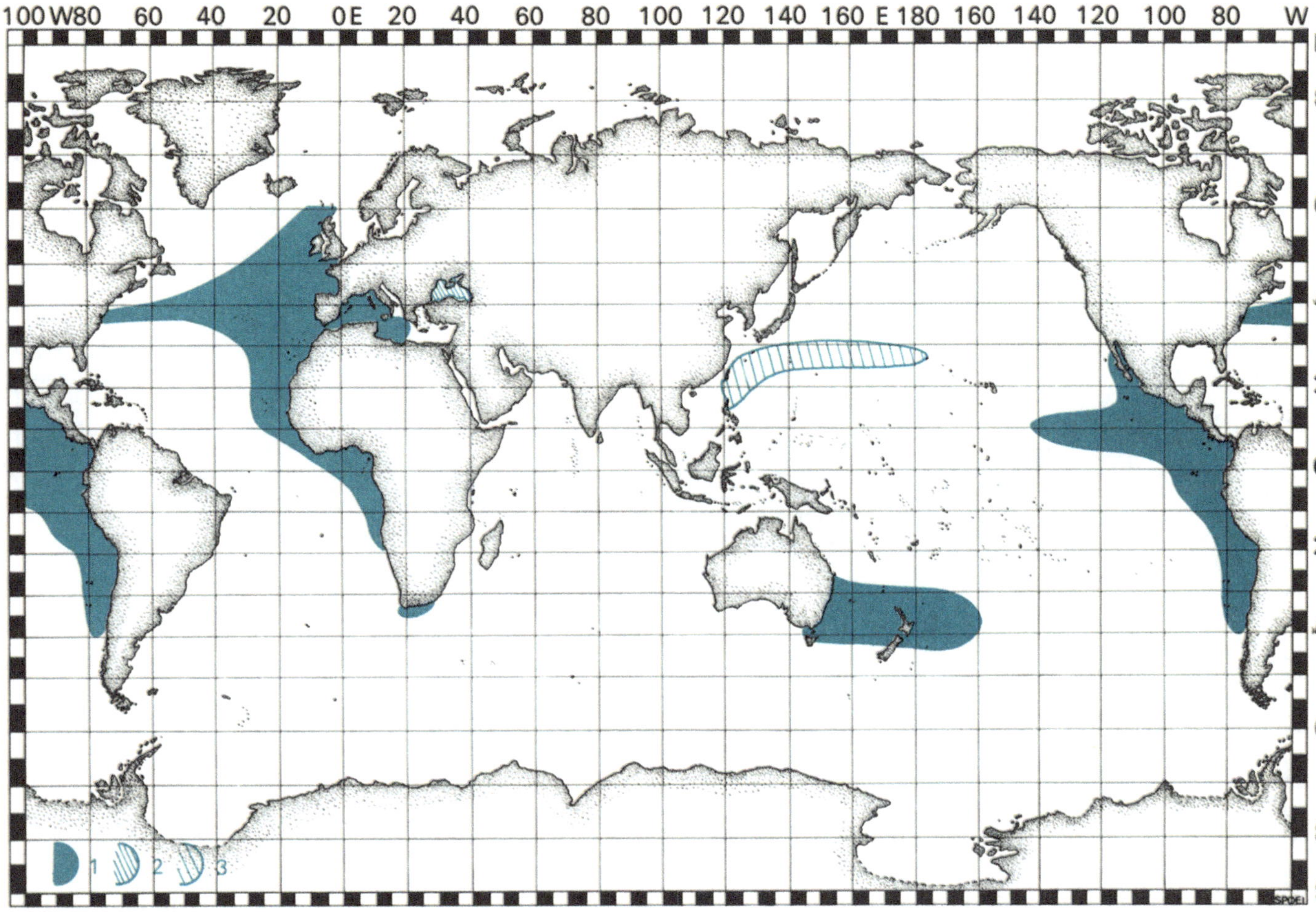
100 W 80 60 40 20 0 E 20 40 60 80 100 120 140 160 E 180 160 140 120 100 80 W
N 80 60 40 20 0 20 40 60 80 S
1
2
3

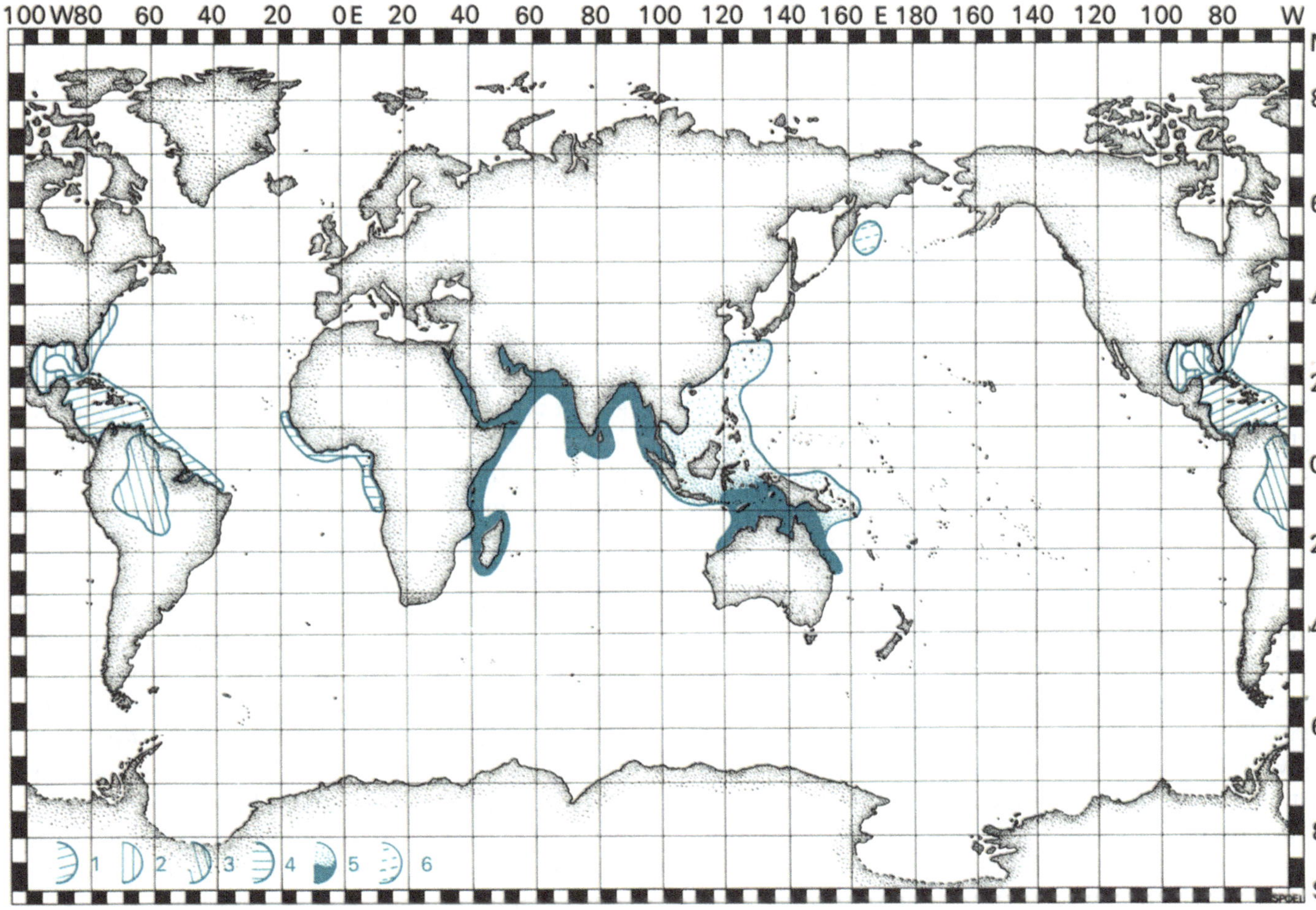
100 W 80 60 40 20 0 E 20 40 60 80 100 120 140 160 E 180 160 140 120 100 80 W
N 80 60 40 20 0 20 40 60 80 S
1
2
3
4
5
6

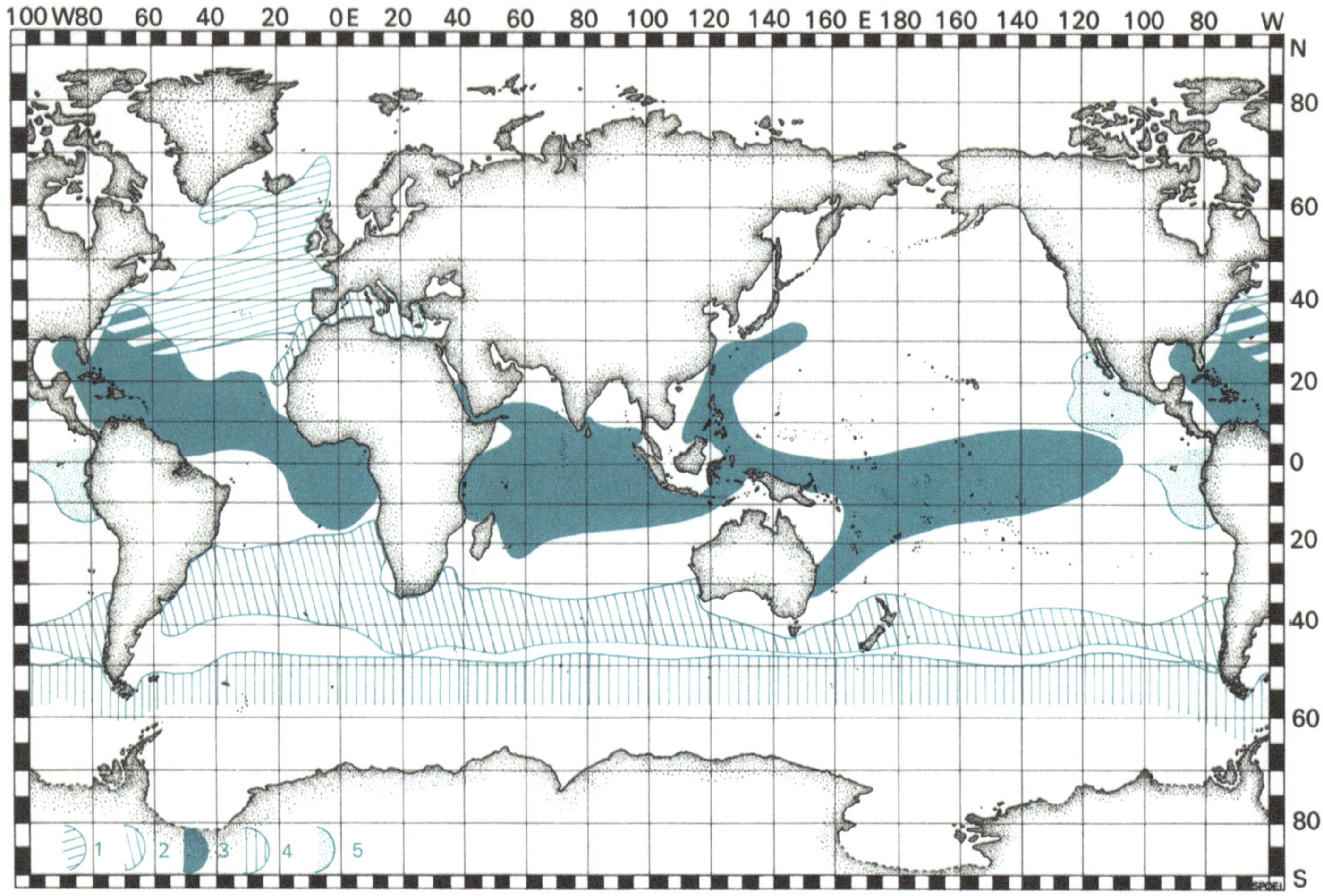

manatus latirostris), a LITOPENAEUS SETIFEROIDES type (fig. 131) (*T.m. manatus*), a EUPHAUSIA HANSENOIDES type (fig. 120) (*T. senegalensis*), an ACETES JAPONICOIDES type (fig. 103) (*Dugong dugong*) and a CALANUS PLUMCHROIDES type (fig. 136) (*Rhytina gigas*) and a distribution comparable with that of the freshwater species *Acetes paraguayensis* (fig. 105) (*Trichechus inunguis*). The freshwater species of *Trichechus* is, however, certainly not of a Pacific origin as is the case with *Acetes*, and *Rhytina gigas* is to be considered to have derived from the Indo-Malayan populations and not, as in *Calanus plumchrus*, from a Subarctic relative.

Mesopelagic and bathypelagic patterns

The meso- and bathypelagic patterns differ only in detail from the epiplanktonic ranges, as shown in the following maps. Only a few special meso- or bathypelagic types are proposed, but it is very well possible that in the future epiplanktonic distributions of the same type will be found. Most of the typical meso- and bathypelagic patterns found, are represented by micronektonic species, which points to the fact that they are probably more typical of nekton than of plankton.

In fig. 146 mesopelagic fish distributions are given which were already depicted for epiplankton. *Stomias boa ferox* shows the DIACRIA ATLANTICOIDES type (fig. 95), *S. b. boa* shows the FOWLERINA ZETESIOIDES type (fig. 120), combined with the EUPHAUSIA LUCENSOIDES type; *S. affinis* shows the EUPHAUSIA PSEUDOGIBBOIDES type (fig. 77), *S. gracilis* shows the EUPHAUSIA TRIACANTHOIDES type (fig. 86) and *S. atriventer* shows the SAGITTA BIERIOIDES type (fig. 117) of distribution. Some of the patterns are slightly larger or different in shape, compared with the given examples, but these are no principal differences. The distribution of *S. b. boa*, however, combines two plankton patterns, which in plankton are never combined and the *boa* pattern is more likely to resemble a *Peraclis*-like distribution.

Like the mesopelagic representatives of the genus given in fig. 146, the four *Stomias* species given in fig. 147 show normal types of distribution also found in plankton. *S. nebulosus* shows the HYDROMYLES GLOBULOSOIDES type (fig. 112), *S. brevibarbatus* shows the EUPHAUSIA AMERICANOIDES type (fig. 109), though the southern part of this range is not occupied by *S. brevibarbatus*; *S. lampropeltis* shows the EUPHAUSIA HANSENOIDES type (fig. 120) and *S. columbrinus* shows the CAVOLINIA AFFINOIDES type (fig. 53) of distribution.

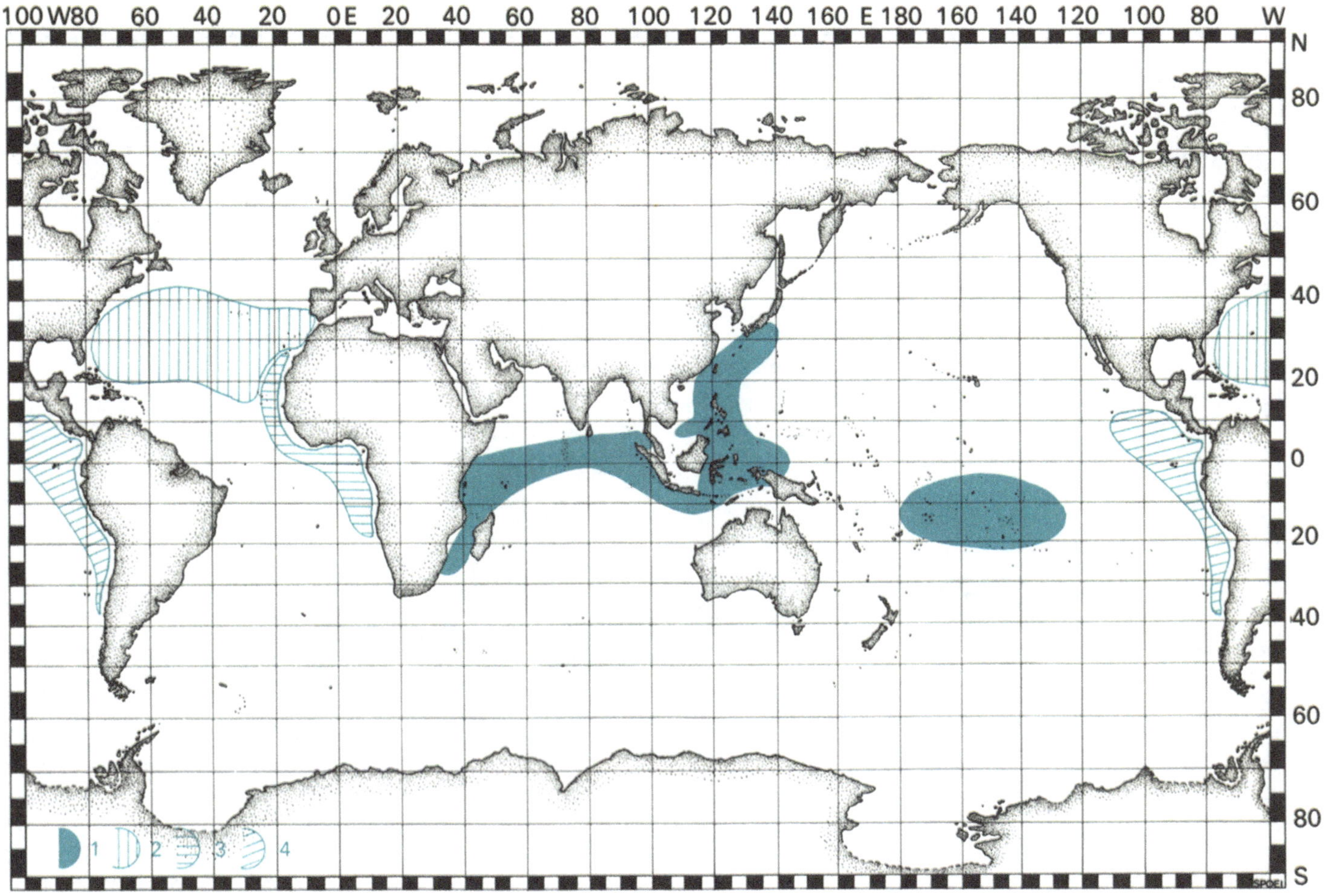

FIG. 146 Distribution of *Stomias* (see also fig. 147).
1 = *S. boa ferox*, 2 = *S. b. boa*, 3 = *S. affinis*, 4 = *S. gracilis*,
5 = *S. atriventer*.
FIG. 147 Distribution of *Stomias* (see also fig. 146).
1 = *S. nebulosus*, 2 = *S. brevibarbatus*, 3 = *S. lampropeltis*,
4 = *S. columbrinus*.

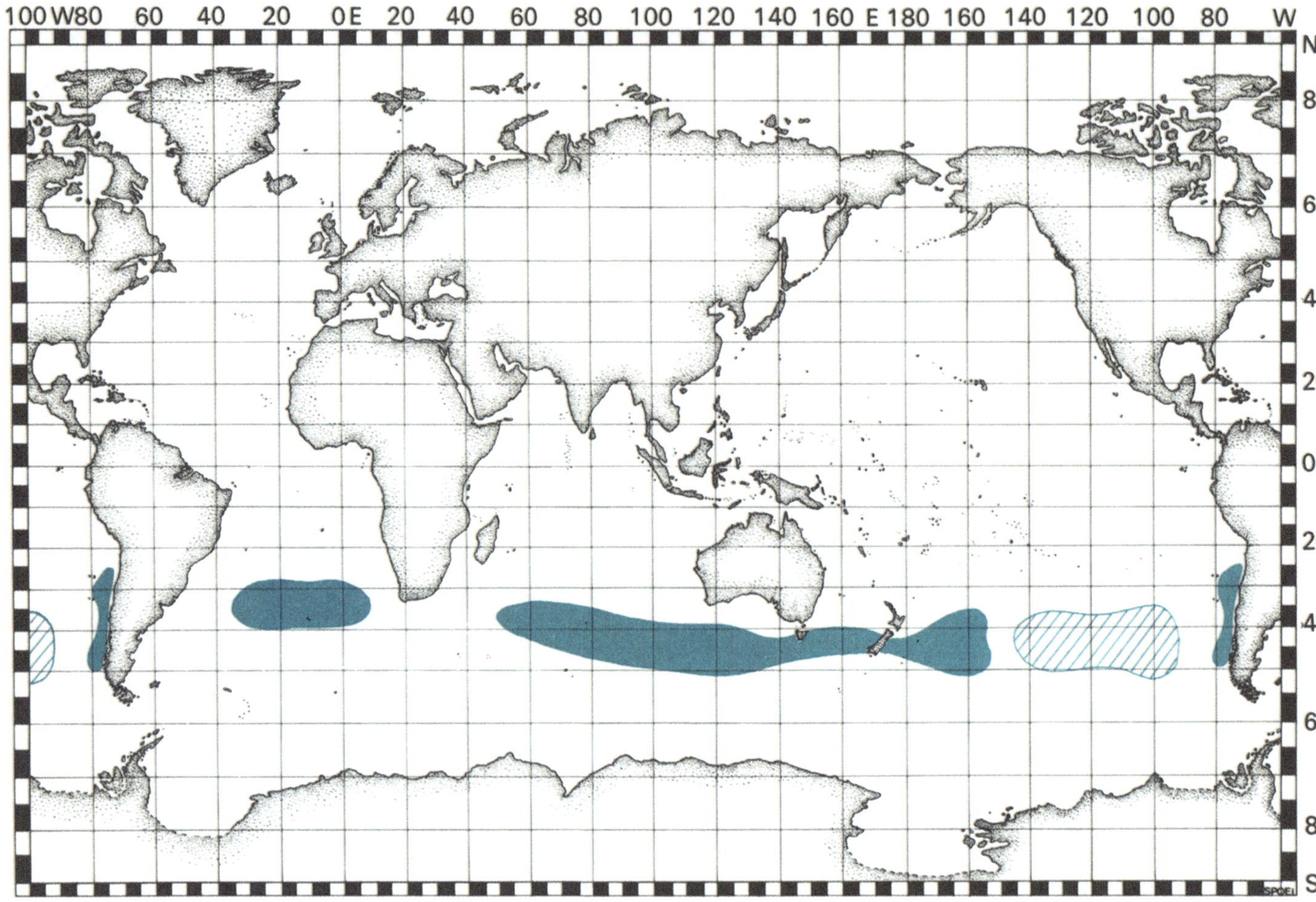

ELECTRONA PAUCIRASTROIDES type

The pattern given in fig. 148 resembles the EUPHAUSIA LUCENSOIDES type (fig. 84), but it is formed by two, probably vicariating, species in the Transitional belt. *Electrona paucirastra* occurs in the whole belt except for the part in the open Pacific Ocean. This pattern is very closely related to the THALIA LONGICAUDOIDES type (fig. 122). The distribution of *Symbolopharus* spec. (fig. 148) of the SAGITTA ATLANTICOIDES type (fig. 118) shows another example of the separate position of the Pacific Ocean east of about 140° W.

The wide bathypelagic distribution of *Cyclothone pseudopallida* (fig. 149) is of the STYLOCHEIRON MAXIMOIDES type (fig. 72). The great distances to the coast-lines compared with the plankton ranges do not constitute an essential difference as they are only caused by the depth of occurrence of this species.

CYCLOTHONE ATRAROIDES type

The pattern of *Cyclothone atraria* (fig. 150) slightly resembles the GLOBOROTALIA HIRSUTOIDES type (fig. 38), but this species penetrates much further to the south and into the Pacific in the east. It is a common phenomenon that deep-water species occupy larger ranges than comparable epiplanktonic ones. The typical penetration into the Antarctic of *C. atraria*, however, did induce the creation of a special type.

C. microdon (fig. 150), shows the CALANUS PACIFICOIDES type (fig. 79), as its range is likewise wider than that of *Eucalanus bungi bungi*, a phenomenon also found in *Diomedea nigripes* and *D. albatrus* (fig. 139).

PERACLIS APICIFULVOIDES type

The mesopelagic distribution of *Peraclis apicifulva* (fig. 59) shows some resemblance with the pattern in fig. 150. It is a taxon which is found in the Atlantic and W. Indian Oceans and in the E. Pacific, separated from the W. Pacific populations.

FIG. 148 Distribution of *Electrona paucirastra* (black) and *Symbolopharus* spec. (hatched).

FIG. 149 Distribution of *Cyclothone pseudopallida*.

FIG. 150 Distribution of *Cyclothone atraria* (black) and *C. microdon* (hatched).

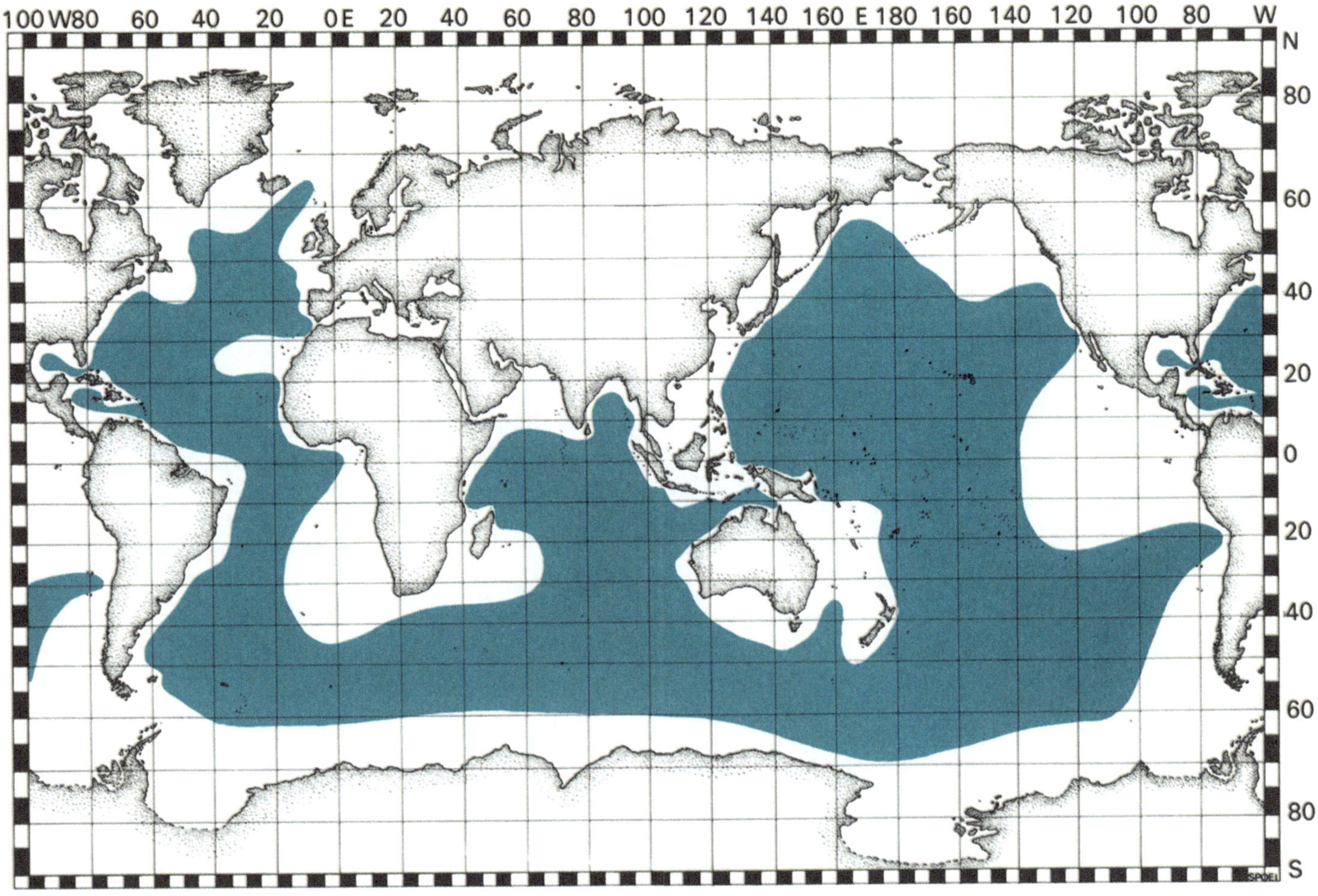
100 W80 60 40 20 0E 20 40 60 80 100 120 140 160 E 180 160 140 120 100 80 W
N
80
60
40
20
0
20
40
60
80
S

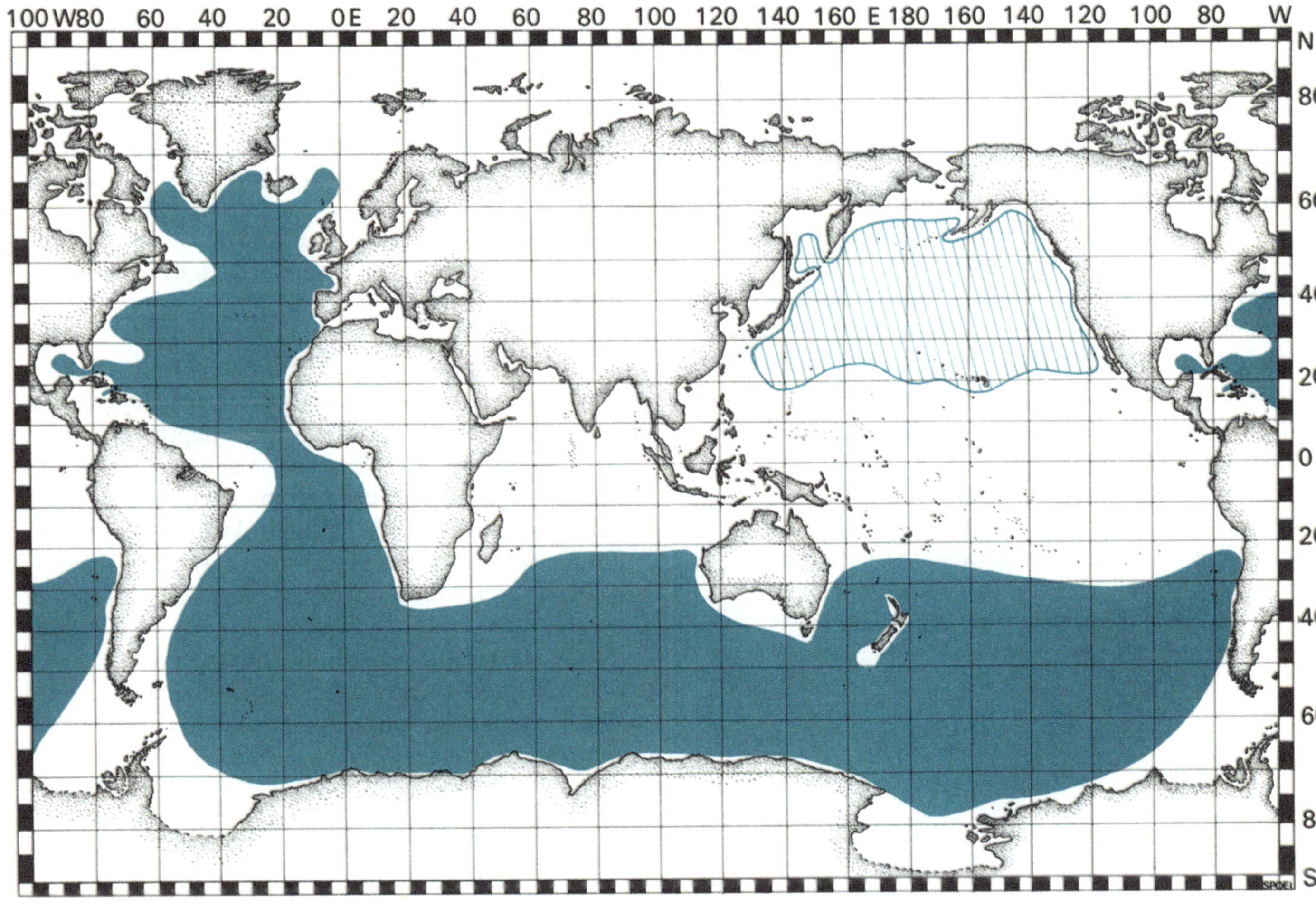
100 W80 60 40 20 0E 20 40 60 80 100 120 140 160 E 180 160 140 120 100 80 W
N
80
60
40
20
0
20
40
60
80
S

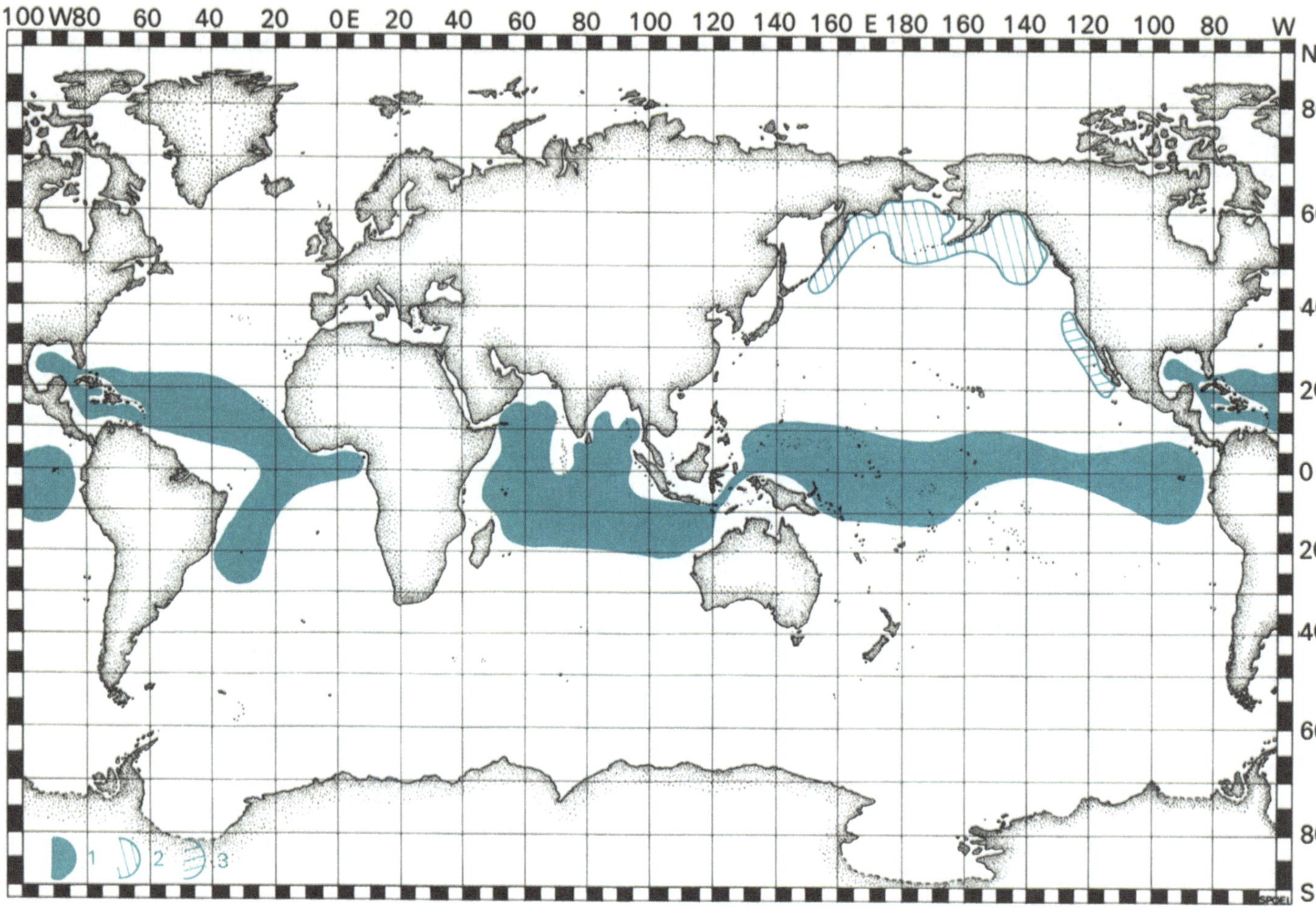

PERACLIS RETICULATOIDES type

The distributions of *Peraclis reticulata* and *P. moluccensis* (fig. 59) show the same tendency as the preceding species but they show a continuous Atlantic-Indo-W. Pacific population and a separated E. Pacific population.

PERACLIS VALDIVIAOIDES type

The pattern of *Peraclis valdiviae* (fig. 60) has much in common with the GLOBOROTALIA HIRSUTOIDES type (fig. 38), but it also occurs around New Zealand, as in the LIMACINA HELICOIDOIDES type (fig. 152). In addition, it shows an Indo-Malayan population, which makes it probably a separate pattern.

PERACLIS BISPINOSOIDES type

Peraclis bispinosa and *P. triacantha* (fig. 60) are both mesopelagic endemics of the Atlantic Ocean characterizing the present type, also found in epiplanktonic species.

The three bathypelagic fishes of the genera *Cyclothone* and *Oneirodus* given in fig. 151 show patterns of distribution comparable with those in plankton. *C. obscura* shows the EUPHAUSIA PSEUDOGIBBOIDES type (fig. 77), *O. bulbosus* shows the LIMACINA ACUTOIDES type (fig. 82) and *O. acanthias* shows the DESMOPTERUS PACIFICOIDES type (fig. 111) of distribution.

LIMACINA HELICOIDOIDES type

The distribution of *Limacina helicoides* (fig. 152) is of a very special type as it occurs at great depth only in the Atlantic and near New Zealand. A good explanation for this pattern has never been given. Although it is not found exclusively in the bathyplankton but also in the abyssal benthic realm, as shown by e.g. *Onchnesoma steenstrupi* (cf. Murina, 1977), it is considered a separate type.

Clio chaptali (fig. 152) shows the EUPHAUSIA PSEUDOGIBBOIDES type (fig. 77), though it occurs in a disjunct range, but this may be due to insufficient sampling of this deep-living species.

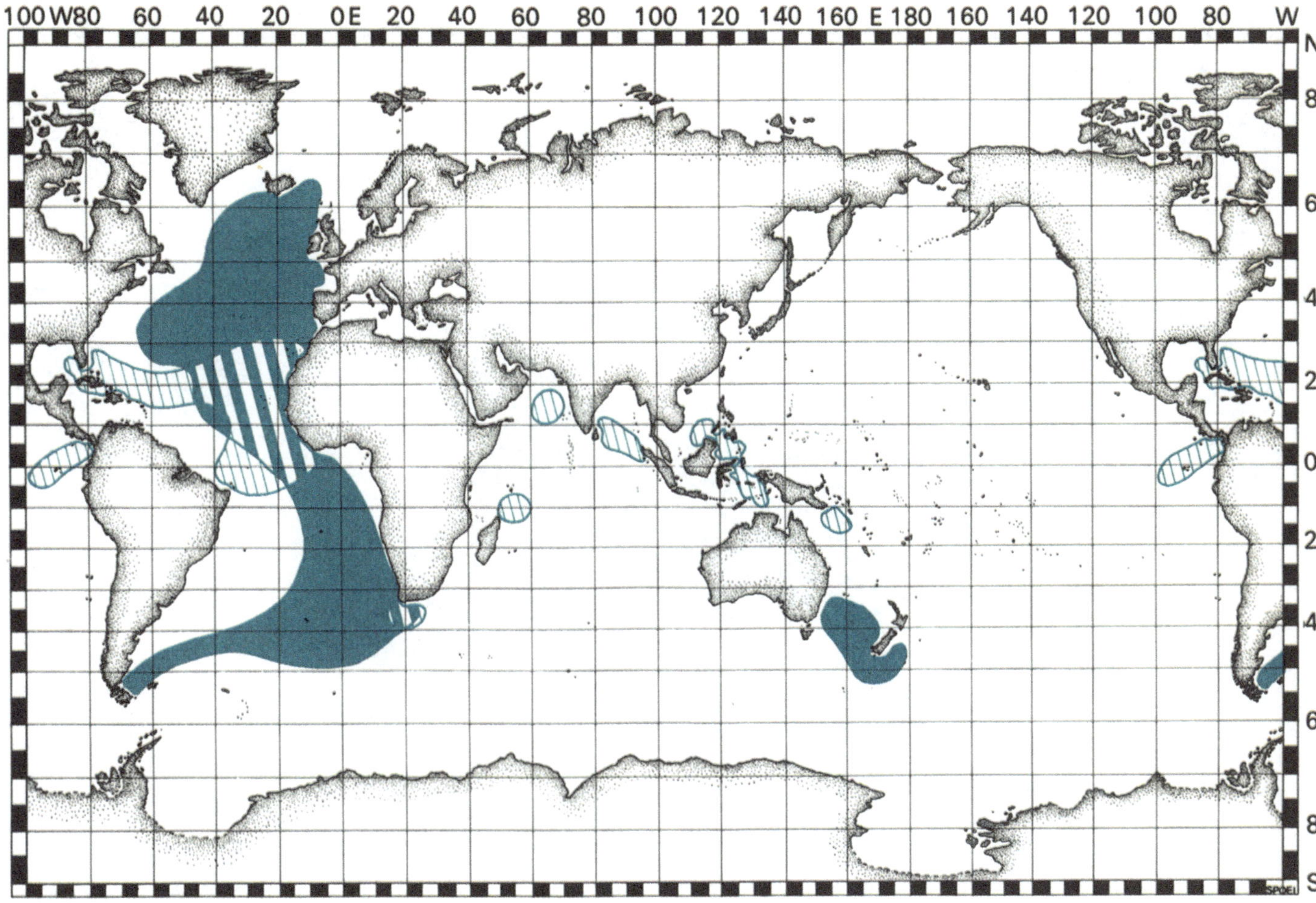

FIG. 151 Distribution of *Cyclothone obscura* (1), *Oneirodus bulbosus* (2) and *O. acanthias* (3).
FIG. 152 Distribution of *Limacina helicoides* (black) and *Clio chaptali* (hatched).

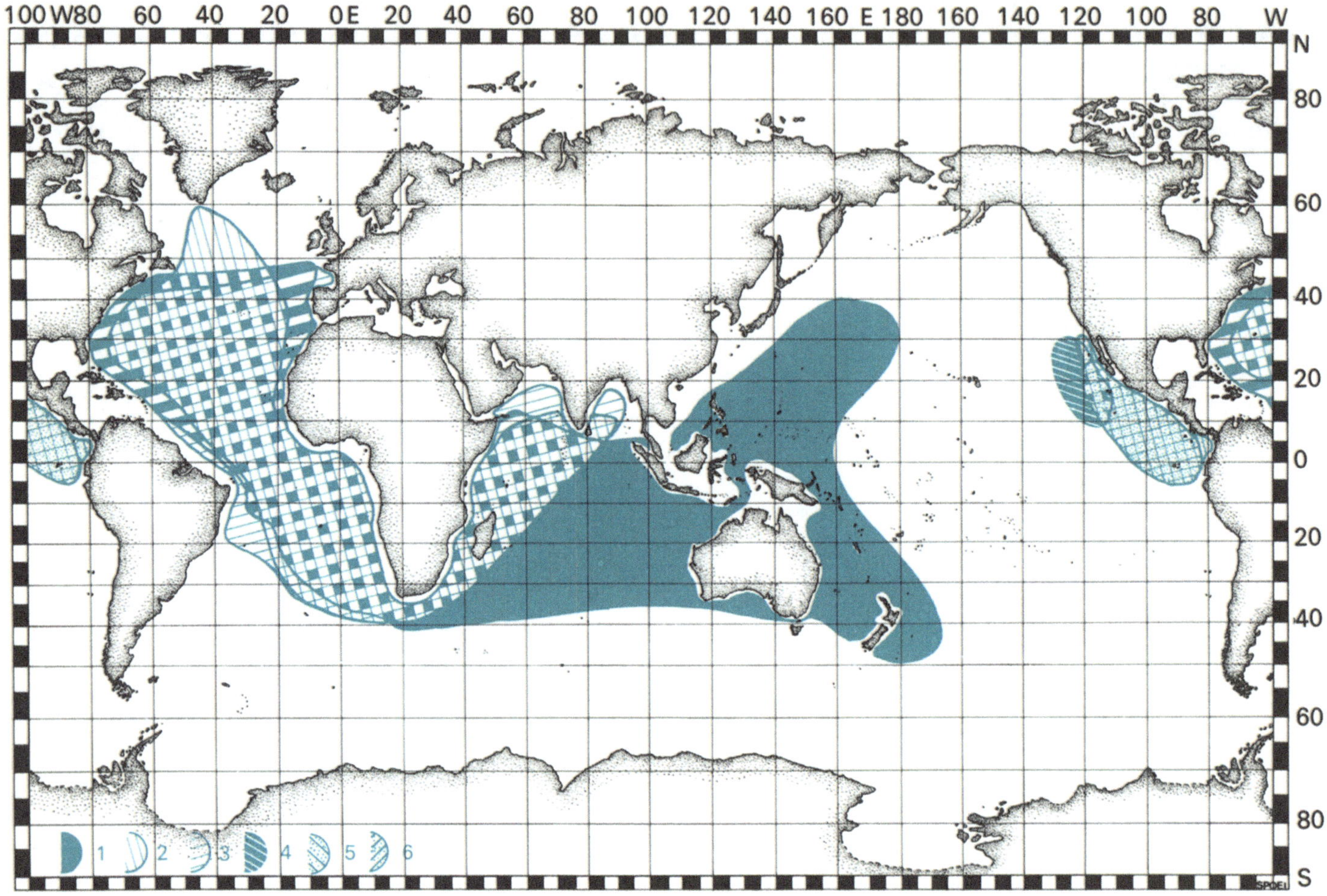

FIG. 153 Distribution of abyssal-benthic molluscs. 1 = *Abra profundorum*, 2 = *Poromya tornata*, 3 = *Limopsis p. pelagica*, 4 = *Abra californica*, 5 = *Poromya perla*, 6 = *Limopsis p. dalli*.
FIG. 154 Distribution of *Bathyteuthis abyssicola*.
FIG. 155 Distribution of *Todarodes sagittatus* (1), *Moroteuthis robusta* (2) and *Spirula spirula* (3).

The three mollusc species *Abra profundorum*, *Limopsis pelagica pelagica* and *Poromya perla* are abysso-benthic taxa with an Atlantic-Indo-W. Pacific distribution (fig. 153) between 50°N and 50°S. They each have a close relative in the E. Pacific. Exactly the same patterns are not found in plankton, but there are evident resemblances with the LIMACINA HELICOIDOIDES type (fig. 82) and the PERACLIS RETICULATOIDES type (fig. 59). The patterns are here also given to show that the separate position of the E. Pacific is not only reflected in littoral neritic epiplanktonic animals, and in bathyplankton, but also in bathybenthos in different taxonomic groups, as shown by the deepsea Holothurians (Hansen, 1967).

The bathynektonic cephalopod *Bathyteuthis abyssicola* (fig. 154) shows a distribution of the LIMACINA HELICOIDOIDES type (fig. 82). In all probability this species also originates from the Atlantic Ocean and shows some isolated populations in the N. Indian Ocean and in the E. Pacific, as *L. helicoides* does near New Zealand. Essential differences between this nektonic range and the planktonic range of this type are not found. Other similarities between plankton and nektonic cephalopods are mentioned in relation to the figs 61, 74 and 155.

The nektonic cephalopod species *Todarodes sagittatus*, *Moroteuthis robusta* and *Spirula spirula* also show patterns comparable with planktonic ranges. *Todarodes sagittatus* (fig. 155) shows the PERACLIS BISPINOSOIDES type (fig. 60), *Spirula spirula* (fig. 155) shows the PERACLIS APICIFULVOIDES type (fig. 59) and *Moroteuthis robusta* (fig. 155) shows the CALANUS PACIFICOIDES type (fig. 79) of distribution.

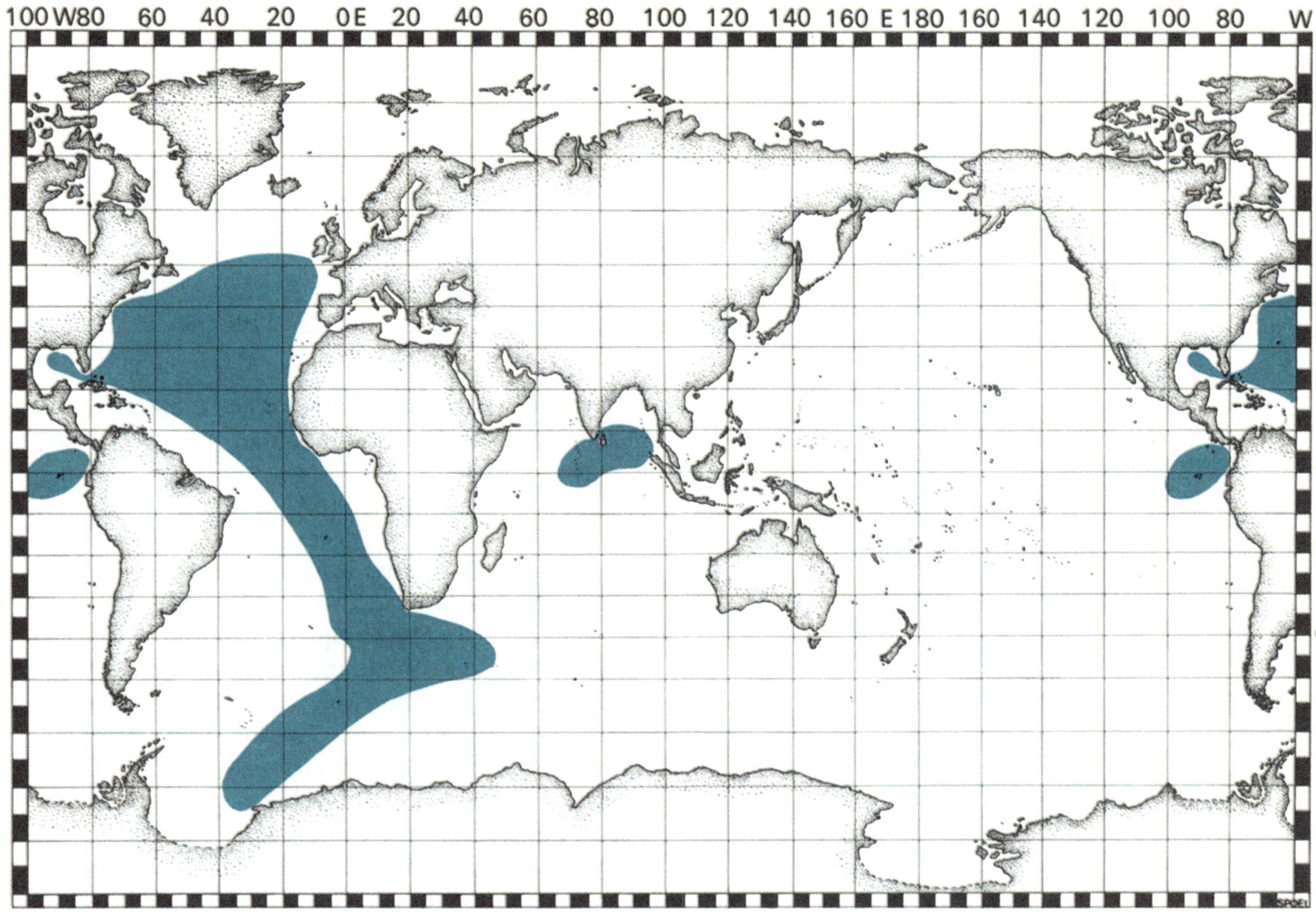
100 W80 60 40 20 0E 20 40 60 80 100 120 140 160 E 180 160 140 120 100 80 W
N 80 60 40 20 0 20 40 60 80 S

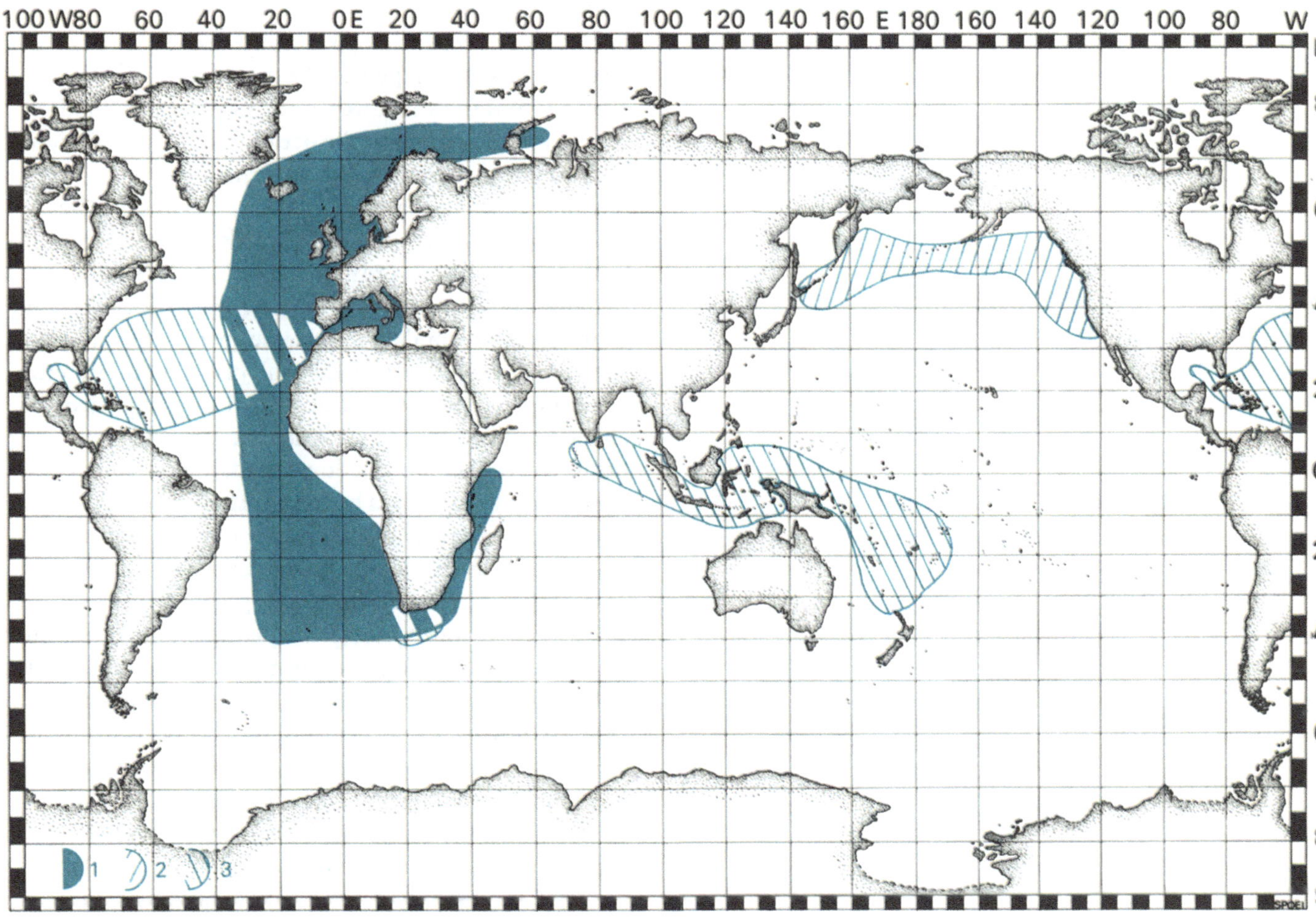
100 W80 60 40 20 0E 20 40 60 80 100 120 140 160 E 180 160 140 120 100 80 W
N 80 60 40 20 0 20 40 60 80 S
1
2
3

4 Vertical distribution

Vertical migration and distribution—Tropical submergence and vertical variation

The vertical distributions of plankton are even less known than the horizontal ones. The use of accurately functioning opening-closing nets has been introduced only recently at any scale and the data available are far too scarce to discriminate between different vertical patterns or migration types. In most species some kind of vertical movement of populations is found. This movement may be diurnal, ontogenetic or it may be caused by feeding-migration or by an up- and downwards following of the isotherms. Consequently, the vertical patterns are seldom stable in space and time, and it seems impractical to distinguish special types of vertical distribution.

The more deeply living taxa, and deeper distributions, can in general be considered to have evolved from the shallower ones. The primary food supply for all animals in the ocean, but for a very slight minority near thermal vents, is formed by phytoplankton in the surface layers. As a result, the foodchain runs vertically downwards so that deeply living organisms can only have developed after the shallower taxa. In recent groups a reverse development, giving rise to secondary epipelagic species, may have been possible, as this process took place in the presence of a complete foodchain. But still, in all groups deepsea characters have proved to be adaptations and not primitive, so in general a downwards succession of evolutionary stages will be found. In some groups, such as the Nemerteans, only deepsea forms are known to exist, which live at depths below c. 1000 m. In these cases it is accepted that the development in the group was also downwards, not in the plankton, however, but along the benthos. A great number of groups are known to have developed from littoral ancestors which have creeped downwards as benthos, especially at the higher latitudes, to give rise to deepsea forms. Therefore, one can conclude that deepsea distribution patterns are younger than the shallower ones. Maybe also the downward migration, either ontogenetic, diurnal or seasonal, can be considered adaptive or younger in evolution. For comparative zoogeographic studies one should therefore use the shallowest distributions of migrating taxa; in diurnal migration this is usually the nocturnal situation, and in ontogenetic migrators usually the distribution of the younger life stages. It is ironical that, chiefly for practical reasons, the day-time distribution of adults has been given most attention.

Vertical migration and distribution

Vertical migration is the phenomenon that organisms perform a shift in depth of occurrence at a fixed period or at several periods in their lifetime. This results in shifts in depth of a population, or of the abundance in a population. Four ecological types of migration are distinguished, viz.: the diurnal, the ontogenetic, the seasonal and the feeding migration. Vinogradov (1970) divided these four into seven categories (cf. fig. 156) but this division is slightly confusing and does not present a complete picture. Table II gives the more complete summary of migration types distinguished.

TABLE II
Types of vertical migration.

Migration:	Period:	Depth range'):
Diurnal Shallow-	days	0- 50 m
Diurnal Epi-	days	0- 200 m
Diurnal Epi/meso-	days	0- 500 m
Diurnal Meso-	days	100- 1000 m
Seasonal	months	0- 500 m
Ontogenetic Epi-	months/years	0- 300 m
Ontogenetic Meso-	months/years	100- 1000 m
Ontogenetic Wide-	months/years	200- 5000 m
Ontogenetic Deep-	months/years	700- 4000 m
Feeding	days	100- 8000 m
Feeding Abysso-	days	3000-10000 m

') average ranges influenced by hydrographic and latitudinal conditions.

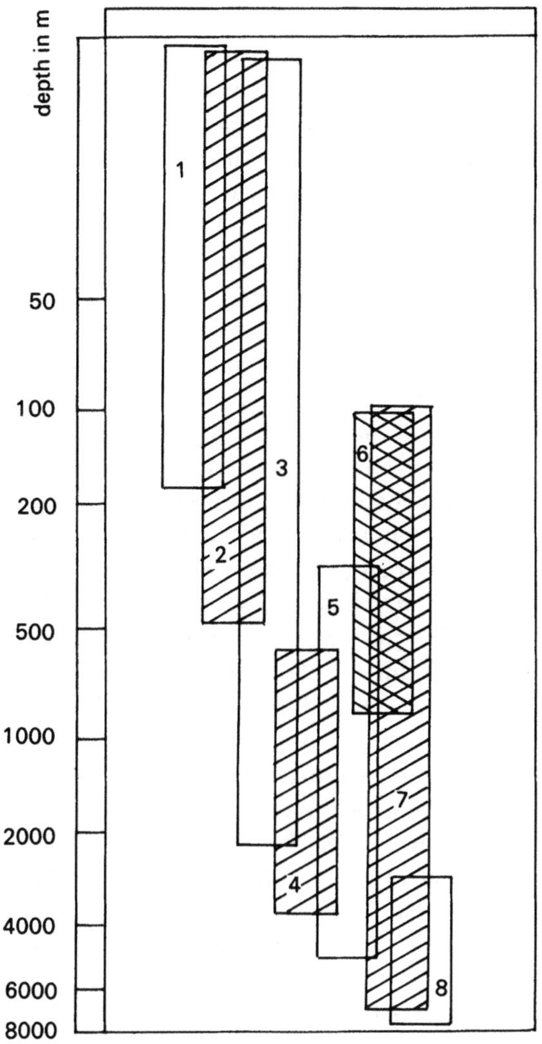

FIG. 156 Vertical migration ranges as summarized by Vinogradov (1970). 1 = diurnal migr., 2 = diurnal + seasonal migr., 3 = seasonal + ontogenetical migr., 4 and 5 = ontogenetical migr., 6 = diurnal + ontogenetical migr., 7 = irregular feeding migr., 8 = abyssal fauna migration.

All these vertical migrations are ecologically determined; light- and temperature-triggered when shallow migrations are concerned, and triggered by food supply when it concerns migrations below 500 m.

It is questionable whether all these eleven migration-types, distinguished chiefly by the Soviet school, really all exist. First of all, the differences between seasonal and ontogenetic migrations are vague, as one can never be certain whether a special life-stage which occurs in its special season, migrates to a particular depth, or whether the ecological conditions determine, in each season, the depth of the population irrespective of the life-stage of the organism. Secondly, it is discutable whether real migrations exist below 1000 m depth.

Light penetration in the ocean is restricted by absorption to maximally 500-1000 m (fig. 157), and turbidity or large plankton blooms may restrict the penetration of light to some few metres only. From figs 156 and 157 it is clear that any migration triggered by light is excluded below 1000 m. Temperatures below 1000 m are very stable and the vertical temperature gradient is here only small, so that influences of temperature on diurnal migration are excluded as well. Angel et al. (1982) found no diurnal migration below this depth.

As many ecological conditions change with the seasons, especially when light and temperature are considered, seasonal differences in vertical migration are expected to occur. Such differences are also found in latitudinal sequence; the intensity of vertical diurnal migration varies with the latitude. 'Maximum intensity was observed at 30°N 23°W and decreased both to the north and to the south. At about 60°N 20°W there was little migration by ostracods, and at 10½°N 20°W only one species performed a migration of the normal pattern and several species underwent reverse migration. Some species that were nearly ubiquitous, like *Conchoecia imbricata*, changed from being non-migrant or weakly migrant species at the extremities of their ranges, to being strong migrants towards the centres.' (Angel & Fasham, 1975). That diurnal vertical migration in one and the same species is also different in different water masses is demonstrated by the migration of *Diphyes dispar*, *Abylopsis eschscholtzi* and *Chelophyes appendiculata* in the N.E. Indian Ocean (fig. 158).

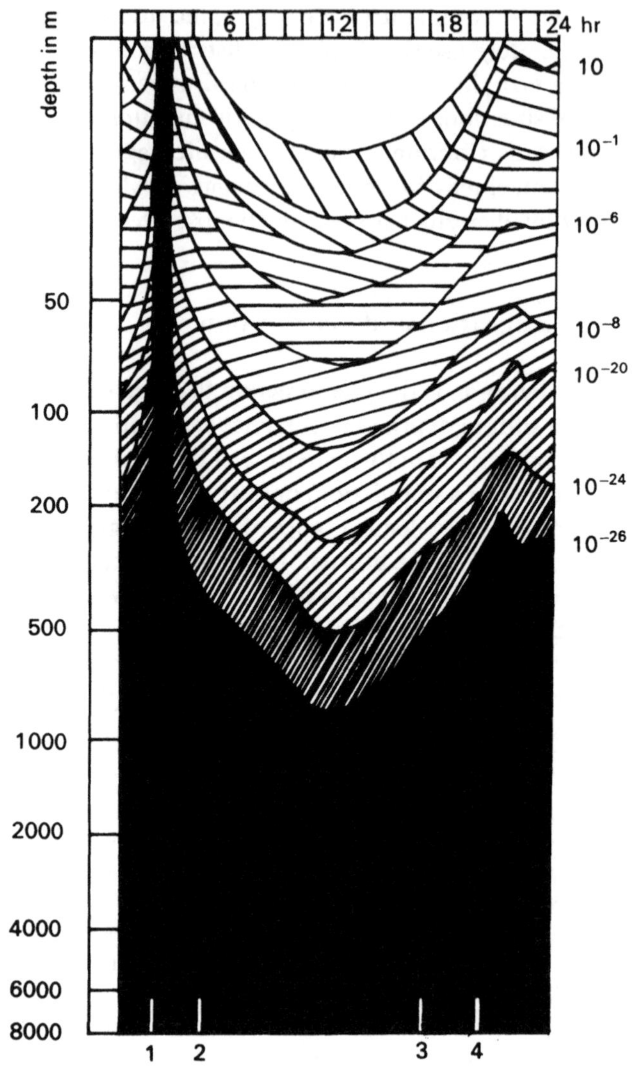

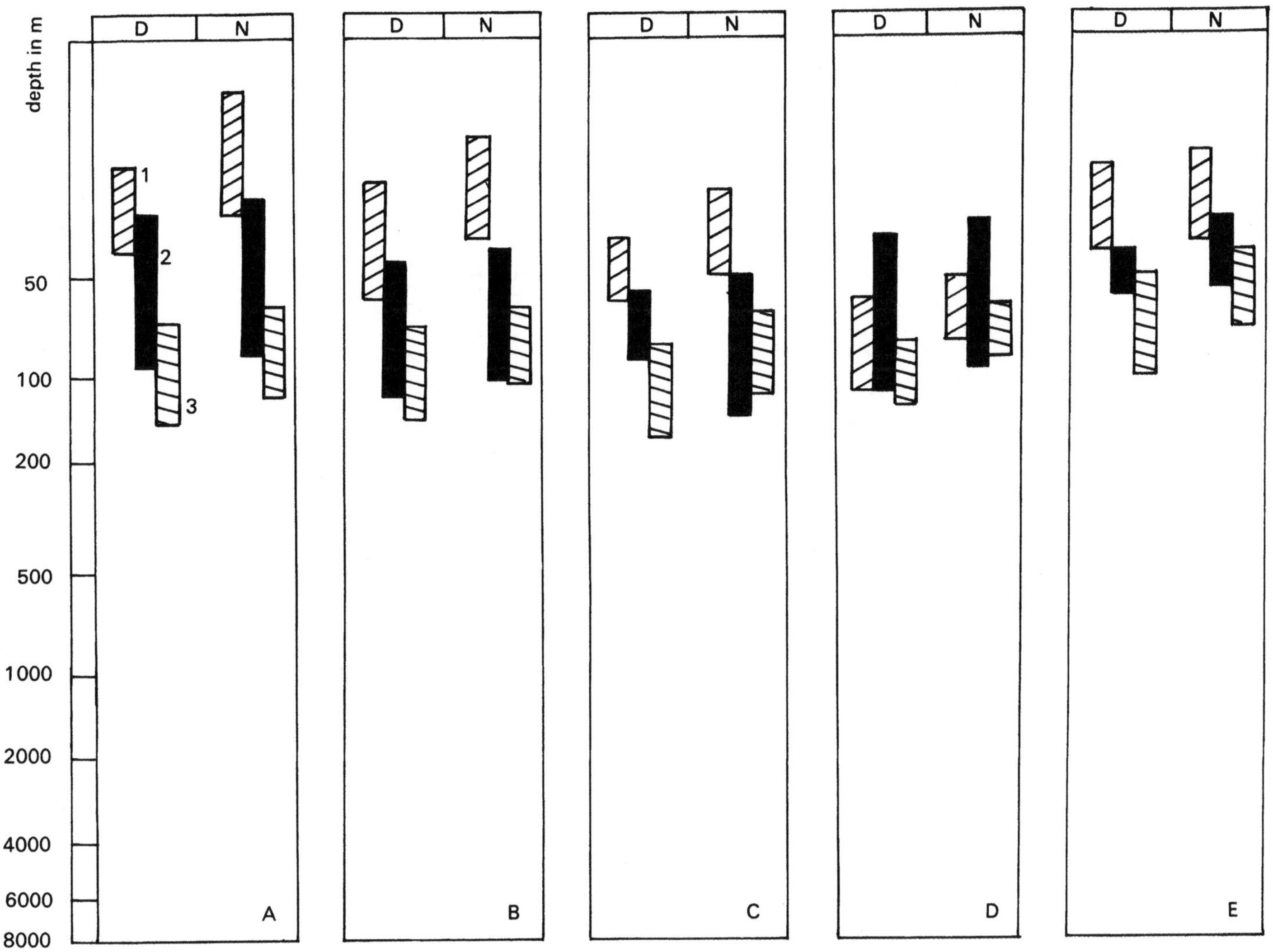

FIG. 157 Light penetration in the ocean with 9 isolumes (mod. after Moore, 1958). 1 = moon rise, 2 = sun rise, 3 = moon set, 4 = sun set.
FIG. 158 Diurnal migration of *Diphyes dispar* (1), *Abylopsis eschscholtzi* (2) and *Chelophyes appendiculata* (3) in the Gulf of Aden (A), Equatorial waters (B), S. Equatorial water (C), the Arafura Sea (D) and in W. Australian waters (E) (after Musaeva, 1976). d = day, n = night.

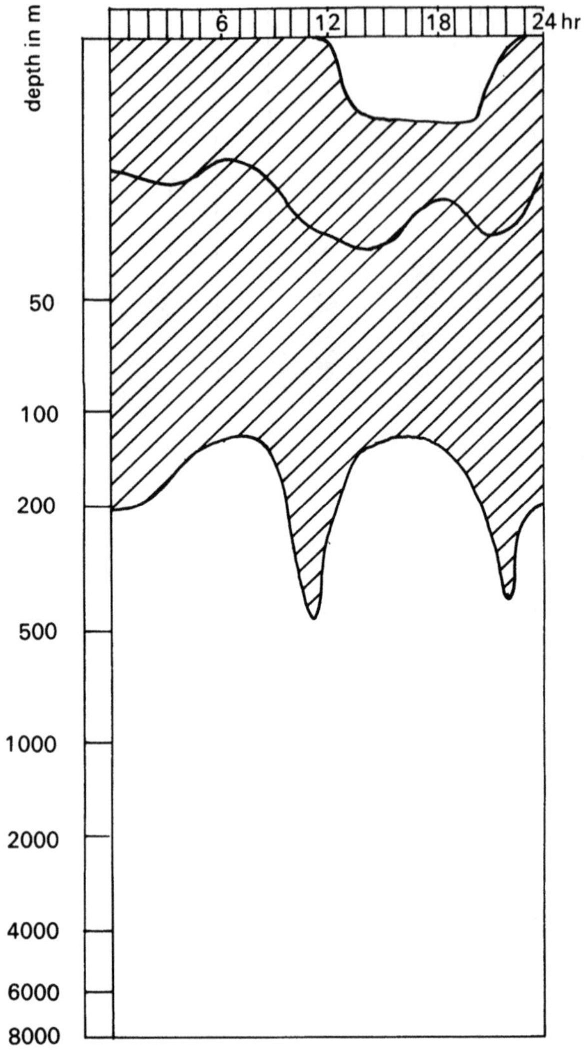

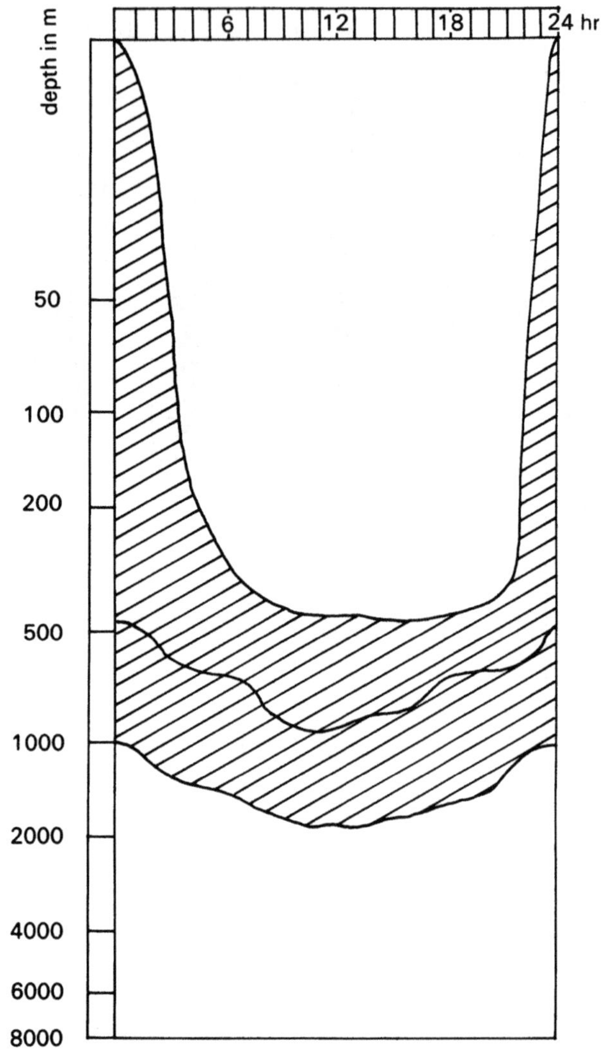

Diurnal vertical migration is evidently strongly influenced by ecological conditions, and when one finds completely different diurnal migration patterns it is difficult to determine whether zoogeographical or ecological differences are concerned.

The migration of *Calanus plumchrus* (fig. 159) in the upper 200 m, showing only a disappearance between 12 and 20 hrs from the uppermost layer, and that of *Gennadas elegans* (fig. 160) going down to 1000 m and disappearing from the upper 200 m layer between 4 and 20 hrs, are typically different, but does this difference depend on the taxa concerned, or the areas investigated (north of 40°N in the Pacific and at 30°N in the Atlantic, resp.)? The migration of *Clio pyramidata* in the Atlantic (fig. 161) resembles much the diurnal migration of *Gennadas elegans* in the same area. When studying more migration patterns in different areas one gets the impression that the migratory behaviour is largely determined by local influences.

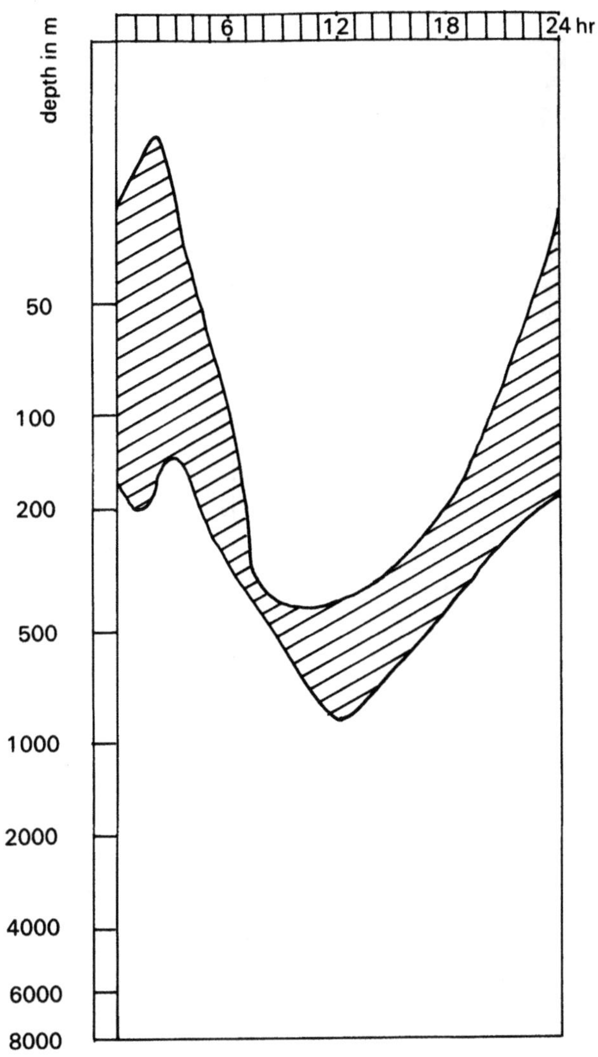

FIG. 159 Diurnal vertical migration of copepodid V of *Calanus plumchrus* (after Vinogradov, 1970).
FIG. 160 Diurnal vertical migration of *Gennadas elegans* (after Moore, 1958).
FIG. 161 Diurnal vertical migration of *Clio pyramidata* in the Bermuda area.

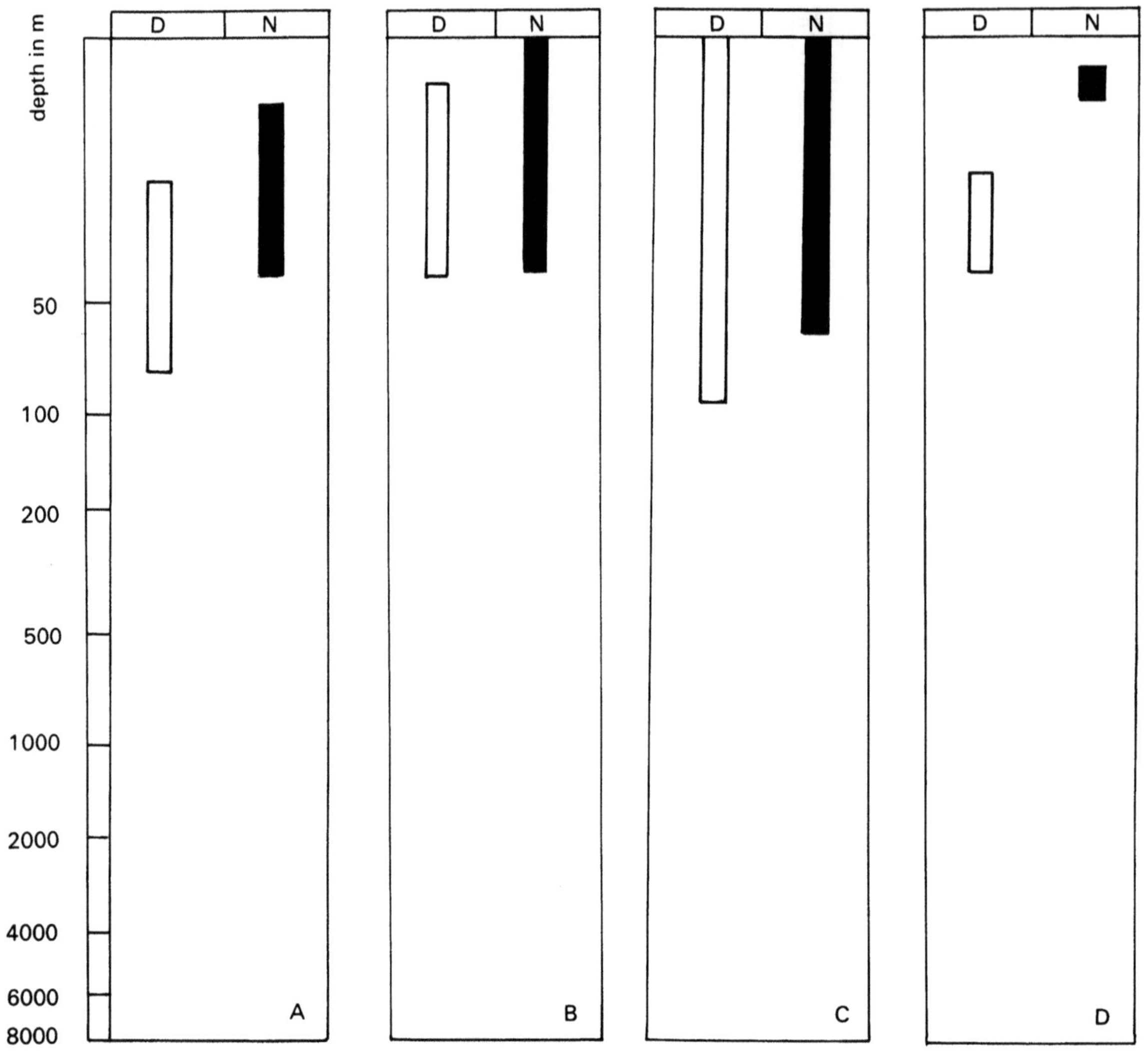

As each taxon occupies its own ecological niche, different species usually show different migration types within one water mass, as shown for *Diphyes dispar*, *Chelophyes contorta* and *Lensia subtiloides* in the Sula Sea (fig. 162). Usually different life stages of one taxon, each with its own ecological demand, show different diurnal migration (fig. 163).

This probably implies that there is real ontogenetic migration, as the different diurnal migrations brought together into one year cycle give rise to an ontogenetic migration (fig. 164). These calanoids show ontogenetic migration in the upper 1000 m. Deepsea plankton as represented by e.g. *Limacina helicoides* and *Clio polita* also shows ontogenetic migration (of the type 3 in fig. 156), the adults living below 1000 m and the juveniles in the surface plankton. Usually the ontogenetic migration consists, as in the two pteropods mentioned, of a movement of the animals downwards with increasing age. Species with such strong ontogenetic or seasonal migration (such as *Calanus plumchrus*) usually show little or no diurnal vertical migration and species with little or no seasonal migration (such as *Metridia pacifica*, fig. 163) show strong diurnal migration.

This rule, however, does not always hold good, as temporary or local hydrographical features, such as more especially temperature stratifications, may strongly affect the migration patterns. It is therefore impossible to give a better division into types of vertical migration than that of table II, to which we can only add the terms double migration and reverse diurnal migration. Double migration is the migration in which, during one cycle, one depth stratum is crossed four times. Double diurnal migration is exemplified by *Creseis virgula* and *Diacria quadridentata* in some parts of their ranges. Stubbing (1938) found this type, characterized by a strong upward migration only at dawn and dusk, for the first time in the Arabian Sea. Reverse diurnal migration is the phenomenon of the organisms living in the upper layers in the day time and in the lower levels at night. Reverse ontogenetic migration means that the younger stages live below the adult stage.

It should be stated that vertical migration has different influences on horizontal distribution. One of the advantages of vertical zooplankton migration is the possibility it provides to move horizontally within a current system, with a minimum of energy-loss. Hardy

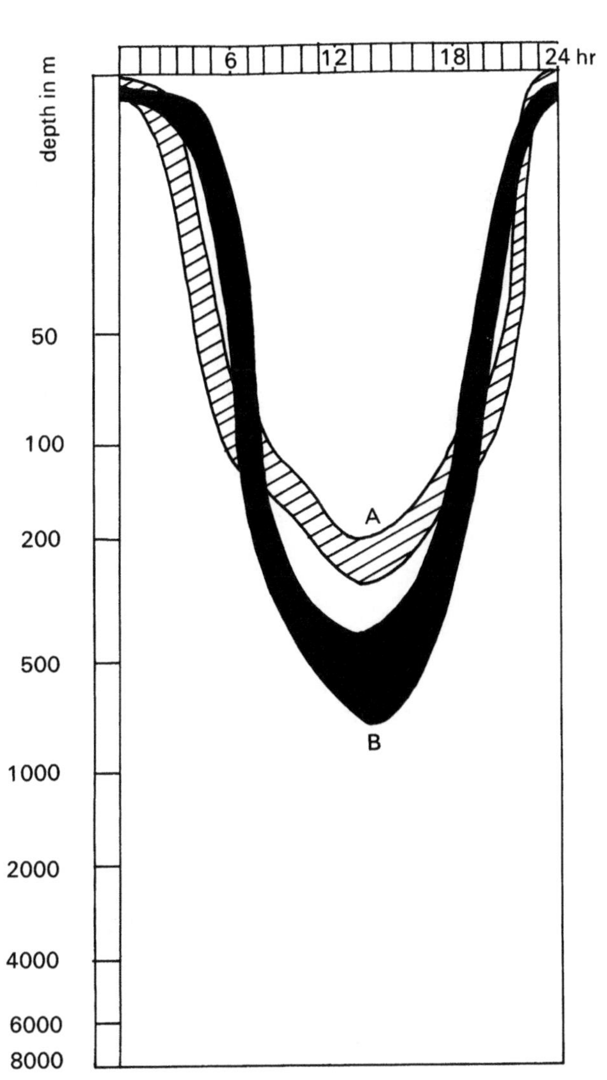

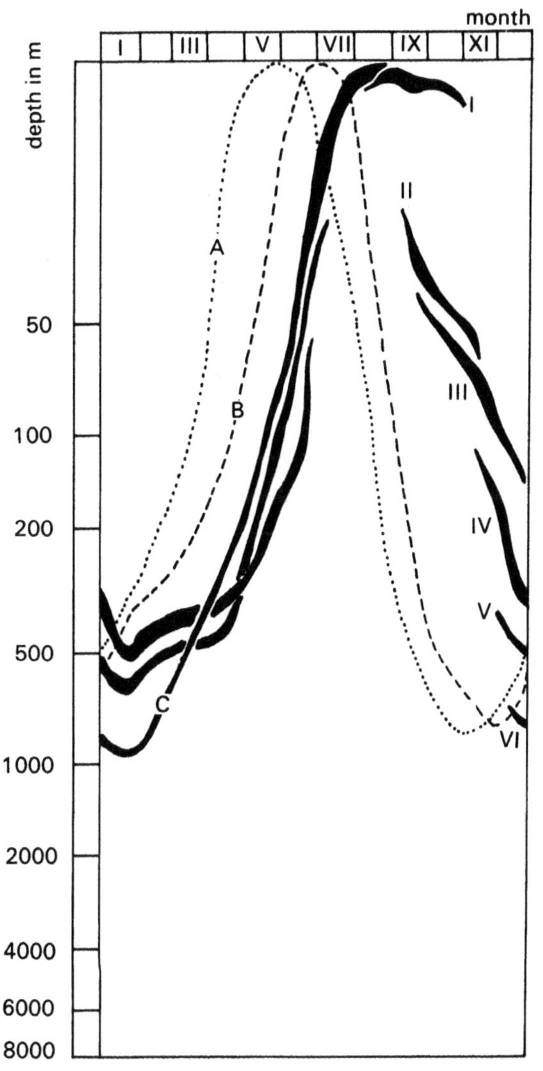

(1935) suggested a relation between diurnal vertical migration and horizontal movement of planktonic animals; in 1970 Miller found a confirmation of this relation.

The vertical movements, however, are mainly of importance to maintain a position in a restricted area. It has been pointed out already (p. 127) that all vertical migrations are ecologically determined. This explains the opposite reactions of two species to the same phenomenon as exemplified by the vertical migratory behaviour of *Calanus pacificus* and *Euphausia pacifica* in relation with the same seasonal upwelling in the California Current (Koslow & Ota, 1981). Two types of vertical migration are distinguished, viz. one with a long, and one with a short periodicity.

The vertical migration types with long periodicity are important because of the possibility to transport specimens over large horizontal distances. Feeding migration and diurnal migration, of short periodicity, are of restricted importance, though their high frequency may give them a greater value for dispersal. The ontogenetic and seasonal migrations, both with a long periodicity, were called long-term migrations by Longhurst (1970). He stated: 'Since most, though not all, plankton species which show these sorts of migrations have several generations each year, it is not useful to separate ontogenetic from seasonal migrations, since so frequently they are one and the same thing.'

FIG. 162 Diurnal vertical migration of *Lensia campanella* (A), *Diphyes dispar* (B), *Chelophyes contorta* (C), *Lensia subtiloides* (D) in the Sulu Sea (after Musaeva, 1976). d = day, n = night.

FIG. 163 Diurnal vertical migration of copepodid I (A) and copepodid VI (B) of *Metridia pacifica* (after Vinogradov, 1970).

FIG. 164 Diurnal vertical migration of three interzonal calanoid species (A, B, C) with seasonal ontogenetic vertical migration of the copepodid stages I-VI (after Vinogradov, 1970).

In the Antarctic waters the long-term migrations of *Euphausia superba* and *Calanoides acutus* are similar. In winter time the *C. acutus* population lives south of the Antarctic Convergence below 250 m. During the summer, most of the advanced copepodids and all the youngsters concentrate in the surface layers until the autumn, when the whole population sinks again.

In spring, the metanauplii of *Euphausia superba* are

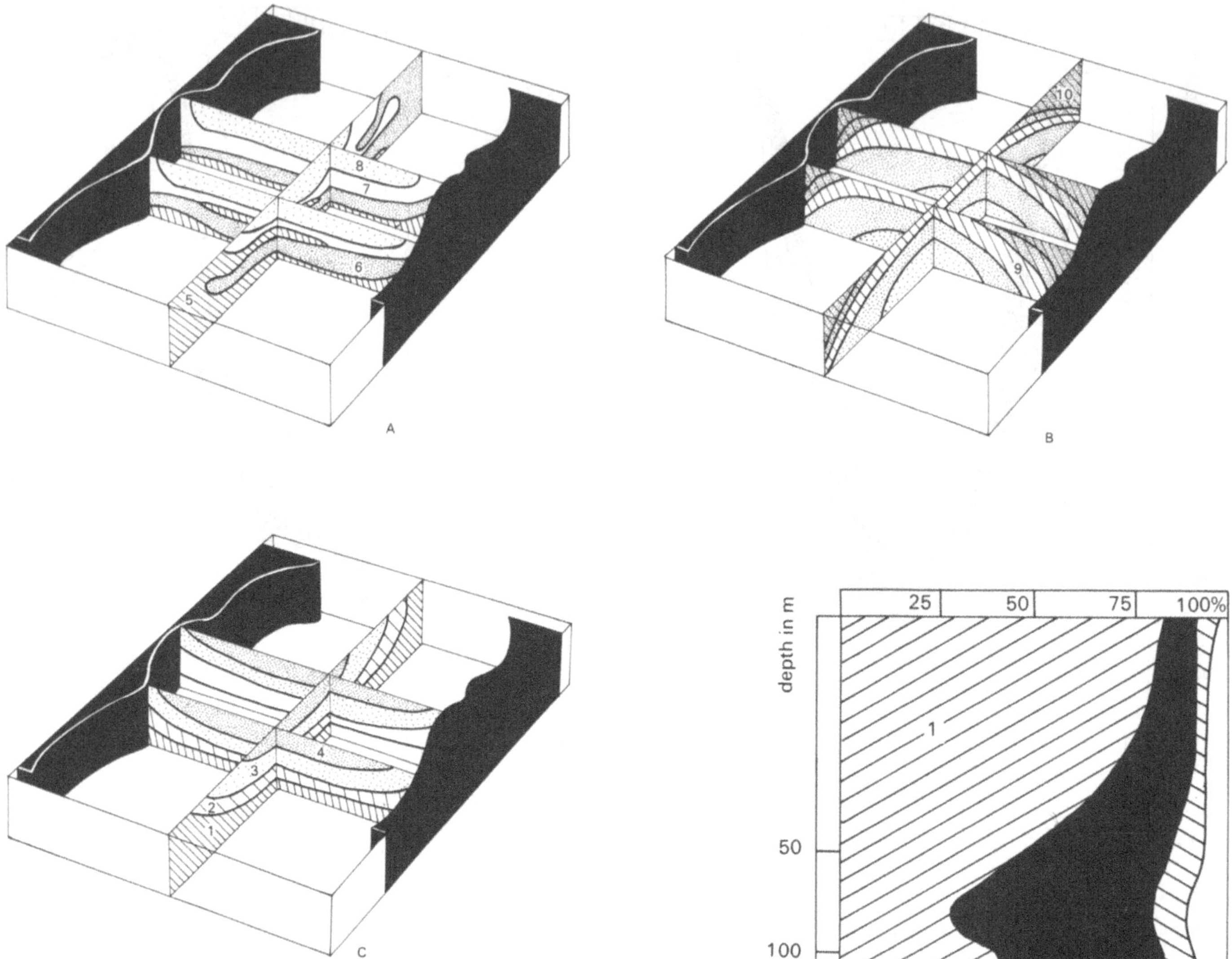

FIG. 165 Ocean structure in the vertical plane, based on water-masses (A), on productivity (B) and on zoogeography (C). 1 = deep and polar, 2 = deep, 3 = Transitional, 4 = Tropical, 5 = Antarctic, 6 = Arctic, 7 = Intermediate, 8 = Tropical, 9 = low productive, 10 = high productive.

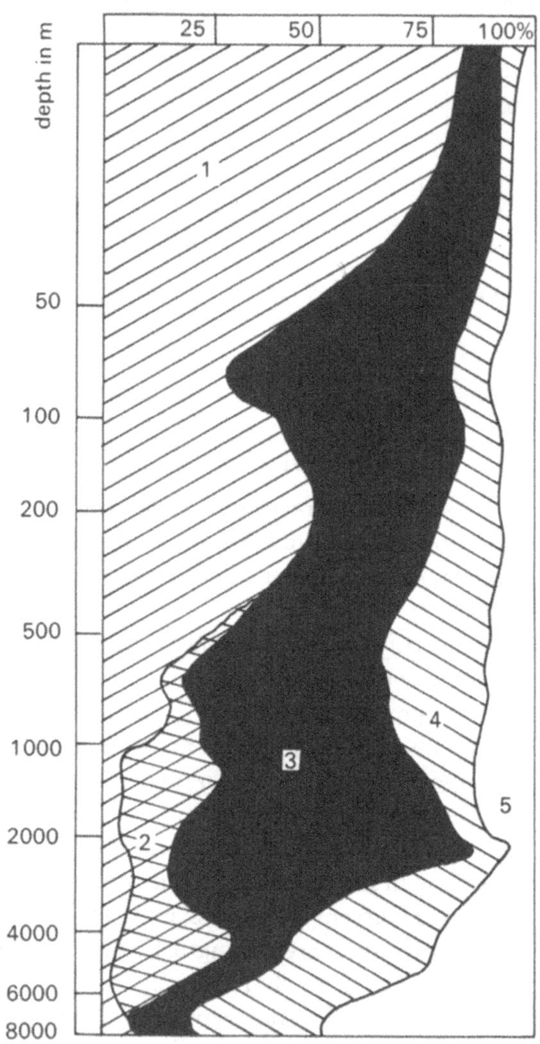

FIG. 166 Relative abundance of feeding specialisms. 1 = filter-feeders, 2 = filter-feeders in a non-feeding stage, 3 = carnivores, 4 = euryphages, 5 = radiolarians and other groups (after Vinogradov, 1970).

very abundant at a depth of more than 500 m. The young furcilias show strong diurnal migration. In summer the older furcilias remain in the surface layers. They live epiplanktonically and are carried northward as far as the Antarctic Convergence by surface currents. During the autumn the euphausiids sink into a current which brings the population back to the south (Vinogradov, 1970).

Most of the well-known long-term migrations occur at high latitudes. These vertical migrations have not been found at low latitudes, except for places with substantial seasonal changes, such as coastal upwelling areas. Long-term migration seems mainly to occur where it is important to limit the area of distribution. The very implications are only fragmentarily known.

The high-frequency migrations, such as diurnal and feeding migrations, are accepted to have more influence

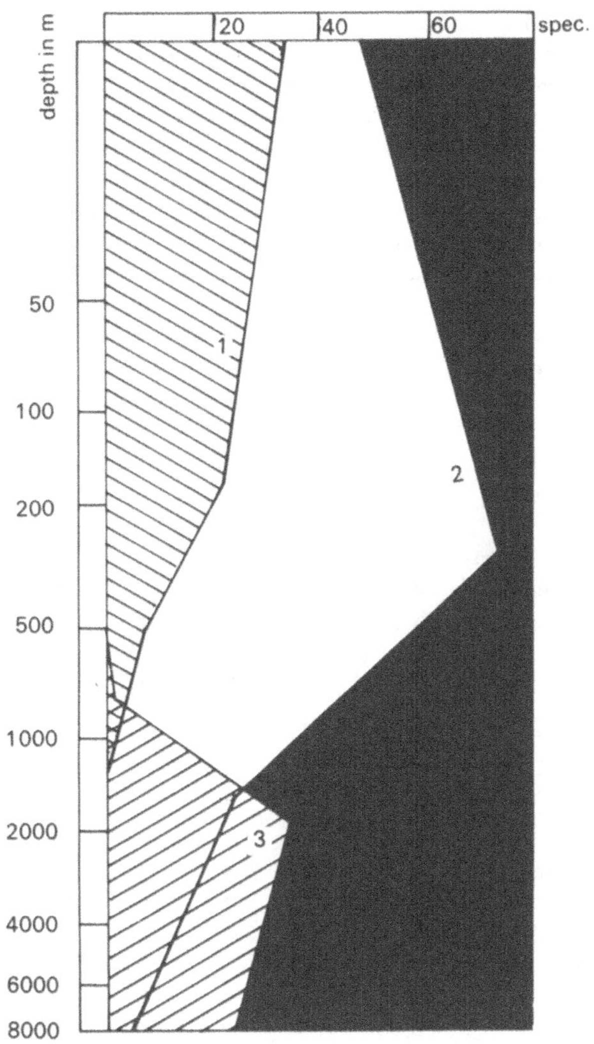

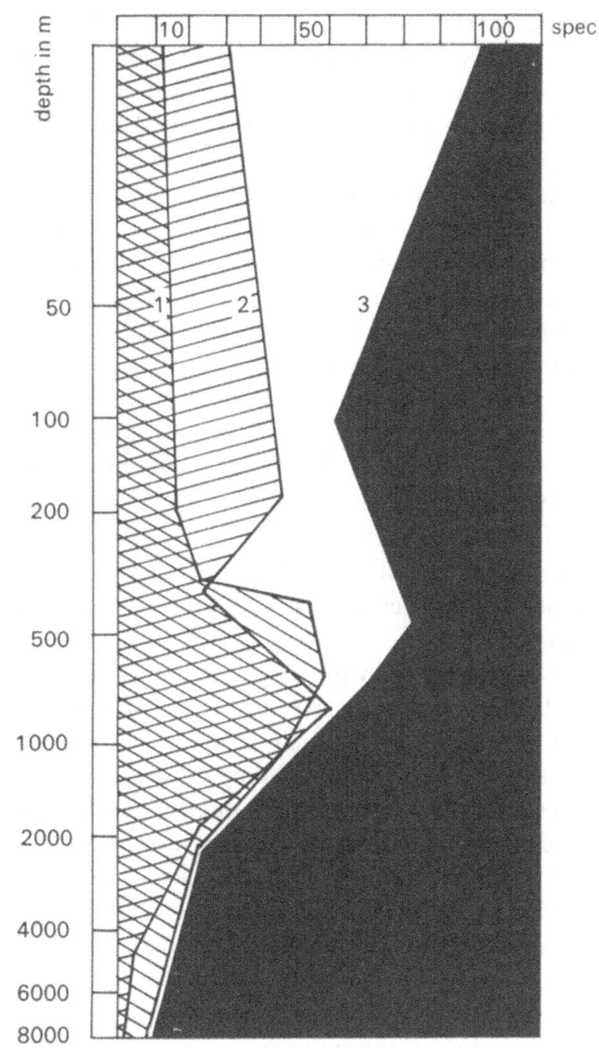

FIG. 167 Vertical differences in diversity (number of species) per group. 1 = Heteropoda, 2 = Euphausiacea, 3 = Nemertines.
FIG. 168 Vertical differences in diversity (number of species) per group (after Vinogradov, 1970). 1 = Calanoida, 2 = Ostracoda, 3 = Radiolaria.

on horizontal dispersal than the long-term migrations.

The vertical structure of the ocean is partly determined by temperature and salinity (cf. figs 25-35). Water masses can be defined by these two parameters. Fig. 165A gives these water masses diagrammatically for a hypothetic ocean, and an image of piled bowls emerges. In fig. 165B the ocean structure is given as based on fertility which results in a picture, the reverse of that of fig. 165A.

Zoogeography, however, seems more dependent on water mass configuration than on productivity, as shown in fig. 165C based on rough vertical patterns; this configuration again resembles that in fig. 165A.
It is indeed the Tropical-Subtropical surface waters and Arctic and Antarctic deep and Intermediate waters which shape the vertical distribution.

Besides the water masses, it is the feeding ecology which is strongly related to vertical distribution. The surface waters are dominated by largely phytophagous filter-feeders, the mesopelagic realm consists chiefly of carnivorous taxa and euryphagous taxa while detritus-feeders dominate the depths (fig. 166).

Feeding ecology is partly adaptive, but also distinctly determined by evolution as a 'group-special' character.

Salps can hardly become catching animals and nemerteans cannot develop into filter-feeders. As a consequence, each group shows its maximal diversity at a particular depth-level as shown for a chiefly bathy-, a meso- and an epipelagic group in fig. 167 and for three epi-meso pelagic groups with different preferences for the epiplanktonic zone in fig. 168.

Van Soest (1979) studied the relation between horizontal N-S distribution and vertical distribution (table III) and concluded that epipelagic and mesopelagic organisms are restricted in their distributions by strong latitudinal barriers, less strong barriers are often surpassed but still limit the distribution of many epipelagic organisms. Mesopelagic and bathypelagic organisms show an increased tendency to have a cosmopolitan N-S distribution. It can be inferred from the data presented that most if not all holoplanktonic groups have their origin in surface waters of the lower latitudes of the oceans. The vertical structure of the ocean presented in the different figures makes it necessary to define the concepts epi-, meso- and bathypelagic and some others more precisely in relation to this structure.

Pleuston consists of animals that float or move on the surface of the water, such as e.g. *Halobates*, *Velella* and *Glaucus*.

Neuston is formed by the plants, bacteria and animals that live in the uppermost centimetres of the water. The epipelagic zone is usually taken to comprise also the neuston, and to be determined by the degree of light penetration. In relation to light intensity under water, one distinguishes the euphotic zone, in which photosynthesis is possible, the disphotic zone in which visible light penetrates but where photosynthesis is not found, and the aphotic zone which is dark. Some scientists take the underside of the euphotic (± 80 m), others the underside of the disphotic zone (± 200 m) as the lower limit of the epipelagic realm. The seasonal thermocline (varying in depth between 10 and 250 m) is regarded as important by others. For zoogeographic studies it seems impossible to use isotherms or thermoclines to limit the epipelagic realm. Epipelagic biota also exist in polar water throughout the year, living at temperatures much lower than the average global temperature at 200 m depth. It seems most practical to take the layers in which visible light is present as epipelagic zone, a technical problem is that light measurements at any depth are seldom made. The mesopelagic realm is the zone of twilight, the zone in which very small, but still measurable light quantities penetrate. In clear ocean water the maximum depth of measurable light penetration is 1200 m. Two problems arise here: 1) such deep measurements are rarely made; 2) mesopelagic animals are usually more temperature-sensitive than light-sensitive.

The consequence of taking a temperature as the lower limit of the mesopelagic realm is the absence of this realm from polar areas, and moreover, many different temperatures (usually 10°C or 5°C) are used for the lower limit. It is known that organisms considered to be mesopelagic live at greater depths in the Tropics than at higher latitudes and that bathypelagic animals live at very shallow depths in the Subpolar regions.

TABLE III
Distribution of epipelagic (e), mesopelagic (m) and bathypelagic (b) organisms, grouped according to the higher taxa and shared ranges. The numbers represent the totals of species; the distributional categories are: wide (i.e. occurring at all latitudes), moderately wide (occurring between 50°N and 50°S), restricted to high latitudes (higher than 50°N or 55°S), restricted temperate latitudes (50°-30°N, 30°-55°S) and restricted to low latitudes (30°N-30°S) (after Van Soest, 1979).

Animal groups		N-S distribution wide	mod. wide	restr. high lat.	restr. temper. lat.	restr. low lat.
Foraminifera	e	0	12	5	1	10
	m	—	—	—	—	—
	b	—	—	—	—	—
Siphonophora	e	—	35	3	—	18
	m	2	45	2	2	6
	b	4	6	—	—	2
Medusae	e	1	16	6	2	13
	m	—	8	2	—	2
	b	9	6	3	—	—
Tomopteridae	e	—	4	2	—	—
	m	11	14	3	—	—
	b	3	—	—	—	—
Mysidacea	e	—	3	5	3	12
	m	3	13	17	10	14
	b	12	7	1	1	4
Euphausiacea	e	—	17	—	7	12
	m	1	1	4	3	9
	b	—	—	—	—	—
Hyperiidae	e	—	4	8	1	10
	m	1	1	—	—	—
	b	—	—	—	—	—
Phronimidae	e	—	8	—	—	—
	m	—	—	—	—	—
	b	—	—	—	—	—
Copepoda	e	3	64	—	—	26
	m	21	74	1	—	12
	b	3	7	—	—	2
Euthecosomata	e	1	21	2	—	7
	m	—	—	—	—	—
	b	—	1	—	—	—
Pseudothecosomata	e	—	7	—	—	5
	m	—	4	—	—	—
	b	—	3	—	—	—
Gymnosomata	e	—	17	2	8	11
	m	—	3	—	1	—
	b	—	2	—	1	1
Heteropoda	e	—	21	—	—	12
	m	—	—	—	—	—
	b	—	—	—	—	—
Chaetognatha	e	—	13	2	4	2
	m	5	2	1	—	—
	b	—	—	1	—	1
Copelata	e	—	32	13	3	14
	m	—	—	—	—	—
	b	—	—	—	—	—

(table III continued)

Doliolida	e	—	5	—	1	8	
	m	—	—	—	—	—	
	b	—	—	—	—	—	
Pyrosomatidae	e	—	5	—	2	2	
	m	—	—	—	—	—	
	b	—	—	—	—	—	
Salpidae	e	—	21	5	—	17	
	m	—	—	—	—	—	
	b	—	—	—	—	—	
total	e	5	305	53	42	181	Σ 586
	m	44	165	30	16	43	Σ 298
	b	31	34	5	2	11	Σ 83
total percentage	e	1%	52%	10%	7%	31%	Σ 60%
	m	15%	55%	10%	5%	15%	Σ 31%
	b	39%	41%	6%	3%	13%	Σ 9%

For zoogeographic purposes it is probably better to give the more restricted definition of the concept as listed with fig. 169. The use of these terms, though modified, is necessary as the use of water masses to describe vertical distributions is not yet possible because of a lack of sufficient biological and hydrographical data.

In the epiboreal pelagic realm there is usually a strong mesopelagic influence, and in the epipolar and mesopolar pelagic realms usually evident bathypelagic influence is found. This rough subdivision must suffice for the time being.
The zonation proposed in fig. 169 is similar to that proposed by Casey (1971) for the Pacific Ocean's Radiolaria (fig. 170), so that the subdivision and terminology of fig. 169 are proposed to represent the faunal zones.

The pleuston fauna is limited to the warmer parts of the oceans between the Subtropical fronts. Both in the pleuston and in the neuston one cannot speak of vertical distribution. For the epiplankton it should be possible to distinguish vertical patterns. Many ecological studies have elucidated the stratified occurrence of populations and species. To distinguish vertical zoogeographic patterns, however, is difficult, if not excluded. In the tropical-subtropical parts of the ocean, the main thermocline is found at the lower limit of the epipelagic zone or below it. In the colder parts, polar to about 40° latitude, there is no thermocline in winter, and in the spring and summer a thermocline develops that sinks to

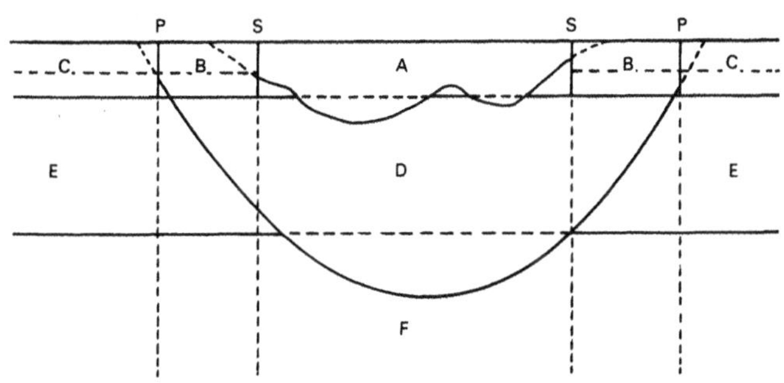

FIG. 169 Vertical cross-section through a hypothetical ocean from north to south giving the biotopes as discussed in the text.
A = epipelagic:
between Subtropical fronts, surface and 15 °C isotherm
B = epiboreal-pelagic:
between Subtropical and Polar fronts, surface and 200 m depth
C = epipolar-pelagic:
polewards of Polar fronts, between 0 and 200 m depth
D = mesopelagic:
below 200 m or below 15 °C isotherm and above 5 °C isotherm
E = mesopolar-pelagic:
below 200 m or below 5 °C isotherm and above 1000 m
F = bathypelagic:
below 1000 m or below 5 °C isotherm
P = Polar front, S = Subtropical front.

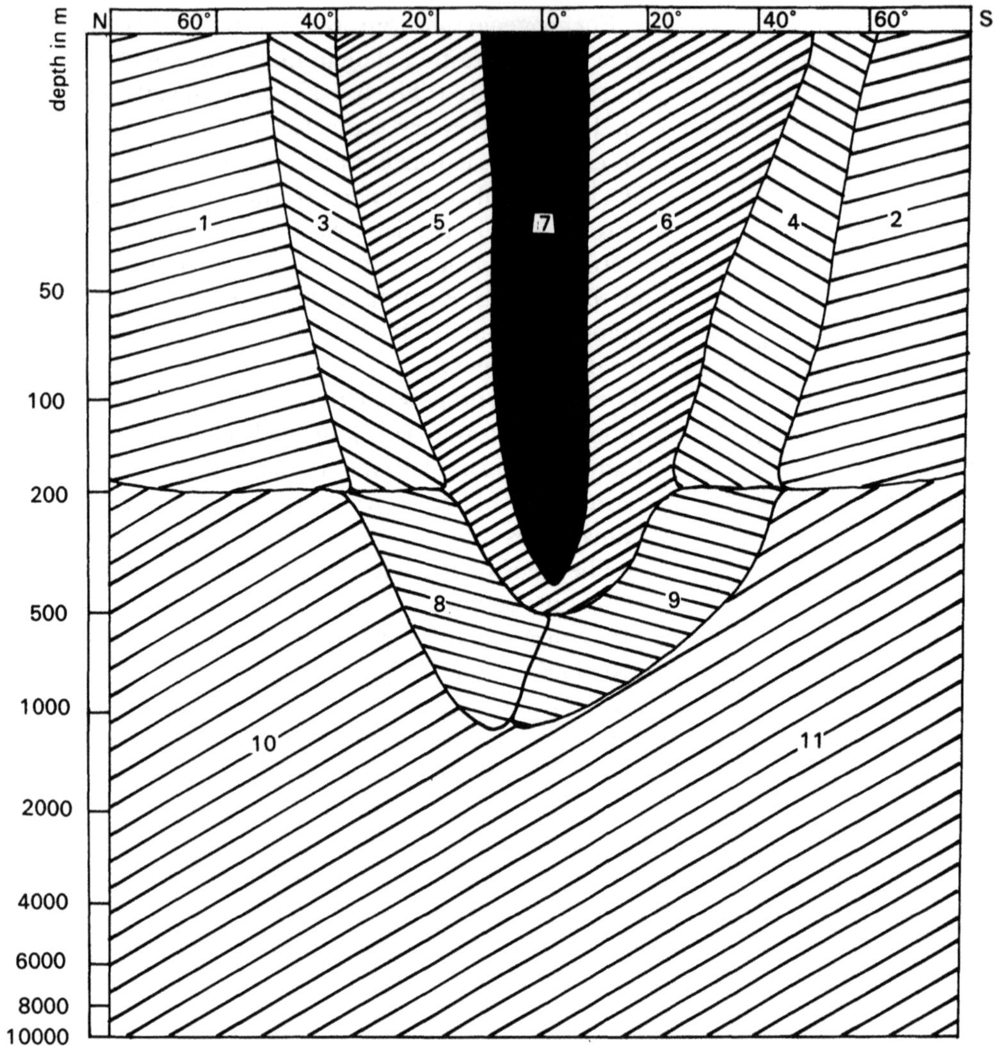

a depth of 50–150 m. These differences in temperature structures have a great influence on the zoogeography. In the Tropics there is a continuous vertical area available throughout the year and this area is strongly delimited by the thermocline. In the colder region there is, in winter, a vertical continuum caused by the absence of a thermocline and by strong mixing even with waters from below ca 200 m. In summer, however, each year a stratum above the summer thermocline developes with a special productivity and ecosystem. For the Tropics no subdivision in the vertical of the epipelagic realm is proposed. For the higher latitudes, the epipelagic realm can be divided into a summer-surface biotope, occupied by specific taxa absent in winter. In winter, the lower living epipelagic taxa also occupy the shallower strata. In fig. 169 the epipelagic realm is therefore given without subdivision and the epiboreal pelagic realm with a temporary subdivision (broken line). The species living in the summer-surface biotope are frequently lower-latitude taxa which only penetrate the higher latitude in summer. This is not a penetration into an expatriation area, as completely self-supporting, successfully breeding populations occupy these areas.

The mesopelagic realm gives more possibilities for a subdivision, the Soviet scientists in general propose the subdivision into 'upper interzonal' and 'lower interzonal' species. The term interzonal is not synonymous with mesopelagic as it comprises all animals which spend part of their lives in the surface zone and another part in the deepsea (Vinogradov, 1970), but they are taken separately from deepsea organisms. Upper interzonal are species moving up- and downwards roughly above 500 m. The lower interzonal species move below approx. 500 m. The use of the concept interzonal by e.g. Vinogradov is caused by the fact that it was accepted that most if not all plankton of deeper layers is in constant motion. When one considers the mesopelagic realm as a faunal zone, however, it is clear that different taxa have a different preference for specific depth layers. Due to migration and latitudinal differences it is, however, not possible to distinguish special vertical ranges within the mesoplankton.

The bathypelagic realm, comprising the largest section of the vertical profile of an ocean, is never considered to be composed of different zoogeographic areas. Insufficient sampling or real absence of vertical barriers below 1000 m may explain this.

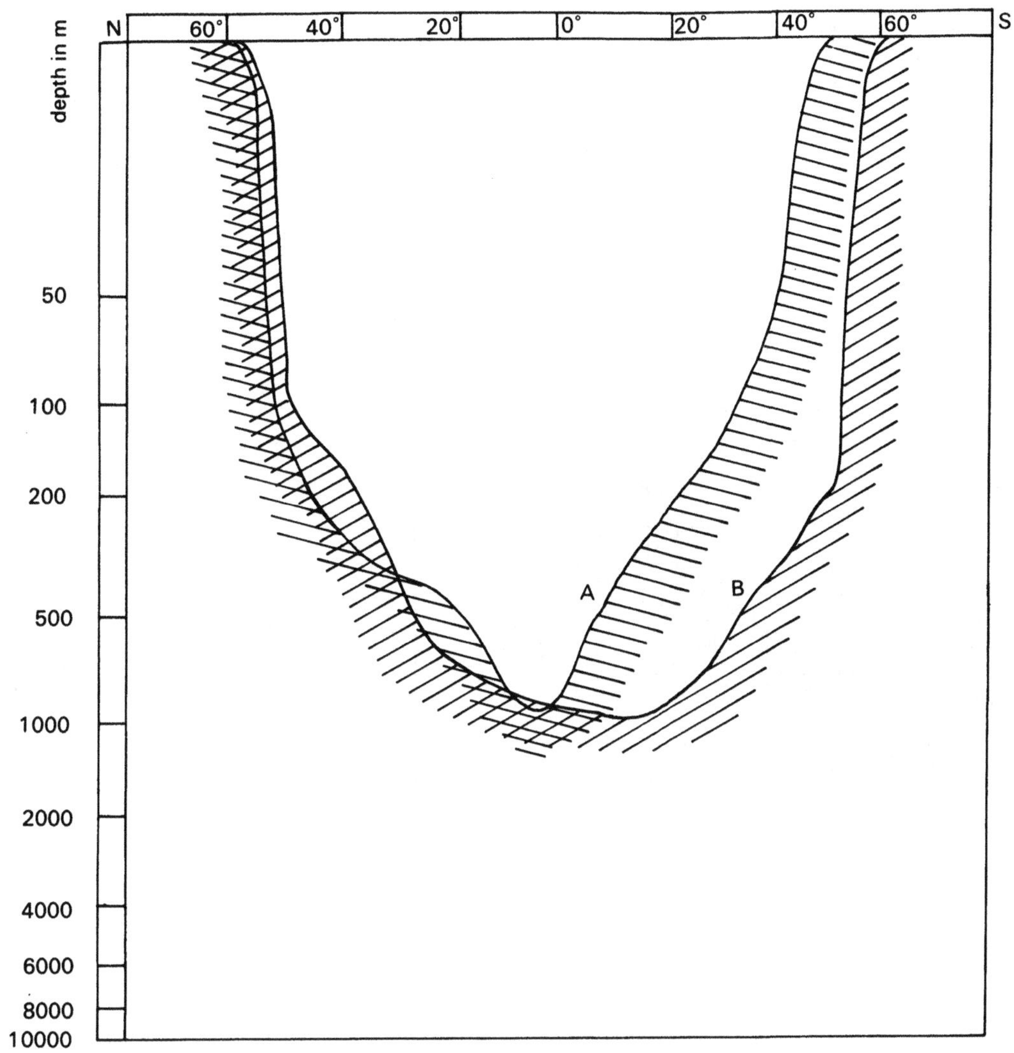

Tropical submergence and vertical variation

The bowl-shaped vertical pattern of faunal zones (figs 165, 169), provides the wide-spread species, occurring in different water masses, with the possibility to develop a bowl-shaped distribution. *Eukrohnia hamata* (fig. 171) clearly shows this pattern over the epipolar-mesopolar and bathypelagic realms, and is a representative of a large group of taxa with this pattern. It is a cold-water species (polar waters) and a bathypelagic species (deep layers) which avoids the warmer surface waters at lower latitudes, a phenomenon called 'tropical submergence'. *Sagitta planctonis* also shows tropical submergence (fig. 172) but the pattern is more complex as the species is represented by two forms (Pierrot-Bults, 1972), the form *planctonis*, showing the real tropical submergence and the form *zetesios*, living in the warmer surface layers. This polytypic species clearly shows that tropical submergence is not necessarily an escape from warm water but can also be an adaptation to special conditions.

When the disappearance from the surface waters occurs at about the polar or subpolar fronts, the term tropical submergence is not adequate. Still, there are numerous species, such as *Argyropelecus olfersi*,

FIG. 170 Biogeographic zones for polycystine radiolarians of the Pacific Ocean (after Casey, 1971). 1 = Subarctic shallow zone, 2 = Antarctic shallow zone, 3 = Transitional shallow zone, 4 = Subantarctic shallow zone, 5 = N. Central shallow zone, 6 = S. Central shallow zone, 7 = Equatorial zone, 8 = Transitional Central zone, 9 = Subantarctic Central zone, 10 = Subarctic Intermediate zone, 11 = Antarctic Intermediate zone.

FIG. 171 Vertical position of upper limit of highest population densities of *Eukrohnia hamata* an the Atlantic (A) and Pacific Oceans (B).

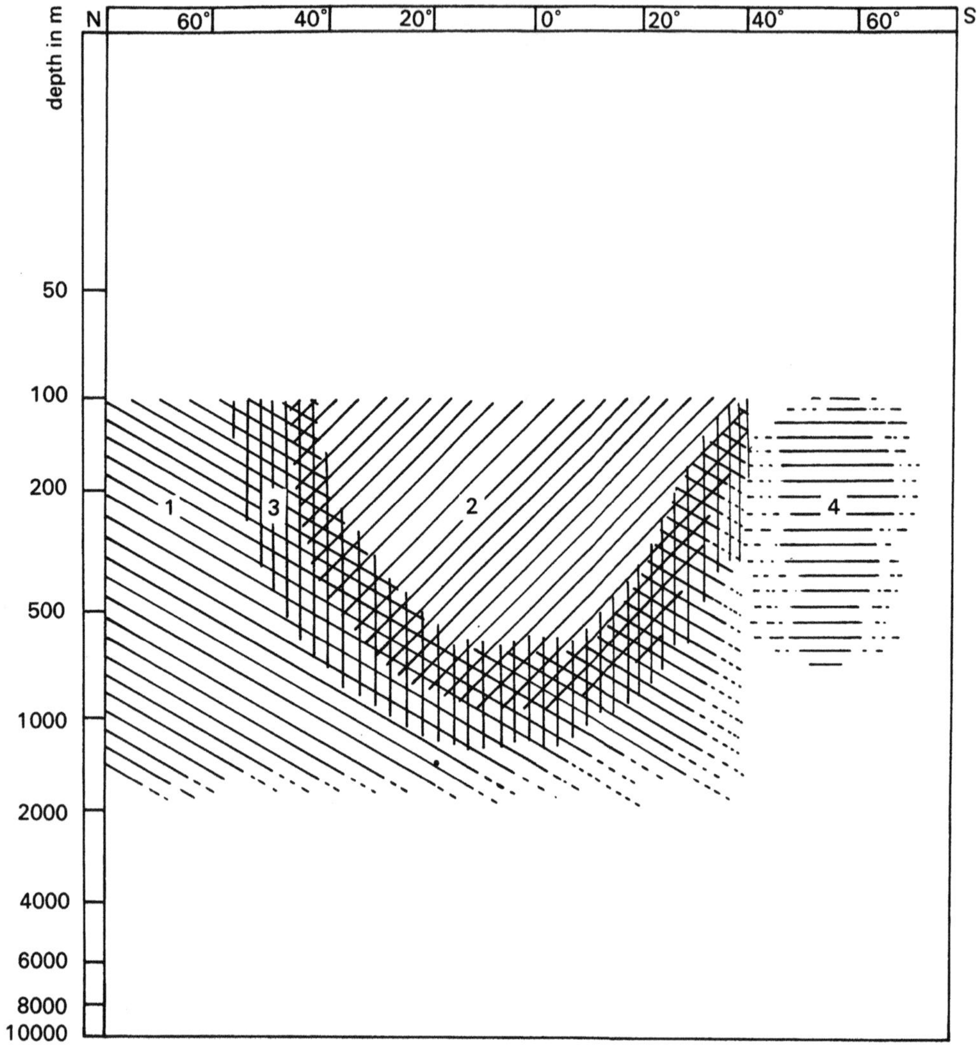

A. aculeata and *Periphylla periphylla*, which live in the whole water column north of 55 °N and gradually become deepsea dwellers in the high Tropics. This diving to greater depth where surface temperatures are higher is also tropical submergence. In *Argyropelecus* it is also associated with vertical variation (Pafort-v. Iersel, 1981). It seems as if temperature is the main factor for this vertical zoogeography and variation.

In most plankton species no vertical variation is found and this is explained by taking vertical migration as a force which mixes the populations. In *Sagitta planctonis* and *Argyropelecus*, both with vertical variation, however, vertical migration is obvious. This proves that distribution patterns can exist above one another in the ocean without losing their characters through vertical movements.

FIG. 172 Vertical distribution of *Sagitta planctonis*. 1 = *S.p. planctonis*, 2 = *S.p. zetesios*, 3 = *S.p.* intermediates, 4 = *S. marri* (after Pierrot-Bults, 1976).

5 Some general trends

Historical trends—Geomorphological influences—Climate influences—Sea-level changes—Horizontal patterns—Ecological factors—Speciation around the species-level and distribution—Plankton communities and patchiness

The marine environment is the only uninterrupted, internally moving, three-dimensional environment. Uninterrupted means that, in principle, each organism can reach all places in this environment; internally moving means that inactive organisms are constantly subjected to transport; three-dimensional implies that populations and migratory movements are also three-dimensional. With regard to dispersal these three factors are unique. The marine environment, sea-water throughout, is uniform, relatively stable and isolated from other biotopes. Thus, the selective pressure is rather unvariable in space; and stable means that it is not very variable in time either, while the isolated position means that influences, biotic as well as abiotic, from other surrounding biotopes, do not affect the marine fauna. The dominant ecological features which nevertheless shape the distributions in the oceans, are the currents, the temperature, the food web and the interaction of species, and it is also these features which give shape to the communities. If the taxonomy and dispersal are to be rightly understood, however, the historical zoogeography should be treated first, before focusing on the above-named phenomena.

Historical trends

Distribution is variable in time and in this chapter the possible trends along which present-day patterns developed, will be discussed. Only on horizontal distribution sufficient data are available to allow of some conclusions; the vertical pattern of the deeper layers remains obscure.

In the Cenozoic, only some 53 to 36 m.y. ago, the deepsea circulation changed thoroughly to reach its present configuration and to provide the ecological conditions existing at present (Herman, 1979). The ability to live in the deeper strata is, therefore, at least in most cases, a relatively recent development and adaptation (Pierrot-Bults & Van der Spoel, 1979). Though the deep-sea fauna in its present composition is of recent origin, lack of data, chiefly due to inadequate sampling techniques, prevents us from understanding the vertical dispersal. It is only possible to distinguish between an epi-, a meso- and a bathypelagic realm (cf. p. 118), each with its typical planktonic representatives. Typical epiplankton is: *Limacina inflata*, *Cavolinia longirostris*, *Sagitta enflata*, *Euphausia nana* and *Nyctiphanes capensis*; mesoplanktonic are: *Clio pyramidata*, *Cavolinia tridentata*, *Sagitta planctonis*, *Euphausia mutica*, *Meganyctiphanes norvegica*; bathypelagic representatives from these groups are: *Limacina helicoides*, *Clio polita*, *Sagitta macrocephala* and *Euphausia similis*. Some of the genera are represented by epi-, meso- and bathypelagic species and some genera are only found in two of the three strata. It is, however, not possible to distinguish real mesopelagic or bathypelagic genera, let alone higher taxonomic groups, which is also evident from table II (after Van Soest, 1979). Consequently, in the following discussion it is accepted that more deeply living taxa derived from epiplanktonic forms.

Vertical migration is in general an ecological adaptation related to temperature- and light-tolerance and to feeding- and reproductive behaviour. Research with sound scattering has proved a very regular diurnal vertical migration in the upper 600 m. (Schalk, M. S.), but findings based on sound scattering cannot be related to taxonomic groups and the existence of migration in deeper layers is dubious (Angel et al., 1982). Phenomena in the seasonal, ontogenetic and feeding migrations are fragmentarily known, as discussed above (p. 128). Any conclusions as to evolution and development of vertical migration is so far excluded, the only hypothesis which can be put forward, is that of closely related taxa, the migrating taxa are further evolved than the non-migrating ones.

As most plankton species do not fossilize, data on time variability of distribution are even more rare than data on vertical variability. The geological history of ocean morphology, sea-level and climate gives some indications as to how the present patterns came about. In the eight geo-cladograms below, a summary is given of the general trends of zoogeographic developments, as derived from geological theories and zoogeographic data. All these geo-cladograms are so arranged (fig. 173) that the branches 1, 2, 3 etc. are more primitive than the branches A, B, C etc.; the continuing line goes in the direction of the most evolved group. The indication # means that along that branch a certain amount of extinction is expected, while the arrows mark branches where migration is supposed to have occurred.
As a general assumption one can accept that cold-water species derived from warm-water taxa (Pierrot-Bults & Van der Spoel, 1979) and cold-water species are usually ancestral to deep-water taxa. After the Cretaceous the climate became more and more severe, so that first a shallow-temperate and later, e.g. in the early Oligocene, a cold-water fauna could develop. The original cold-water faunas will have been separated by the Tropics. This separation occurred gradually, and when the taxa were still rather eurythermous they could, of course, make contact through the Tropics more easily than at the end of their evolutionary adaptation to the cold-water areas. During the cooler periods of the Ice-Ages a last possibility for these north-south contacts will have existed. That the splitting of the cold-water fauna into separate northern and southern hemisphere faunas is of rather recent origin can also be concluded from the close relation of some Antarctic and Subantarctic taxa with Arctic and Subarctic taxa.

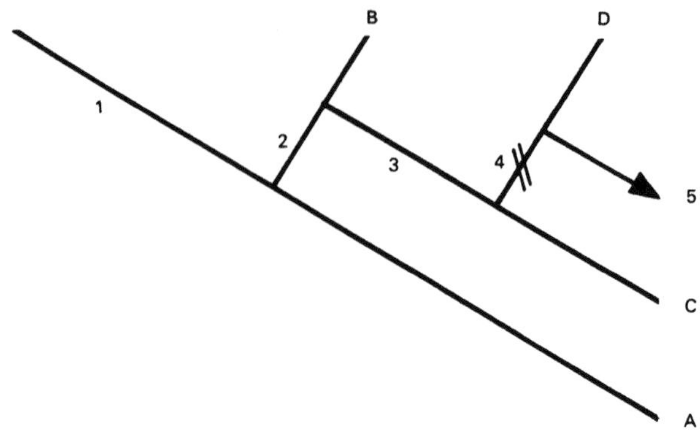

FIG. 173 Cladogram composition; for explanation see text.

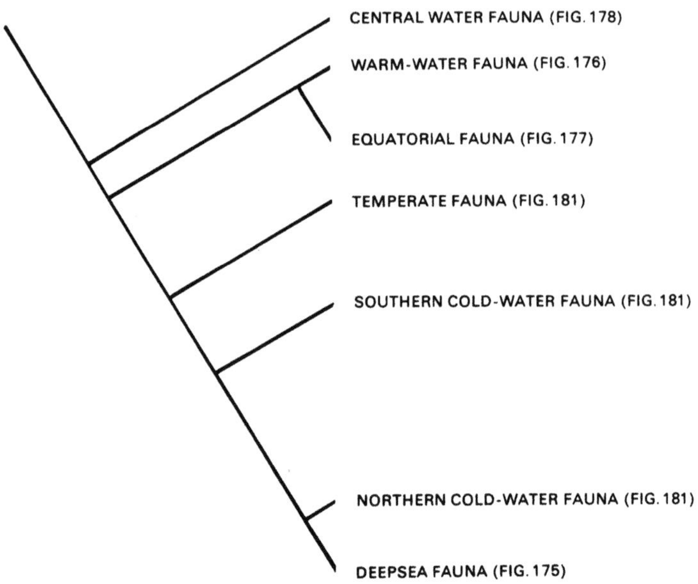

FIG. 174 Geo-cladogram of the major planktonic faunas.

After the establishment of the cold-water faunas, deepsea taxa could develop. This development can be represented by the following geo-cladogram (fig. 174).

As the North Atlantic was the first deepsea basin to develop, many authors accept that the deepsea fauna first developed here and dispersed afterwards. For Peraclididae (fig. 186) this dispersal is given and in fig. 175 it is represented as a geo-cladogram.

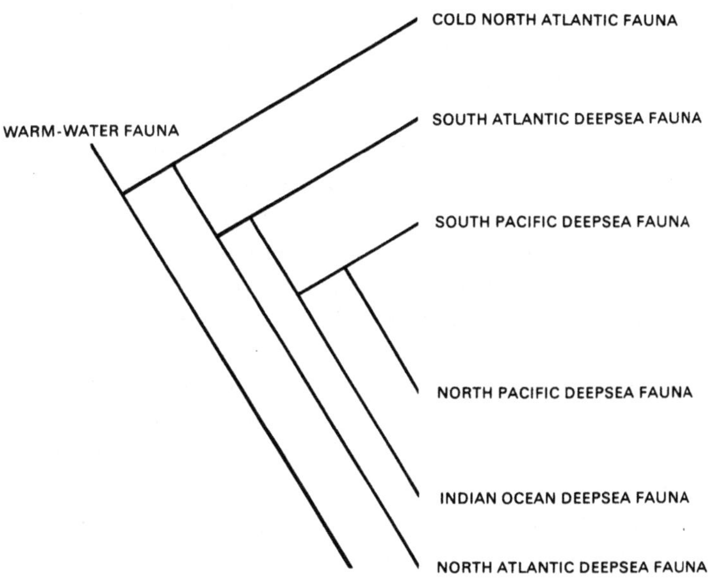

FIG. 175 Geo-cladogram of the deepsea faunas.

The Central water taxa (p. 68) are warm-water species which established their distribution probably in relatively early times, so that in fig. 174 a first splitting is proposed of warm-water taxa into Equatorial, Central water and general warm-water taxa. From the maps it is evident that two belts of warm-water taxa are to be distinguished, besides the Central water and Equatorial taxa, viz. the 40°-40° belt and the 30°-30° one; in addition, some taxa show a special East Pacific distribution.

As a diagram for speciation of general warm-water oceanic plankton the geo-cladogram in fig. 176 can be added to the preceding figure. A time-axis is hard to indicate in the geo-cladogram of fig. 174, but for the present one (fig. 176) the branching points can roughly be placed in geological periods. The development of the 30°-30° distribution out of the 40°-40° distribution (branching point 1) can be placed in the Eocene-Oligocene, when global temperatures dropped. The branching point 2 is of later date, when the Atlantic warm-water belt became separated from the Indo-Pacific one; this may have been in the early Pleistocene.

During the Ice-Ages there were several successive temperature drops, so that splitting point 2 may have occurred also more than once (Pierrot-Bults & Van der Spoel, 1979). The isolation of the East Pacific (point 3) is probably even more recent as to conclude from the taxonomic level of the taxa endemic to the E. Pacific.

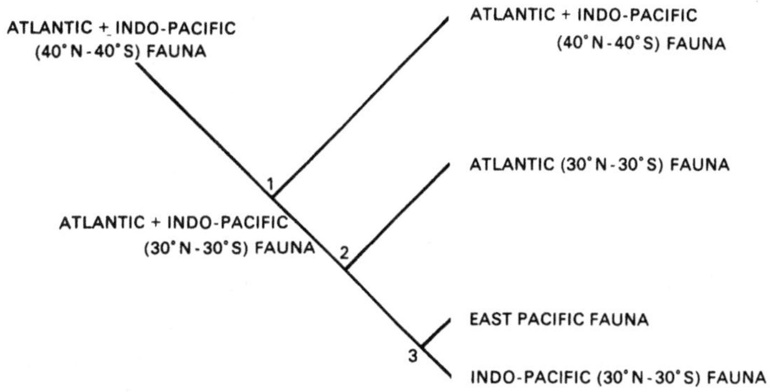

FIG. 176 Geo-cladogram of the warm-water faunas (splitting points 1-3 are discussed in text).

For the more restricted equatorial distributions the diversification is very simple, as it is a mere splitting of ranges. The original Equatorial fauna living in the circum-world Paleozoic ocean first split off a fauna ancestral to the Atlantic and Mediterranean ones (splitting point 1). For some taxa, the Indo-Malayan-Australian land masses formed a barrier so that the

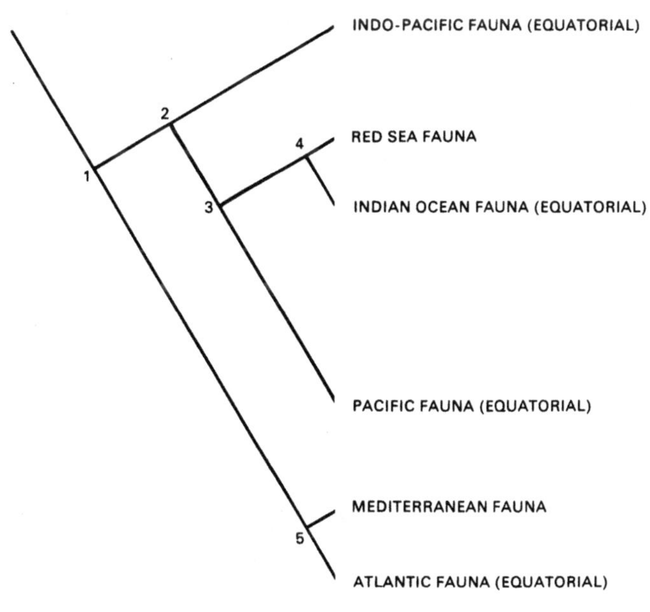

FIG. 177 Geo-cladogram of the more restricted equatorial faunas (splitting points 1-5 are discussed in text).

Indian and Pacific faunas became partly split (splitting points 2 & 3) and later the Red Sea and Mediterranean became more isolated basins (splitting points 4 & 5).

The very old patterns of Central water species are considered to have developed by migration (⟶) and extinction (#) (cf. p. 68). The influences splitting the Central water species off from the warm-water faunas cannot be found in any geological phenomenon; probably the Central water fauna formed directly under ecological influences as an adaptation to oligotrophic Central waters. If this adaptation has existed already for so long, it would also explain the strong connection of this fauna with Central waters up to the present day. In this geo-cladogram the development of the present Central water pattern is considered to be dependent on migration into the Atlantic and Indian Oceans after the major change of the tectonic plates by continental drift. Another factor is the extinction (of the Tethys fauna between 4 and 5; of the N. Atlantic fauna between 5 and 7; of the S. Atlantic fauna between 7 and 9 and of the Indian Ocean fauna after 9) due to temperature drops during the Ice-Ages (see fig. 178).

To conclude from the phylogeny and zoogeography of the genera *Acetes*, *Peraclis*, *Sergestes*; from the phylogenetic relations of East and West Pacific taxa; from the Atlantic-Tasman Sea patterns and from many others phenomena, such as current patterns at higher latitudes and deep water circulation, the pelagical, and more especially the distant-neritic fauna, shows a strong tendency to eastward dispersal after the Miocene (Van der Spoel, 1983) (fig. 179).

In the following geo-cladogram (fig. 180) this eastward dispersal (⟶) forms the basis for the relations indicated for the neritic taxa. The faunas of the Panama Passage and South American inland sea are extinct. In fig. 180 they are considered to be ancestral to all Atlantic distant neritic taxa. This is a simplification as faunal elements could also have migrated to the Atlantic through the Indian Ocean via the Agulhas area.

The Temperate faunas are circumglobal in the southern hemisphere and rather different in the N. Pacific and N. Atlantic. There may have existed connections between Temperate faunas of the northern and southern hemispheres, but as these contacts were at the most incidental, they cannot be reflected in a geo-cladogram. In addition, the frequency of these contacts, if any, will have been much lower than the contacts between the Arctic and Subarctic faunas, as the blockage by the American continent was much stronger and continuous in the Temperate belt compared with the high latitude belts. The development of the Temperate fauna out of the warm-water fauna was simultaneous in the different oceans and hemispheres.

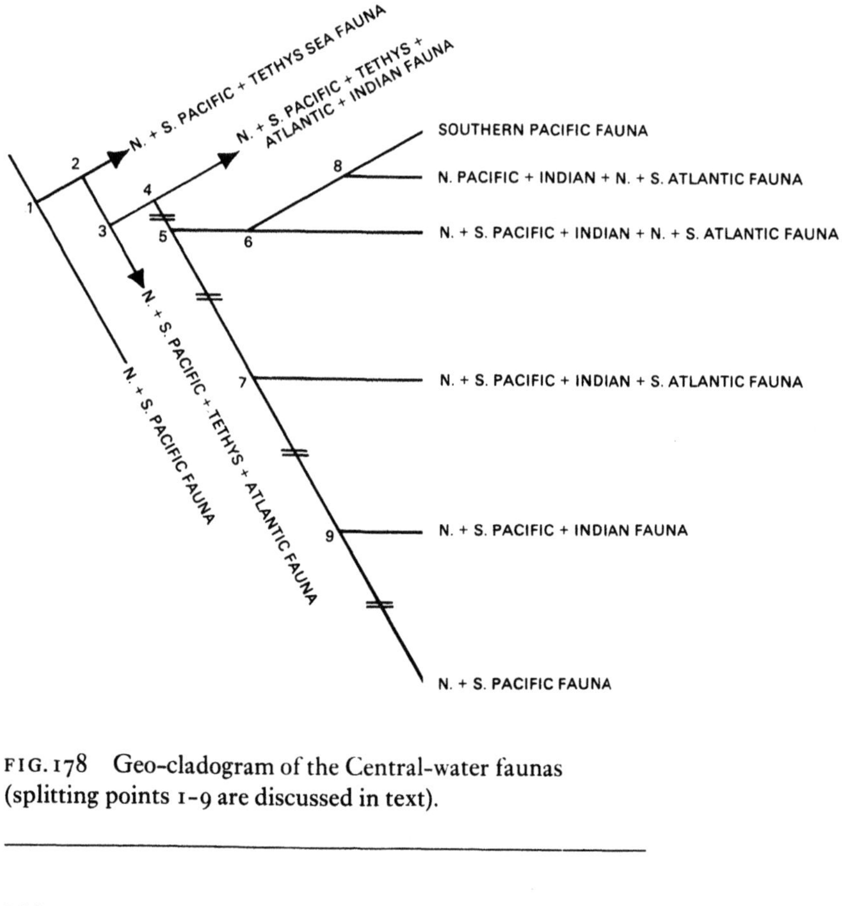

FIG. 178 Geo-cladogram of the Central-water faunas (splitting points 1–9 are discussed in text).

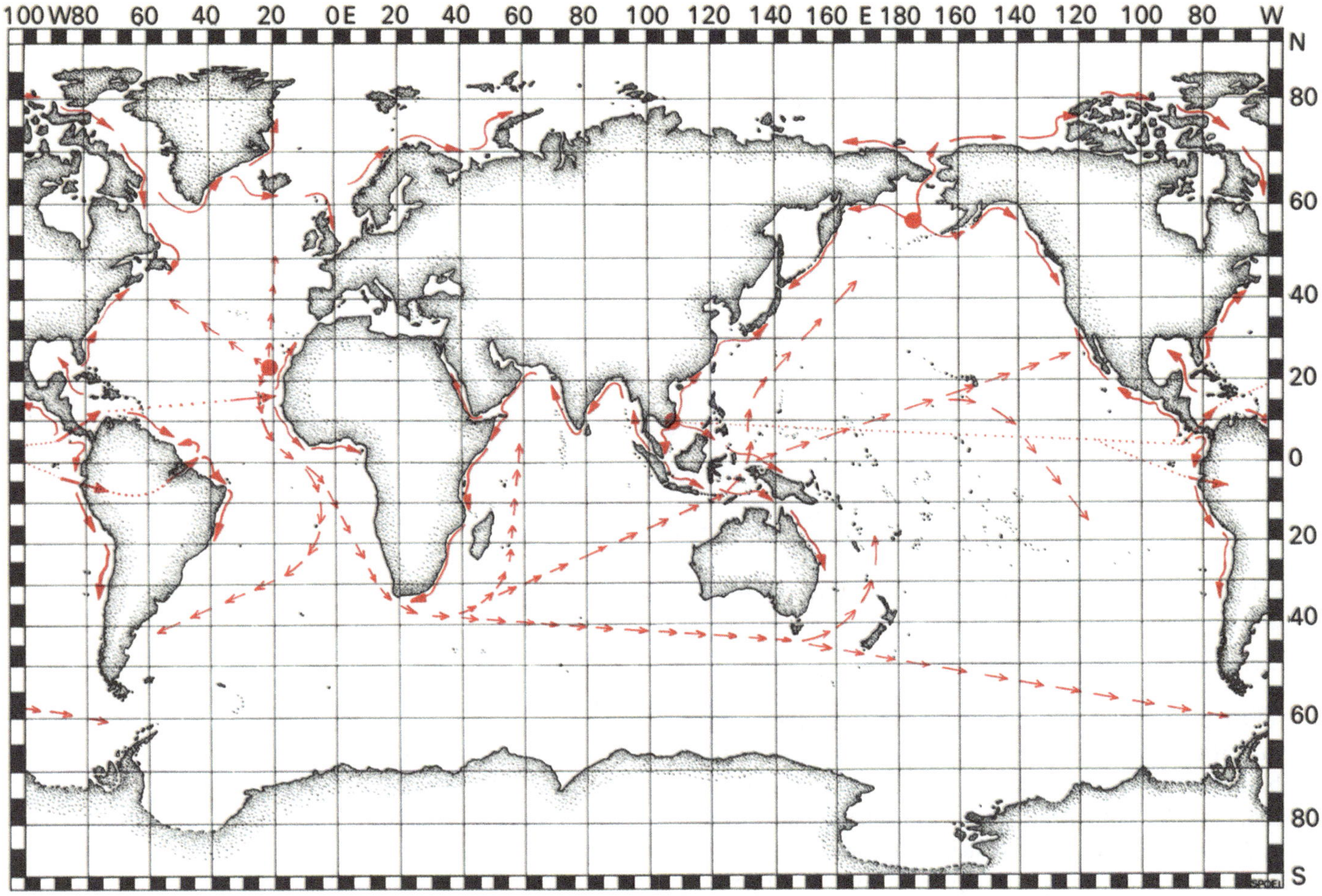

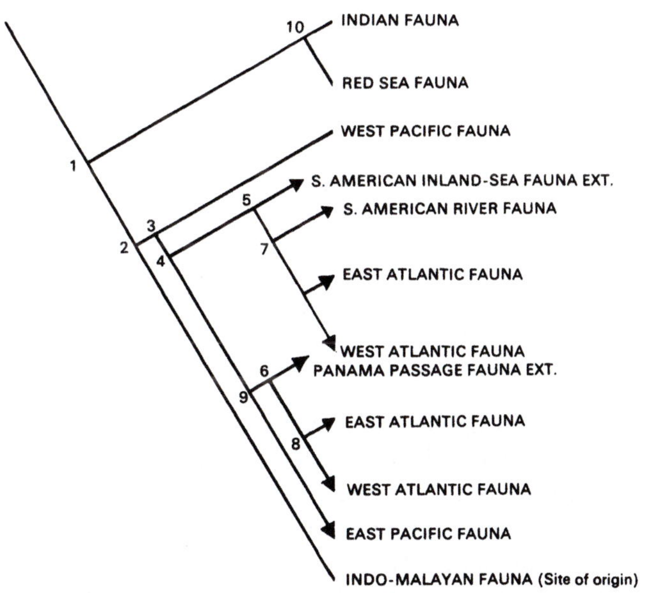

FIG. 179 Dominant routes of dispersal for neritic taxa (arrows) and taxa of the Dakar Centre (broken arrows).
FIG. 180 Geo-cladogram of (distant) neritic faunas (splitting points 1–10).

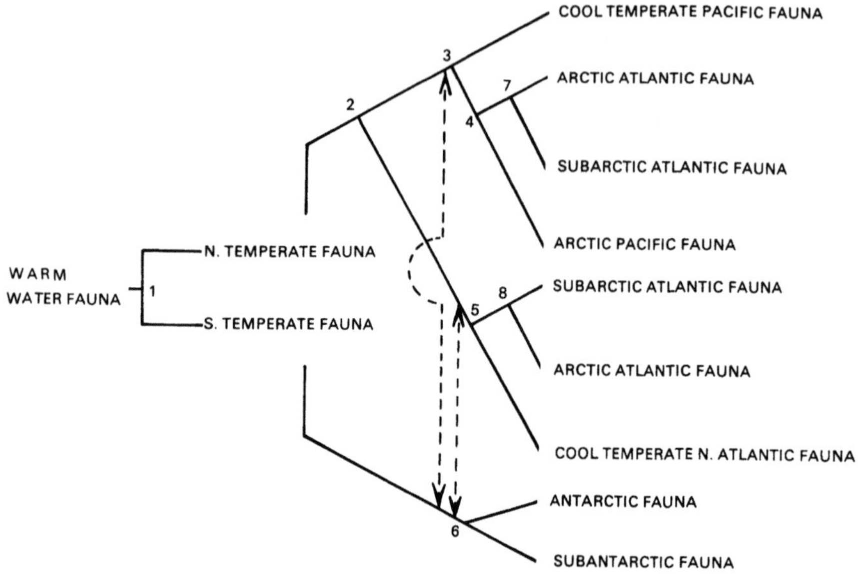

FIG. 181 Geo-cladogram of high-latitude faunas (splitting points 1–8 are discussed in text).

The faunas of the higher latitudes have also developed in a latitudinal sequence, but their relations are slightly better understood and can be, tentatively, depicted (fig. 181). Accepting the presence of separate northern and southern Temperate faunas, developed out of the warm-water faunas (splitting point 1), one can follow a separate line of development in the north and in the south. The mentioned north-south connections, e.g. during the Ice-Ages are indicated by broken lines in fig. 181. In the southern hemisphere the development is considered to have gone via a cold-water fauna, which split into an Antarctic and a Subantarctic fauna probably fairly recently, after the development of the Antarctic Convergence.

For the northern hemisphere a separation of N. Pacific and N. Atlantic is considered to have occurred relatively long ago (splitting point 2). For some taxa this may be correct, but rather recent contacts between N. Atlantic and N. Pacific (splitting point 4) are known (Dunbar, 1979). These connections only favoured a west to east penetration of chiefly Arctic elements, as Dunbar (1979) stated that the 'Pacific Subarctic' does not exist (splitting point 3) whereas the Subarctic and Arctic Atlantic faunas also split off from the cool Temperate N. Atlantic (splitting point 5). The splitting of Subarctic and Arctic elements (splitting points 7 & 8) will have gone in the same way as the splitting of Subantarctic and Antarctic faunas (splitting point 6): the coldest faunas will have developed from less cold ones, or a previously cold-water fauna split into a cold and a very cold fauna.

In Chapter 2 a general trend for the evolution of recent species-groups was discussed (p. 42) and the geographic cladogram for many recently developed species-groups can be given as in fig. 182. In this figure

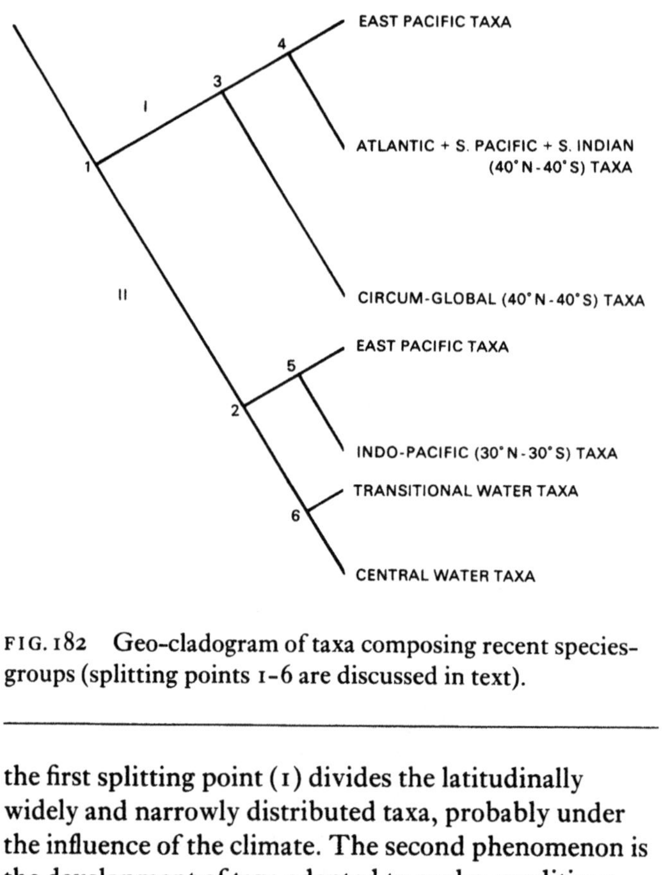

FIG. 182 Geo-cladogram of taxa composing recent species-groups (splitting points 1–6 are discussed in text).

the first splitting point (1) divides the latitudinally widely and narrowly distributed taxa, probably under the influence of the climate. The second phenomenon is the development of taxa adapted to cooler conditions (splitting point 2) in the second group (II) and the splitting (3) in the first group (I) of the more endemic taxa from the circumglobal ones. For both groups a development of East Pacific taxa is proposed (splitting points 4 & 5). The group most adapted to cooler conditions (II) may develop a Central water and/or a Transitional taxon.

All geo-cladograms deal with 'older' developments, and therefore only hold for 'original faunal elements'.

However, in recent times too, ranges can develop and be occupied. The trends found in recent dispersal are discussed below (p. 153).

The geo-cladograms should be compared with taxonomic cladograms. However, such cladograms are not available, except for the five cladograms given below which cannot demonstrate all the developments proposed above. On the other hand, the geocladograms do not provide a complete picture of the events in the past. The most important restriction to be made for the geo-cladograms presented, is that they concern groups which have a fixed type of distribution, e.g. an 'Equatorial' or 'Central water' type. If, however, this type changed during its development or if some members of the group changed their ecological preferences, combinations of two or more geo-cladograms have to be used to explain distributions.

The cladogram of *Eucalanus* (fig. 183) for example, can only be explained when recognizing the fact that within the group one line is oceanic, while the other line comprises neritic taxa; moreover, the oceanic species developed partly as warm-water, partly as Temperate water taxa.

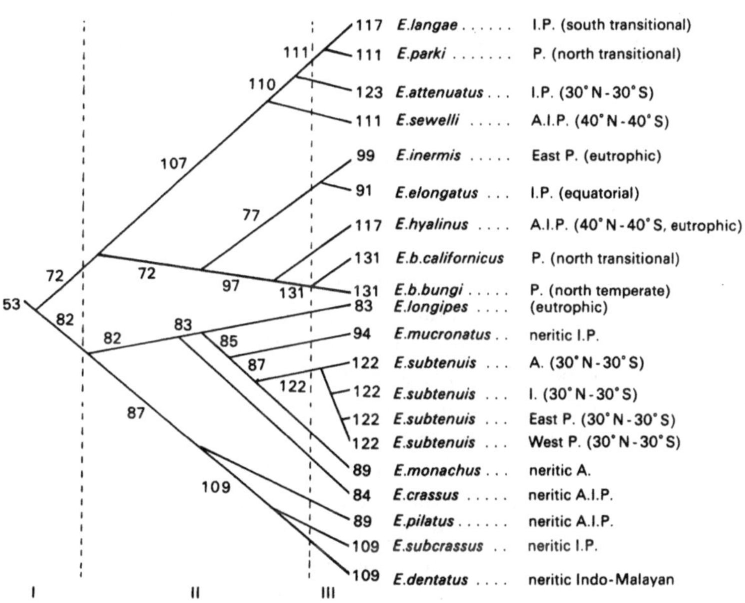

FIG. 183 Cladogram of *Eucalanus* based on the number of pores in the integument of females, and zoogeographic data. The numbers given along the branches represent maximum numbers of pores shared. A = Atlantic Ocean, I = Indian Ocean, P = Pacific Ocean.

The cladogram of *Pontellina* (fig. 184) follows exactly the geo-cladogram for warm-water plankton (fig. 176).

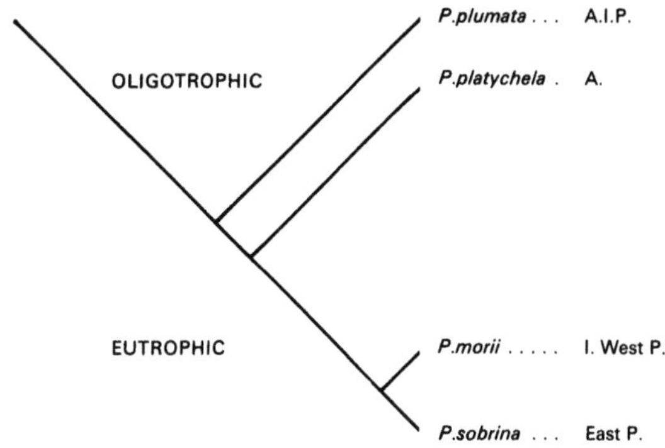

FIG. 184 Cladogram of the species-group *Pontellina plumata*. A = Atlantic Ocean, I = Indian Ocean, P = Pacific Ocean.

The cladogram of fig. 185 for the *Calanus helgolandicus* group covers part of the geo-cladogram for high-latitude faunas (fig. 181).

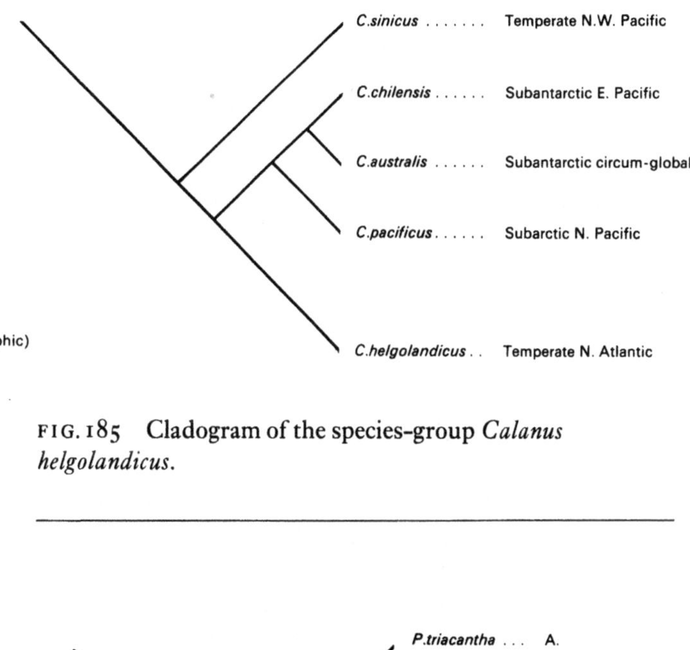

FIG. 185 Cladogram of the species-group *Calanus helgolandicus*.

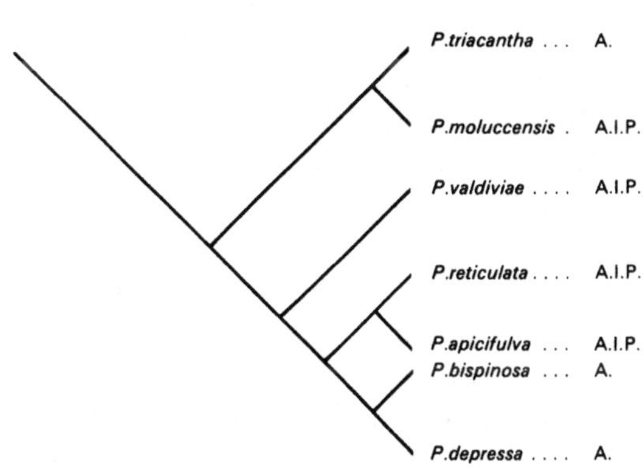

FIG. 186 Cladogram of the genus *Peraclis*. A = Atlantic Ocean, I = Indian Ocean, P = Pacific Ocean.

The cladogram for the *Diacria quadridentata* group (fig. 187) fits in well with the geo-cladogram for warm-water species and for Equatorial species. The Central water populations in the diagram should not be explained as real Central water dispersals, but as recent ecological adaptations to these water masses.

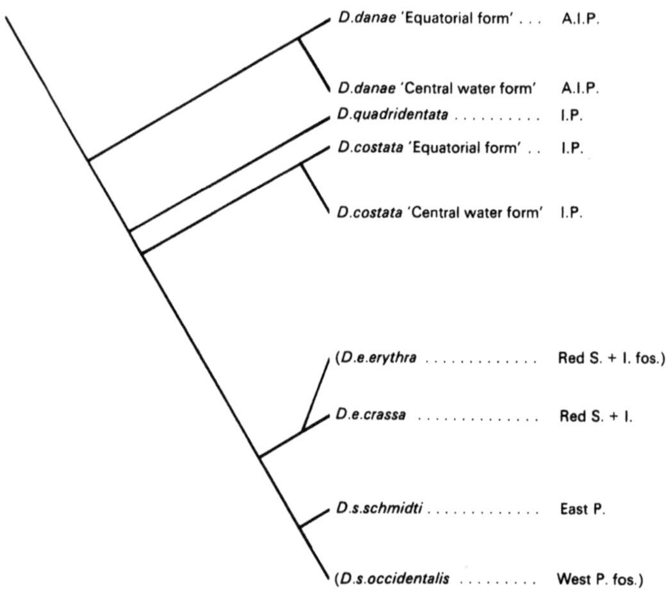

FIG. 187 Cladogram of the species-group *Diacria quadridentata*. A = Atlanta Ocean, I = Indian Ocean, P = Pacific Ocean.

Geomorphological influences

The geo-cladograms given above explain distributional developments on the basis of geological events and ecological behaviour of taxa. It has been stated (Pierrot-Bults & Van der Spoel, 1979) 'that diversity in recent planktonic groups largely developed after the Mesozoic' and the fauna from which this development started was a circumglobal warm-water fauna and a shallow Tethys Sea fauna. But this only concerns oceanic plankton, neritic faunas will have existed already on a large scale at the time.

A subdivision of the oceanic warm-water fauna can have existed also in or long before the Mesozoic, as the splitting off of Central water taxa (fig. 178) can have occurred already in the early Cretaceous (Van der Spoel, 1983) when the first displacements of continental plates had taken place.

For the most important geological influences affecting the patterns we have to look in post Cretaceous times, in which the Tethys Sea disappeared, the Panama Passage closed, the North Atlantic became connected with the Arctic, the Indo-Malayan archipelago formed a barrier between Pacific and Indian Oceans and the Red Sea and the Mediterranean Sea started to develop. In table IV these important geomorphological features are given.

It is not possible to give any exact indication of the period in which faunas split and occupied specific ranges, by comparing the given geo-cladograms with the geological time-table in table IV. The only indication which such a comparison can yield, is the maximum age a specific development can have. The same problem is met with when representing zoogeographic developments as presented by Pierrot-Bults & Van der Spoel (1979) (see fig. 188).

Only the maximum age of a zoogeographic phenomenon may be indicated, but even this has to be done with restrictions as different taxa behave differently and groups with e.g. only little inclination to dispersal can have taken up an isolated distribution long before geomorphological isolations established themselves.

Consequently, isolated distributions can have existed in the Atlantic long before the Panama closure. Arctic or Antarctic taxa can have existed long before the cooling period started, as warm-water taxa which adapted to cold water in their geological history. The cladograms, table IV and fig. 188 can, however, be used in comparison to explain the development of faunas.

TABLE IV
Major influences affecting the dispersal of plankton.

Mill. Years B.P.:	INFLUENCES
0.006-0.005	Climate optimum, smallest extension of cold-water faunas, optimal development of Equatorial and Subtropical faunas.
0.008-0.006	Warming up of N.W. Atlantic Ocean.
0.012-0.010	Warming up of Central Atlantic Ocean.
0.015-0.013	Warming up of N.E. Atlantic Ocean, cold-water fauna splits definitively into northern and southern cold-water faunas.
0.03	Würm glaciation, sea-level drop, Subtropical Convergence north of 31° S, largest extension of cold-water faunas which make N-S contacts, last strong split of Atlantic and Indo-Pacific warm-water faunas.
0.12	Riss glaciation, sea-level drop, Subtropical Convergence north of 31° S.
0.19	Mindel glaciation, Subtropical Convergence north of 31° S.
0.28	Günz glaciation, sea-level drop, Subtropical Convergence north of 31° S.
0.40	Donau glaciation, sea-level drop, Subtropical Convergence north of 31° S.
1	Biber glaciation, sea-level drop, Subtropical Convergence north of 31° S, first splitting of the warm-water faunas of Atlantic and Indo-Pacific Oceans.

(table IV continued)

2-1	Outflow of Amazonas starts to enter the Atlantic Ocean, last possibility for Pacific-Atlantic dispersal of euryhaline neritic taxa.
2	Last Alpine folding, start of northern glaciation, first development of Polar fronts, formation of the E. Pacific biotope.
3.5	Closure of Panama passage, final separation of Tropical-Subtropical Pacific and Atlantic Oceans.
5	Opening of connection between Atlantic and Mediterranean, full development of circum-Antarctic current system.
7-5	Mediterranean closed off and dried out.
15	First development of circum-Antarctic currents.
16	Tethys Sea split into eastern and western bassin, isolated Atlantic warm-water fauna starts to develop.
20	Outflow of Amazonas into pacific Ocean becomes blocked.
25	Oldest W. Mediterranean sediments formed.
30	Oligocene sea-level drop.
38	Glaciation of Antarctic continent and adjacent seas.
45	Arctic basin develops.
50	Arctic Ocean connects with Atlantic and Pacific Oceans, Australia drifts northwards and disconnects from the Antarctic continent, development of deepsea circulation and deepsea fauna.
65	End of global sea-level rise, start of global temperature decrease.
70	Oldest E. Mediterranean sediments formed.
80	End of major ocean-floor spreading.
85	Atlantic and Indian Oceans make contact south of Africa.
90	First opening of southern Atlantic basins, start of exchange between north and south Atlantic faunas.
140	First opening of Indian Ocean.
150	Opening of Tethys Sea into Atlantic Ocean, begin of major ocean-floor spreading, first dispersal of Pacific faunas towards the Atlantic Ocean.
200	Start of continental drift, N. Atlantic basins start to develop.

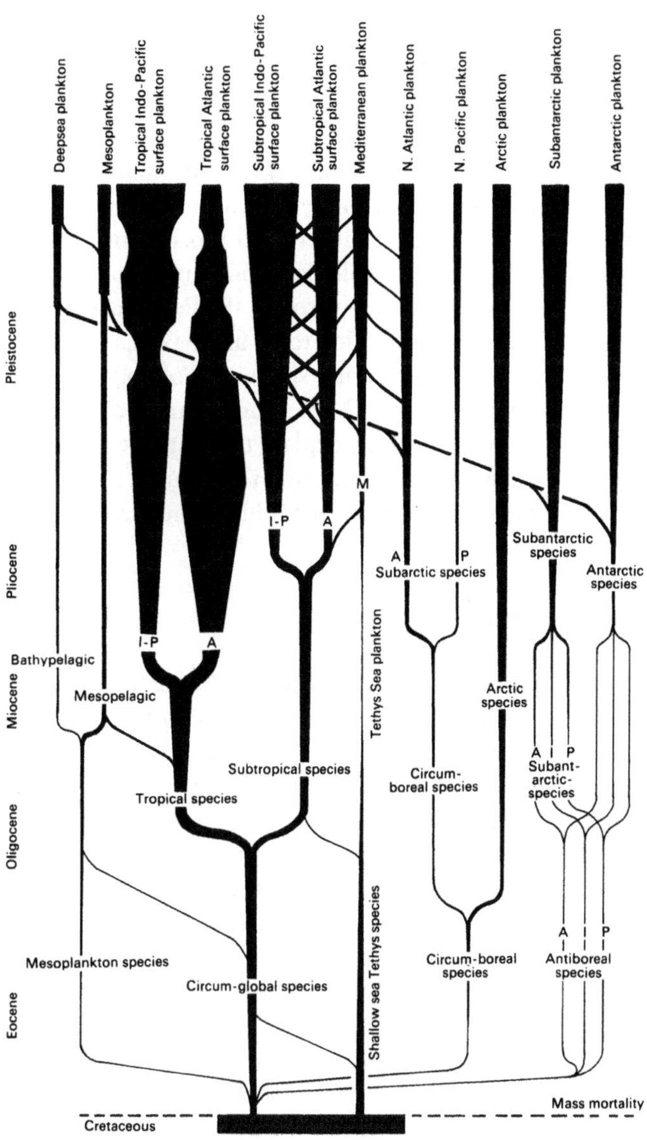

FIG. 188 Development of the planktonic faunas during the Coenozoic based on climatological and geological phenomena (after Van der Spoel and Pierrot-Bults, 1979). A = Atlantic Ocean, I = Indian Ocean, M = Mediterranean Sea, P = Pacific Ocean.

The influences of the Tethys Sea mentioned only for Central water patterns, should not be exaggerated. For distributions of terrestrial and littoral taxa the Tethys is of great importance, but for planktonic and certainly for oceanic taxa its importance is restricted.

The Tethys Sea was a relatively shallow sea connecting the recent Caribbean, Mediterranean, Caspian Sea, the North Indian area, the Indonesian seas and the Coral Sea. Dating back to the times before the shifting of the continents started it was a 'continental' sea forming a band over the archaeic continent. The essential effect of the Tethys Sea is the connection it once formed between the Caribbean and the Indonesian area and the Coral Sea. The Obic Sea was a northward extension of the Tethys, connecting Arctic waters and Black Sea. For bathyplanktonic organisms the Tethys Sea was of no importance but for the distribution of epiplanktonic animals it probably provided the first opportunity to pass to the newly formed basin. The areas with clear seismic belts and Alpine structures coïncide almost exactly with the area of the Tethys as it was laying right over the plate borders characterized by heavy folding. This implicates that at the moment the continents started their migration, the bottom of the Tethys became strongly disturbed. The Alpine mountains rather quickly cut off the connections between the oceans for planktonic animals. Distributional patterns reflecting the influences of the Tethys can therefore only concern taxa present before or at the moment of the beginning of the continental drift.

The connection between the North and South American continents is a rather recent formation and influences of the Pacific holoplankton fauna on the Atlantic one and vice versa through the Panama area was possible. The only restriction to be made is that the passage may have been rather shallow, probably of the type of the passage through the Indonesian Archipelago at the moment, which may explain the restricted number of distributions reflecting the influence of the pre-Miocene passage through Middle America.

Climate influences

The climate has a direct influence on the surface waters (fig. 190) and an indirect influence, frequently with a large time-lag, on all deeper strata. The major climate changes were already discussed in the preceding paragraph together with geomorphological changes, and in table IV. A variety of smaller climate fluctuations will also have affected distributions in the past. Warming up or cooling of only a few degrees Celcius, for example, can shift the borders of ranges over hundreds of miles; a small decrease of eastward winds off the coast of Peru can enforce the El Niño effect and can make species disappear over larger areas, and likewise a change in monsoon force over the Indian Ocean can alter the distribution pattern in large parts of this ocean.

Water on earth occurs in three phases, viz. as gas, as liquid and solid. The ratio of these phases is of great influence on the climate and in its turn, depends on the climate. The greatest changes of these phases occur when the sea-level drops and ice ages occur. It is therefore necessary to pay some more attention to these phenomena.

There is an important difference between small-scale fluctuations and the climate changes giving rise to the Ice-Ages or to sea-level drops. Ice ages show a form of inertia; once the ice sheets are formed they tend to stay put, and once gone they tend to stay away, while the time needed for the transition between the two stages is short. The smaller fluctuations are mere small oscillations around an average condition usually without a long-lasting stable phase.

The influence of the Ice-Ages on the marine environment was not the same at all geographic positions. Naturally, the temperature drops were more clear-cut at high latitudes than at lower latitudes and ice coverage only occurred near the Poles, but there were also local differences at the same latitude. For example, the sea-surface temperature fluctuated during the Pleistocene over all the oceans, but high surface temperatures in the S.E. Pacific were at least partly synchronous with low temperatures in the Atlantic, as Ericson and Wollin stated in 1970. These authors also mentioned that the Pleistocene climate records in the Pacific vary from area to area, though this variation was smaller than the variation between the Pacific and Atlantic Oceans.

As a general phenomenon can be accepted that warm and cold periods succeeded each other, though not synchronously, in all basins. The temperature drops differed from one region to another: temperature during the last glaciation in the Caribbean was 7-8 °C below present, in the Gulf of Mexico 6-7 °C below present, off Dakar 4-6 °C below present and in the Mediterranean 11 °C below present (Emiliani et al., 1975). The fluctuations of the temperature-determined distributions were therefore neither synchronous in all basins, nor everywhere of the same scale in all seas.

The temperature effect of the Ice-Ages can be considered to have determined the latitudinal shifts in distribution and at a few places, extinction of species which could not escape as e.g. High-Tropical species trapped in the cooling Atlantic.

Another effect of the Ice-Ages was the sea-level oscillation over 70-100 m. These oscillations caused the alternating closure and opening of some sea straits such as the Bering Strait, and affected the distribution in two ways. First, the oscillations alternately cut off and restored connections between distributions, and secondly, they affected the flow of surface waters which may have had tremendous influence on marine conditions. Closure of passages in the north may have interrupted the outflow of fertile Arctic water into the N. Pacific and N. Atlantic, which may have resulted in a decrease of fertility in the northern basins. Closure of the shallow sea passages in the Indo-Malayan Archipelago put a halt to the flow of Pacific water through the Archipelago, and enlarged the return-flow of Equatorial waters in the Pacific. The total sea-surface decreased with lower sea-levels and the simultaneous extension of the ice-cover. The smaller the open surface, the smaller the amount of evaporation and as a consequence also precipitation decreased.

Thus, the total of evaporation, precipitation and freezing showed strong oscillations. As circulation of water in the ocean is determined by the earth's rotation, by the wind and by the conditions of surface waters, as determined by evaporation, precipitation and freezing, the circulation in the Ice-Ages will have differed strongly from the present conditions. Especially the Polar fronts, but probably also the Subtropical and Tropical fronts were absent during glaciations, as can be deduced from figure 189 in which the currents which decreased in strength during glaciations, are dotted.

It is very well possible, that due to lack of current, the Subtropical Convergence (SC) was not present, which resulted in upwelling in the Equatorial area becoming a downwelling during the Ice-Ages' cold periods. Speciation related to Subtropical and Polar fronts and cold-water, especially beltshaped patterns, are therefore of a relatively recent origin.

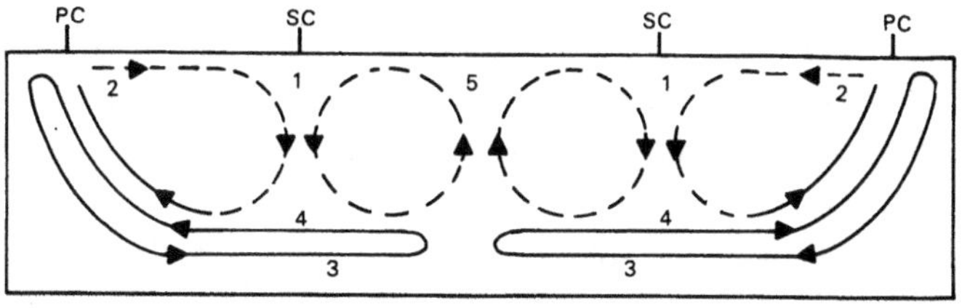

FIG. 189 Hypothetical vertical circulation in an ocean. PC = Polar Convergence, SC = Subtropical Convergence, 1 = downwelling of highly saline water due to evaporation, 2 = currents to the tropics due to melting of ice and precipitation, 3 = downwelling of cold and highly saline water due to freezing, 4 = upwelling to supply the water lost by current no. 3, 5 = upwelling to supply water lost by current no. 1.

The Ice-Ages also affected horizontal surface currents and wind patterns. The current and wind systems will not have differed much from the present situation, but some currents will have had another velocity. The Labrador and Oyashio Currents will have been stronger, whereas the North Atlantic and Kuroshio Currents will have been less important and situated more to the south. This will have caused a stronger east-west diversity in the warmer parts of the ocean and probably a more oblique position of the northern and southern boundaries of several species.

Cyclonic winds entering the high latitudes in an east-polarwards direction are formed near the hydrological fronts between warm and cold water, off the east coast of continents. During periods of cooling, the formation of these cyclonic winds is decreased and their influence on turbulence-determined productivity in the higher latitudes will have been smaller. Production also depends on vertical temperature stratification. The influence of cold water coming from higher latitudes was stronger at lower latitudes during periods of cooling. This will have caused a slower break-down of stable early-spring stratigraphy of the warmer-Temperate water masses as the warming up will have been retarded, so that the pyknocline shifted downwards much more slowly than it does at present. The maximal primary production will then also have been retarded, which made the relatively short summer period still shorter. The area with highest primary production shifted towards the equator by the mere drop of temperature and by the change in hydrological conditions.

The band with high environmental stability was much narrower during cold periods and the slightly changed pattern of currents shows that there will have been a smaller energy transport from the equator to the poles which, in general, causes greater latitudinal provincialism as it will give rise to more different latitudinal belts. This latter phenomenon will have affected speciation and distribution in a reverse way as compared with the other changes caused by Ice-Ages cooling.

Because of the more pronounced character of cold waters and the more isolated occurring of cold-water taxa during cold periods, the Ice-Ages can be taken to be responsible for a great amount of speciation at higher latitudes where cold-water taxa developed from Tropical and Temperate ancestors.

The present-day influence of the climate on distribution has been discussed in Chapter 3. Slight climate fluctuations in the last thousand years have influenced the distribution patterns, but it is not to be expected that they had dramatic consequences; usually, the effects were restricted to minor shifts in the borders of ranges. Such small changes may have important influences on small ranges. For example, neritic species and the oceanic plankton of the narrow Transitional belts (cf. *Coccolithus pelagicus*, fig. 39) may become extinct due to such small changes.

A summary of the situation during the Ice-Ages can be given in the following excerpt of the paper by CLIMAP (1976), concerning the maximum extent of the last Ice-Age, approximately 18,000 B.P.:

—The Gulf Stream slightly shifted southwards off the Carolinas and swept almost directly eastward across the basin to the Iberian Peninsula, the northern limit at 42 °N marked the southern edge of the polar water mass.
—Warm water transport in the North Atlantic (Norwegian, Irminger Currents) was much less intensive.
—The great difference in temperature between 18,000 B.P. and recent of the North Atlantic north of 42 °N was also found in the north-west Mediterranean, where the temperature decrease was 11 °C; in the east Mediterranean there was only a difference of 1 °C.
—Ice coverage stretched from New York-New Foundland-Greenland-Iceland to England.
—In the N. Pacific there was a southward shift and steepening of the temperature gradient in the region between the Kuroshio and Oyashio Currents.
—The Japan Sea was blocked in the south (lowered sea-level) so that temperatures here were low (absence of northward flow of warm water).
—The Bering Strait closed, therefore no Arctic water came into the Bering Sea and there was less cooling in the N. Pacific compared with the N. Atlantic.
—The Antarctic polar front shifted 4° latitude in the

Pacific, 6° in the western S. Atlantic and 3° in western Indian Ocean.
—There was little or no displacement of the Subtropical Convergence, if it was still active, so because of the moving polar fronts and the stable Subtropical Convergence, the area of Subantarctic water was reduced.
—The central gyres showed little or no temperature changes.
—Subpolar and Transitional water masses were reduced in areal size.
—The eastern boundary currents were stronger, and transport reached further southwards off Portugal and N.W. Africa.
—The stronger eastward flow of the Gulf Stream brought tropical species more eastward than today.
—The Benguela Current was stronger and extended more westward, probably also the axis of the current was found more to the west.
—The Agulhas Current decreased strongly, which also decreased the Indian-Atlantic Ocean contact south of 30° S.
—The Mozambique Current was partly absent and turned eastward to parallel the West Wind Drift.
—The cold-water flow northward along W. Australia was stronger than today.
—The cold California Current extended 1000 km further south.
—The temperatures off Mexico were much lower.
—The Chile Current was stronger.
—The Equatorial Pacific and Atlantic Oceans were much cooler.
—The upwelling along the Pacific equatorial divergence (mostly in the East Pacific) was much more intensive.
—There was increased upwelling in the Arabian Sea.
—The surface transport increased everywhere.
—The Red Sea blocked off because of sea-level drops.
—The Indonesian Archipelago (except for the Timor passage) closed because of sea-level changes.
—The temperatures for 18,000 B.C. and Recent are, respectively, south of S. Africa: 13° C and 18° C; south of S. America 1° C and 3° C; south of Australia 7° C and 13° C and in the Strait of Timor 24-26° C and 28° C.

Sea-level changes

Concentrating on the geological periods after the Cretaceous, one should not forget to note that the sea-level was very high during the late Cretaceous (fig. 190). During the Triassic and Jurassic the sea-level fluctuated around a mean also found in post-Cretaceous times, but from the beginning to the end of the Cretaceous the sea-level rose constantly. This means that during the splitting up of the continents, the Tethys Sea and connections between Pacific and Proto-Atlantic and Proto-Indian Oceans were deeper and there were fewer barriers to dispersal. Other remarkable anomalities of the sea-level are found in the end of the Oligocene and in the Pliocene when very low sea-levels existed. These sea-level drops measured 70 to 120 metres.

For the neritic fauna the sea-level changes were not very important, as they did not enlarge the biotopes for neritic taxa. The continental slope bordering the shelf is too steep to create a large shallow area when the sea-level is low. In general, the neritic area will have become smaller as more shallow waters became dry. Interruption of ranges by land becoming dry is for neritic taxa the same as for oceanic taxa. The Oligocene sea-level drop differs from the Pliocene drop, in that the latter came at the same time as the Ice-Ages, causing world-wide cooling of sea-water and extension of sea-ice in the northern hemisphere. Another difference between the two periods is the position of the continents during the low sea-level. During the Oligocene the land masses becoming dry in the Indo-Malayan area did not close the Indo-Pacific connection in the Tropics whereas during the Pliocene drop this area became a barrier between the Indian and Pacific Oceans. The closing of the Bering Strait resulted in an identical barrier in the Oligocene and Pliocene.

During the Oligocene, barriers in the N. Atlantic were possible because of sea-level drops, but during the Pliocene, ice-covered seas blocked these areas.

In recent distribution patterns no traces are found of sea-level changes in the past. It is, however, not only the sea-level itself which changes, but also the land masses can move up- and downwards. Sea-level changes are world-wide; land mass movements, however, are local phenomena. A typical example of land movement is the oscillation of the Scandinavian Peninsula and of the Brazilian plate.

The Scandinavian Peninsula and probably also the Canadian land mass were pressed downwards by the ice-load covering them during glacial periods and they slowly rose when the ice had melted. In the period before the rize and after the disappearance of the ice-cover, the sea had free passage over the continents, so that marine organisms could migrate to the Baltic over Scandinavia and between the N. Pacific and N. Atlantic over northern Canada. These pathways were only open to neritic and littoral taxa.

The Ice Lakes, composed of sea-water and an increasing amount of melt water, pressed in front of the southward moving glaciers over Canada and Siberia, were another way for neritic taxa to migrate from one basin to another, but here only strongly euryhaline taxa could be involved.

The tilting of the Brazilian plate made a passage possible over the South American continent between the Alpine Folding period and the Pliocene. As discussed in relation to the dispersal of *Acetes*, the western side of the Brazilian plate rose during the Alpine Folding and an

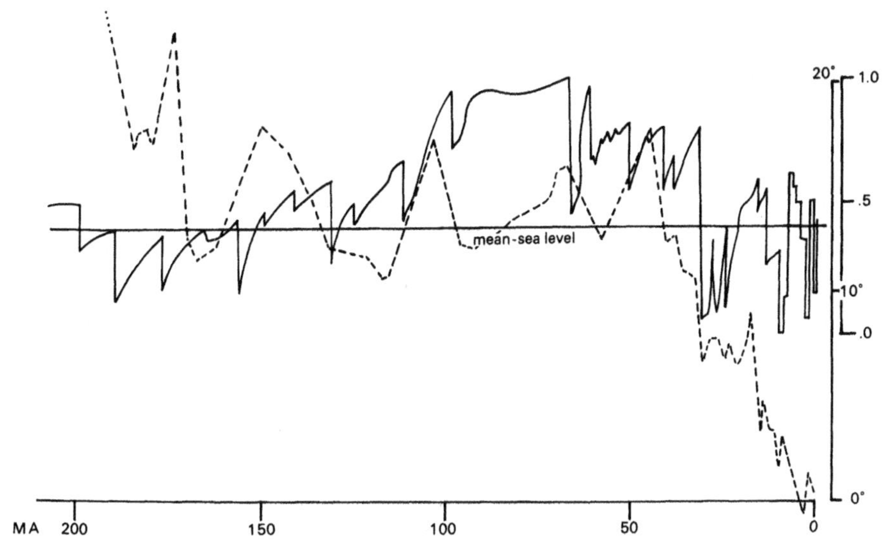

FIG. 190 Temperature fluctuations (broken line) and sea-level changes (continuous line) in the last 200 MA.

inland sea was enclosed which later made contact with the Atlantic as a result of the lowering of the eastern side of the plate. Here also, only strongly euryhaline neritic taxa could use the pathway. In general, sea-level drops and land mass movements only influence dispersal of taxa with strongly neritic characters.

Horizontal patterns

Too little is known on vertical patterns to discuss general trends in vertical distributions. For the horizontal patterns, however, some statements can be made. In Chapter 3, the individual types of distribution were discussed and each provided with a name and it was attempted to group the patterns according to their geographic shape. It is, however, also possible to group at least some of these patterns according to the type of zoogeographic development they had in the past (table V). For many of the patterns discussed, the development cannot be traced due to the absence of ancestral patterns or taxa; such patterns are left out of the grouping given below.

For a large group of patterns it seems evident that they have not changed much; these are the primitive patterns (group A), which are chiefly circumglobal warm-water patterns. In recent groups also patterns of a 'primitive' type can be found, which also present a picture of unchanged distribution. As these taxa, such as pteropods, are known to occupy recent ranges, they are given in a separate group (A′).

Cosmopolitan patterns, of either old or recent origin,

TABLE V
Grouping of zoogeographic patterns according to their history.

Group:	Type:	Examples: (numbers refer to figs)
A	Primitive, not much changed in the past	*S.carinatum* (63), *S.abbreviatum* (64), *S.maximum* (72)
A′	Resembling A but too recent to reflect influences of the past	*L.lesueuri* (47), *C.cuspidata* (48)
A″	Old and younger patterns with Ice-Ages influences	*S.fusiformis* (51), *C.pyramidata* (57)
B	Indo-Pacific primitive patterns (extinct Atlantic populations)	*S.microphthalma* (65), *S.pacifica* (117)
C	Primitive neritic patterns	*A.indicus* (103), *A.serrulatus* (104)
D	Central-water patterns	*E.brevis* (89), *S.suhmi* (90), *S.affine* (91), *T.obtusifrons* (93)
E	Temperate and cold-water patterns	*G.pachyderma* (80), *N.cylindrica* (81), *L.helicina* (82)
F	East Pacific patterns	*D.pacificus* (111), *S.microps* (143)
G	Red Sea and Mediterranean Sea patterns	*D.erythra* (123), *S.setosa* (124)
H	Atlantic patterns	*F.zetezios* (120), *E.americana* (109)
I	Indian Ocean patterns	*D.gardineri* (111)
J	Dakar Centre elements	*Peraclis* (59), *O.carribaea* (61)
K	Indo-Malayan Centre elements	*S.orientalis* (101), *H.globulosa* (112), *L.hanseni* (130)

clearly reflecting only the Ice-Ages' influences, then form a third, still rather original group (A″). The distributions covering the warm waters of the Indo-Pacific (group B) are also considered to be primitive as it seems to be originally circumglobal patterns from which the Atlantic populations were lost or split off.

For the neritic faunas, the patterns around the Indo-Malayan Archipelago seem the most primitive ones (group C). For oceanic taxa the archipelago is not a centre of primitive distributions.

Distributions which have changed strongly in the past, but which still reflect old patterns, are the Central-water taxa (group D). Evidently derived patterns are the Temperate and cold-water patterns, which both developed from warm-water taxa (group E). Patterns which developed by more or less recent isolation are the E. Pacific patterns (F), the Red Sea patterns and the Mediterranean patterns (G), the Atlantic patterns (H) and the Indian Ocean patterns (I). Less recently developed but clearly reflecting the past development are the patterns around centres, such as the patterns around Dakar (group J) and around the Indo-Malayan Archipelago (group K).

Bipolarity was one of the first marine zoogeographical patterns to be described. However, no special attention has been given here to this pattern as real bipolar species do not exist. Such species should be species occurring only in northern and southern cold water without, in recent times, a connection. A taxon with this distributional pattern will tend to show geographic speciation, ultimately breaking down into species. When one considers bipolar 'species' to be typical forms of sibling species, or special representatives of polytypic species, their existence can be explained.

The bipolarity in *Limacina helicina* or *Clione helicina* is a good example of sibling taxa in the high north and south. *Clio pyramidata* forma *pyramidata* and *Clio pyramidata* forma *sulcata* show bipolarity as special representatives of one polytypic cosmopolitan species.

Bipolarity is reported for the following Radiolaria species, however, six species are not found in the North Atlantic Ocean. *Auloceros arborescens* Haeckel, 1887 subsp. *irregularis* Haeckel, 1887; *Aulographonium antarcticum* Haecker, 1908; *Aulodendron verticillatum* Haecker, 1908; *Aulastrum spinosum* Borgert, 1901; *Sagenoscena irmingeriana* Borgert, 1901; *Coelographis antarctica* Haecker, 1907; *Aulacantha lavissima* Haeckel, 1887; *Castanura primitiva* Reschetnjak, 1952; *Castanidium inclinatum* Reschetnjak, 1952. For these Radiolaria, however, it is not clear which type of bipolarity is concerned, probably the bipolarity is not complete in these taxa, as north-south contacts, however small, do exist. All types of tropical submergence (p. 139) can be incorrectly interpreted as bipolarity.

No attention, so far, has been paid to the phenomenon that two taxa occur in a different latitudinal sequence in two oceans. One of the best studied examples of this phenomenon is found in the species *Thysanopoda aequalis* Hansen, 1905 and *Thysanopoda astylata* Brinton, 1975. Both species migrate between 0 and 280 m depth and seem largely dependent on environmental conditions at about 200 m. *Th. astylata* is found in the Equatorial waters of the Indian Ocean and in the Central waters of the Pacific, while *Th. aequalis* lives in the Equatorial waters of the Pacific and in the Central waters of the Indian Ocean. The distribution in the Pacific Ocean is determined by the temperature at 200 m depth which is $< 16\,^\circ$C in the range of *Th. astylata* and $> 16\,^\circ$C in that of *Th. aequalis* (Boden & Brinton, 1957). The distribution in the Indian Ocean seems to be determined by the same ecological conditions, as in this ocean the temperature at 200 m depth is lower in Equatorial regions than in Central regions. This type of latitudinal reversion can be explained also by salinity preferences, as the salinity is higher in Central Pacific and Equatorial (northern) Indian waters than in Equatorial Pacific and Central (southern) Indian waters. The topographically unexpected reversion is thus explained by environmental conditions; conditions which do not affect all species in the same way, as reversion is normally absent.

A typical example of a disjunct distribution is the Atlantic-Tasman Sea distribution. Species endemic to the Atlantic Ocean and Atlantic taxa of groups showing provincialism have rather frequently a typically disjunct distribution in Atlantic and Tasman Sea waters. This type of dispersal is found in epi- as well as bathypelagic species (cf. fig. 152) and it may be explained by transport of taxa around Antarctica from the Southern Atlantic to the Tasman Sea, which is possible as surface and deepsea currents run from west to east around this continent. The disjunction is, however, in contradiction with this type of dispersal. These Atlantic taxa migrated to the Tasman Sea, but probably not by currents, or the populations along the pathway recently disappeared.

The Tasman Sea does not constitute a part of the real Pacific and can be considered of more recent origin than the Pacific, as it does not lie on a Pacific plate. Probably the Atlantic and the Tasman Sea originated in the same geological period as one area. However, for post-Miocene taxa such as *Limacina helicoides* this does not provide an explanation.

In this Atlantic-Tasman Sea pattern, moreover, always valid species are concerned without traces of speciation in the isolated populations so that a very recent dispersal is to be accepted.

To conclude, one can state that it is evident that there are very typical patterns which are difficult to explain such as the bipolarity and the Atlantic-Tasman Sea pattern while a great number of patterns as listed in table V show evident traces of certain geological influences though these influences are not understood in detail.

Ecological factors

The organism can be treated as part of the foodweb, as part of a community or as part of the environment, as it is an animal, feeding in between other organisms, all with their relations to biotic and abiotic phenomena.

The place in the foodweb is of great importance for distribution. The phytophages are restricted to the surface layers, the carnivores can live at all depths, but the detritus-feeders are most abundant in the deeper layers. The manner of feeding: ciliary-mucous feeding, filtering, and catching: either solitarily or in social relations, is related to the place the animal has in the food chain, but it also influences its biogeography. Spectra of particle size and abundance differ with the seasons, the depth, the latitude and the productivity. In the sea, as on land, the whole biological system depends on solar energy, which is first fixed in phytoplankton as assimilation products. In areas with high productivity and at high latitudes, the particle-size spectra are single-peaked, in Temperate waters with moderate productivity they are multi-peaked, and in the Tropics the spectra are smooth and flattened (Conover, 1979). The peaks at high, productive latitudes are due to mass occurrence of one or a few phytoplankton species and these succeed one another during the seasons, so that the spectra differ with the seasons. Feeders in such productive areas with peaked spectra must be able to feed on many different particle sizes, as these peaks succeed each other rapidly, and there should be no specialisation in feeding. In the Tropics, the flattened constant particle spectrum is formed by numerous different phytoplankters occurring in small numbers. The feeders are frequently specialised so that food specialisation is an important factor of the ecological niche in the Tropics. Food specialisation is difficult to realize in mucous-ciliary feeders, but filter-feeders rather easily adapt to special feeding habitats. Mucous-ciliary feeders are therefore expected to dominate the high latitudes, while filter-feeders, with feeding adaptations, are expected to be most dominant in lower latitudes. This expectation only partly agrees with reality. Mucous-feeders are very numerous in the Tropics too, and specialised feeders are also found in the high latitudes. In the Subantarctic, *Euphausia superba*, for example, eats only a small size-fraction of the diatoms and chiefly takes *Fragilariopsis kerguelensis*. But in such specialised feeders of the high latitudes different life-stages have all their own diet, which follows either in time or in depth the productivity of particular phytoplankters.

The mucous-feeders so frequently found in the Tropics differ from the cold-water mucous-feeders. The mucous surface used by warm-water species is much larger than that of the cold-water species, in order to harvest enough particles also when these are more scarce in the less productive areas.

The animals of pray usually live solitarily and when they show shoal forming it is only in high latitudes where they feed on also shoaling phytoplankton feeders. The phytoplankton feeders usually shoal in cold productive areas and are solitary in less productive areas. The differences are compared in table VI.

It is not certain whether a specific behaviour determines the distribution, or whether the occurrence in a particular area determines the feeding, but there doubtless exists interaction. *Clione limacina* shows starvation when it is not sympatric with *Limacina retroversa*. *Euphausia superba* de- and increases simultaneously with *Fragilariopsis antarctica*; *Corolla spectabilis* is confined to the areas with low productivity.

TABLE VI
Types of feeding related to productivity and behaviour.

	Behaviour:	Feeding:	Examples:
Cold productive areas	shoaling	specialized catchers	*Clione limacina*′)
	shoaling	specialized filter-feeders	*Euphausia superba*
	shoaling	not specialized ciliary-feeders (with small mucous web)	*Limacina helicina*
	not shoaling	not specialized catchers	*Spongiobranchaea australis*
	not shoaling	not specialized filter-feeders	*Ihlea punctata*
	shoaling	not specialized filter-feeders	*Salpa fusiformis*
	not shoaling	not specialized ciliary-feeders (with small mucous web)	*Limacina helicoides*″)
Warm less productive areas	not shoaling	specialized catchers	*Sagitta planctonis*
	not shoaling	specialized filter-feeders	Copepoda
	not shoaling	not specialized catchers	*Phronima sedentaria*
	not shoaling	not specialized ciliary-feeders (with large mucous web)	*Corolla spectabilis*
	shoaling	not specialized filter-feeders	*Pyrosoma atlanticum*

′) feeds exclusively on *Limacina helicina*
″) deepsea species

Consequently, the feeding behaviour can give us information on the zoogeography of a species.

The population structure and social behaviour and patchiness, though of tremendous ecological importance, will be dealt with elsewhere (p. 164).

The food chain runs vertically through the water collumn, as the basis of the food pyramid is formed by the phytoplankton in the surface layers, so that all energy or biomass transfer descends through the water collumn. The slower the downward transport, the more diverse will be the more deeply living populations. Strong stratification and absence of vertical migration will favour the development of diversity in the depth. In the epipelagic realm there is often rigid and complex stratification and significant vertical migration; in the mesopelagic realm there is little stratification and strong vertical migration, while in the bathypelagic realm no vertical stratification and little migration is found. Hence, the biomass transfer is quickest in the mesopelagic realm and here the diversity will be the least. As diversity and productivity are related, one may also expect latitudinal differences in diversity at depth. The higher-latitude mesopelagic realm will be the area with the least diversity. Oxygen minima reduce the loss in biomass during sinking and thus affect the diversity by enlarging the food supply below the oxygen minimum. In deeper layers, however, the food supply is always small and organisms always occur solitarily.

The effects of seasonal changes in ecological conditions on distributions can only be understood when reproduction, growth and mortality are known. Each animal group is characterized by a predominant type of reproduction, growth and mortality. The salps, for instance, are characterized by successive sexual and asexual reproduction, very fast growth and mass mortality, pteropods are hermaphroditic and moderately fast-growing, while euphausiids are of separate sex, slowly growing through many developmental stages, reaching ages of several years.

These group-specific characters of development determine the ecological and therefore zoogeographical 'possibilities' of the group. So, salps are especially adapted to follow in their population-size and -growth the quick plankton bloom of diatoms. The heteropods, with much slower reproduction, are more adequately adapted as predators of slowly developing areas such as the Tropics, to give two examples. Therefore, a relation exists between distribution and reproduction of a taxon.

Seasonally fluctuating factors of particular importance to plankton are temperature, illumination and salinity. The distributions discussed in Chapter 3 frequently showed to be determined by temperature and salinity and related to day length (latitudinal illumination). These factors also change with depth, which brings about a clear relation between horizontal and vertical dispersal. It has been mentioned already that light decreases downwards, so that zoogeographic zonation based on physico-chemical parameters from the equator to the poles is comparable with the zonation from the surface to the ocean depth. This phenomenon explains tropical submergence and illustrates the fact that deepsea organisms are related to cold-water taxa. The ecological importance of temperature is for a major part due to the fact that temperature determines the rate of metabolism of the organism. Low temperature induces a low metabolism, slow growth and a long breeding cycle, whereas high temperature favours a high metabolic rate with quick growth, short breeding cycle and short life-span of the organism.

Thus far it is clear that specific types of growth, reproduction, food specialisation, feeding methods and social behaviour all have a special distribution in the different water masses. However, mixing of water masses influences those patterns, and before focusing on the ecological community, this mixing has to be discussed.

Mixing of water masses as a rule decreases the number of species, so primary water masses (not mixed, and formed under the influence of local atmospheric conditions) show a greater diversity than secondary water masses (formed by mixing). Consequently, transitions from one water type to another can form a zoogeographic barrier. Primary water masses have their own typical plankton communities, secondary water masses have their impoverished communities and taxa in these communities, characteristic of the secondary waters, are as a rule of Tropical origin (Beklemishev, 1961). This also points to the hypothesis that species of colder water masses, largely of secondary nature, are derived from Tropical taxa. Mixing is very important for plankton ecology, as transport in general is such an important factor for plankton.

There is a great difference between the communities of the surface layers and the bathy- and mesoplanktonic communities. The deeper ones are stable throughout the year, while the surface communities change strongly with the seasons. As the abundance, and even the absence and presence, of epiplanktonic species is different in the different seasons, the species compositions of epiplanktonic communities change strongly. Sometimes the species compositions of communities change as a result of extinction of species in the area and subsequent repopulation by migration out of other areas, and sometimes repopulation is possible by the hatching of eggs and the development of resting stages. There are therefore three types of communities: the stable communities from the deeper layers and the Tropics, the instable communities changing in species composition by mixing and migration and the instable communities changing in species composition due to breeding cycles of local species. The first type of instable community is typical of mixed secondary surface waters, the second type is

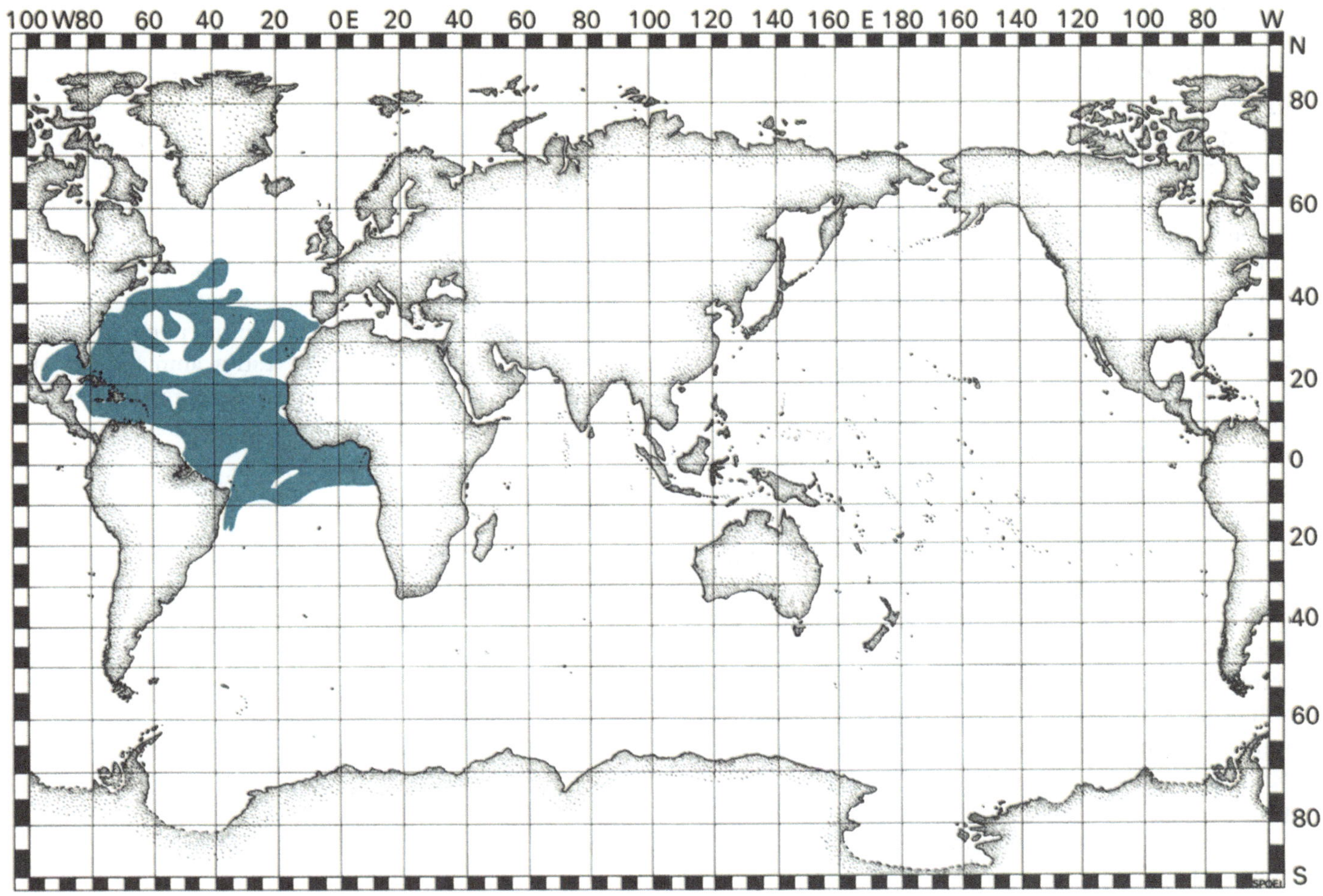

FIG. 191 Distribution of *Planktomya henseni* in the N. Atlantic Ocean.

typical of not mixed, primary surface waters of the high latitudes.

The major influence of currents is the transport possibility they provide to organisms. As a hypothesis one may state that holoplanktonic animals will try to resist the transporting influence, while larvae of benthic invertebrates, dependent on transport for their dispersal, will tend to take as much profit from the currents as possible, as shown by the larval *Planktomya henseni* (fig. 191). For these 'far wandering meroplanktonics' the term teleplanic larvae is sometimes used. For this reason, the dispersal of larvae of benthic organisms by ocean currents will be discussed first, to get an impression of the possible effect of currents on the zoogeography.

For phytoplanktotrophic larvae of Teredinidae Scheltema (1971a) found a transport over long distances along the shores and probably across the Atlantic Ocean. Presence of larvae of coastal species from the western Atlantic was shown over a distance of 50° longitude as a continuous line of records was made from about 65°W to 15°W showing an increasing age of the larvae and a decreasing abundance. According to Scheltema (1971b) the duration of pelagic larval development of benthic invertebrates and the velocities of currents leave no doubt that there is a possibility of exchange of larvae between the eastern and western Atlantic. This means that geneflow between littoral populations of the East and West Atlantic is maintained through the larvae. The Equatorial currents are responsible for a westward transport. The eastward transport is not carried out by the diffuse and seasonal Equatorial Counter Current but by the strong Atlantic Equatorial Undercurrent running below the South Equatorial Current at a depth between 50 and 100 m (Scheltema, 1971b). That not all larvae of marine invertebrates are able to execute reciprocal migration between the littorals of eastern and western coasts of oceans is proved by the faunal differences at the two sides.

Holoplanktonic species always have a longer period during which transport is possible, as they spend their entire life in the plankton. However, this transport by currents does not affect the distribution in all species. In relation to these phenomena, the circulation hypothesis of Damas is of great interest; it states that the distribution of an epiplanktonic marine organism is a product of prevailing hydrographic circulations, biochemical limitations and innate behaviour patterns determining depth range and vertical migration. This hypothesis provides the essential biogeographical principle of the epiplankton, and it shows that besides the current velocity, the time during which an organism is exposed to the current and therefore the behaviour of the species, is of decisive importance for dispersal by currents. Thus, animals bound to a specific current, or depth, e.g. because of their feeding behaviour, are transported further than animals that can freely change their depth, or the current in which they live.

Of minor influence, but still of importance, are the viscosity of the water and the shape of the specimens transported. The sinking rate of an organism increases when the viscosity of the water is high (Tropics) or when the organisms are less well-equipped for floating. In horizontal sense the same phenomena are of influence so that the greatest transport is found in organisms with well-developed floating equipment and when viscosity is low. Cold-water currents will have a greater transporting effect and animals with large 'parachutes' will be transported further. As a secondary effect the important transport of species with well-developed floating equipment causes these to show less provincialism than those without such equipment.

Apart from the surface and deep currents there is a system of cyclonic eddies discovered by the MODE experiment (Hammond, 1974). Eddies are roughly circular flows about 200 kilometres in diameter. They extend vertically down to the bottom of the ocean, although the motion is three or five times stronger above the thermocline. With velocities of 3 to 5 centimetres per second in deep water and 15 to 20 cm per second nearer the surface, an eddy has more kinetic energy than an atmospheric storm and lasts many times longer, up to 80 days or more. Cyclonic eddies are very frequent and close-packed. In the eddies the thermocline shifts vertically. The eddies can be generated by baroclinic instability, by currents, or by radiation of energy from meandering currents. The water in the centre of the eddy is warmer (warm-core ring) than near the periphery, or colder (cold-core ring), depending on the origin of the phenomenon. Eddies strongly influence the community structure and may affect population composition.

Speciation around the species level and distribution

Here we will discuss distributions which have developed recently, or are developing at this moment. As variation in plankton provides most information on this phenomenon, the population has to be treated first.

In plankton a population is the total of swarms or specimens which occur mixed in the period of breeding, so that the specimens have an equal chance to mate (Van der Spoel, 1971). A local population is a swarm in which the specimens can mate optimally. The total of populations of a species covers the range of the species. In planktonic taxa, populations are not stable, as they are usually transported by currents. In all organisms populations may be instable by population dynamics and reproduction but the abiotic transport by water movements provides the instability of planktonic populations with a special dimension.

'Peripheral populations' and 'central populations' as commonly distinguished in terrestrial animals, cannot be used as concepts in plankton zoogeography as spatial relations between populations are instable.

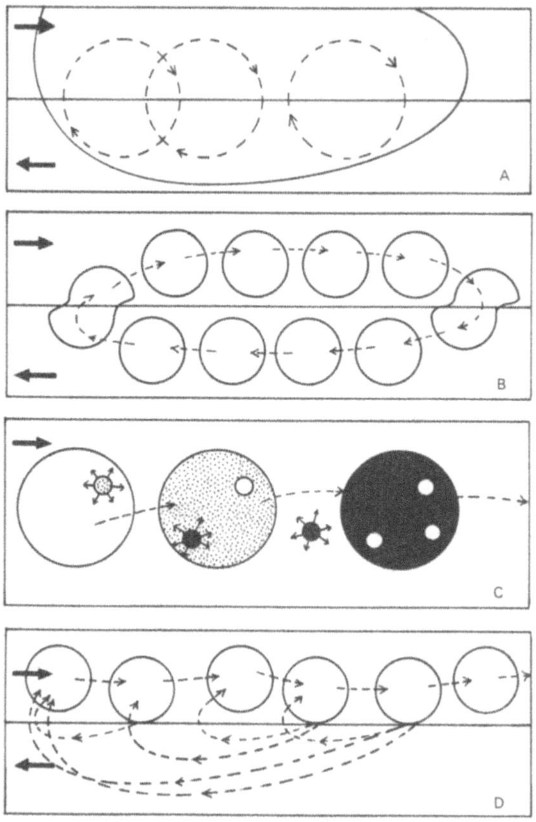

FIG. 192 Examples of mechanisms providing populations with geographic stability. A = specimens execute circular movements with currents, B = entire populations circulate with currents, C = populations shift with currents but some stationary specimens are founders of new populations, D = populations shift with currents but isolated specimens recirculate with other currents and are founders of new populations, thick arrows = currents, thin arrows = repopulation, broken lines with arrows = movements of specimens or populations, lines = borders of populations.

As a taxon is adapted to its special environment, it has to be assumed that more generations have undergone the selective pressure of that environment. As a strong movement of populations would mean a quick movement throughout the range of the species, there has to be a mechanism suppressing the instability, and as possible mechanisms the following can be active:

1) the population can maintain, to a certain degree, its spatial position by the fact that the individual members of the population continually return to their original position by floating in two currents of opposite direction (fig. 192A);

2) the population is transported by currents, but retains its position by continually returning as a unit to its original position by using currents of opposite directions (fig. 192B);

3) the population is transported, but some individual specimens are more stable, maintain there position and are the founders of a new population in the area of the old one which floated away (fig. 192C);

4) the population is transported as a whole, but brought back to its original position by cyclic currents; all or some life-stages of the taxon can execute the whole or part of the cyclic movement;

5) the population is transported with all the specimens, but a few specimens come back by counter currents to the original place and then become the founders of a new population (fig. 192D).

Ontogenetic vertical migration is of crucial importance, as the juveniles may be the only part of the population which is stable or which returns to the original position. Within-population migration is of importance for the phenomena given under 1, 3 and 5, while population migrations are correlated with the phenomena under 2 and 4.

Results of natural selection are different in populations which maintain their positions by different mechanisms. When the populations are more or less stable by circulation of the specimens, adaptation to the selection can differ in the various parts of the range. When whole populations circulate, diversity in one taxon will be restricted and each population must be adapted to a wide range of conditions. When only individual specimens are responsible for the seemingly stable populations, it will always be selection in the adjacent area which determines adaptation in these populations. In the extreme case it is the conditions in the most upstream area which determine the population characters over the whole range.

When the population moves, it will, at some moment, reach the ecological boundary of the taxon where it enters the non-sterile and furtheron the sterile expatriation area (p. 30). When it concerns, however, a population of a polytypic species composed of formae, the crossing of the ecological boundary can result in entering the ranges of another forma of the species. After that crossing, the population will change in characters (phenotypical and probably genotypical) and it will take part in the forma whose geographic area it entered.

All populations having crossed a barrier or boundary will, if they don't die out, develop special adaptations to the other environment. As a rule, however, geographic speciation (sensu Mayr) cannot be the result of such rather continuous population movements.

There can be a great loss of adaptations and variability with the transport of populations out of the range of the taxon, as complete gene-pools of populations become lost for reproduction. This can be considered a special kind of genetic drift which does not only affect peripheral populations, as can be concluded from fig. 192C & 192D. In this figure it is clearly shown that the gene-pool of a population disappears with the population from alle the parts of a range, and only a small amount of the gene-pool, determined by hazard, is left available for new populations to be formed. Genetical drift is in general held responsible for a large amount of variation in smaller populations and in plankton it can also contribute much to random variation as the drift is large and the population in comparison small.

Transport of specimens and populations is determined by the movements in the environment on the one hand and by the behaviour of the species and its morphology and reproduction on the other hand. As the behaviour of a species is chiefly hereditary, the urge to return to an original area or to stay there is also hereditary, and this is one of the characters which is always propitious to the taxon, so that it meets a positive selective pressure. As a consequence, all populations tend to become more stable. The spatial instability due to morphological and reproductive characters is, however, also determined by the phylogeny. This is the reason why at least some taxa seem composed forever of moving populations.

With regard to the definition of the species given by Mayr (1963), the instability of holoplanktonic populations is important. First of all, it should be mentioned that in holoplanktonic animals species are more than groups of interbreeding populations. Also those, usually peripheral, populations which form no longer part of the genetic unit of the species, as they have floated out of the reach of the other, usually more upstream, populations, belong to the species. The species is thus a reproductive community plus the offshoot of this reproduction, which takes no longer, in reality or potentially, part in the communication, without becoming another species.

The concept 'population' as used in the species definition by Mayr cannot be used in plankton either. Mayr (1971) clearly stated that the single 'individual is only a temporary vessel, holding a small portion of the gene-pool for a short time', whereas 'It is the entire effective population that is the temporary incarnation and visible manifestation of the gene-pool. It is in the population that the genes interact in numerous combinations. The continued interaction of the genes in a gene-pool provides a degree of integration that permits the population to act as a major unit of evolution.' Populations which are constantly moving, also through other environments where other selective pressures influence the gene-pool, and which are constantly renewed, are not the major units of evolution as postulated by Mayr. The moving population resembles more the individual holding for a short time a small and constantly altering part of the gene-pool. The concept of populations as given by Mayr holds only good for species with geographically stable populations. This restriction should be kept in mind when studying holoplankton speciation.

The concepts 'peripheral' and 'central' population does not hold for plankton by the mere fact that central populations will become, after a period, depending on the current speed and the size of the range, peripheral and some peripheral ones will become central. In addition, there are two temporary kinds of geographically peripheral populations, the most upstream ones and the most downstream ones. In planktonology it is better to use the terms generating populations and finite populations. The generating populations form the centre where populations develop which will move through the range; they come most close to central populations. The finite populations are the populations near the boundary, bound to disappear (fig. 192D) and most resembling peripheral populations.

The differences in genetical diversity, phenotypical variation and population dynamics found between peripheral and central populations, is not found between generating and finite populations. Generating populations will have, as a rule, low density (animals per volume water), little genetical diversity; there may be more phenotypical variation, as stabilizing effects of homeostatis, heterozis and coadaptation are absent. The finite populations have by the mere fact that they are older than the generating ones, a higher genetic variability, a more stabilized phenotype and a higher density. In addition, very young generating populations have a relatively high number of juveniles while finite populations tend to have only adults.

The fact that populations near the boundaries of the species frequently provide more stable specimens which return to the original locality, makes these populations more important for the genetic community of the species while the populations in the centre of the range are more important in relation to the species as reproductive community.

The wind direction, when it directly influences the currents, as Trade winds do, is also of importance for populations. Fleminger & Hulsemann (1974) found for Tropical species in surface waters that seasonal changes in wind direction may be responsible for the stability of species in their range. These authors stated (p. 111) 'These pools of warm water are subjected to seasonally repetitive changes in the strength of the Equatorial Trade Winds (Wyrtki, 1966, 1967; Taft, 1971). The seasonal changes produce monsoon-like reverses in the circulation of the equatorial segment inhabited by each species' (cf. *Pontellina*) 'This phenomenon apparently provides a sufficiently closed hydrographic circulation to maintain breeding stocks in proximity to suitable nursery grounds and thus ensures continual success of each species'.

It is, however, doubtful that this mechanism of seasonally changing Trade winds is in all parts of the Tropical belt of the same importance. Seasonal changes in e.g. the Tropical West Atlantic are negligible when compared with the changes in the Tropical Indian Ocean. Seasonal current changes of any importance are only found in the Tropical and Subtropical waters. This makes clear that the stimulus to bring populations back to their breeding grounds forms an extra possibility in the Tropics to develop longitudinal variation.

So, the isolation by continents will not be the only explanation for longitudinal variation of species restricted in their latitudinal distribution, also the seasonal circulation changes are responsible for this phenomenon.

The great flexibility and geographic instability of populations in plankton makes it sensible to pay special attention to the concept 'forma' before discussing the subspecies. The forma can be considered identical with the 'dependent subspecies' as distinguished by Huxley (1942). The forma is a taxon composed of populations taxonomically different from other such populations only in the centres of their ranges (due to different selective pressures in these centres) and shading off into one another (due to gene-flow). Hence, it is impossible to indicate an exact geographic boundary between formae.

Formae can be found in horizontal as well as vertical successions. The speciation, in which the forma occupies an important place, has been described for Chaetognatha by Tokioka (1974: 406) 'There have been known in many planktonic animal groups such as chaetognaths, pelagic tunicates, molluscs, crustaceans, protozoans, etc. a number of pairs, trios or groups of species that are closely related to one another, but respectively living in water masses different but adjoining one another. When any twins of such pairs are neritic or eurythermal, it is worthy to check again the validity of these species. The complete absence of any intermediate forms cannot be the prove to separate the species, because the threshold effect will bring about the perfect separation in morphology.'

Groups of species, trios and twins are not the species groups or the superspecies as defined by Mayr (1971), which he considered to develop by isolation; the groups of Tokioka (1974) develop by ecological adaptation, which, in some cases, results in an isolation. 'The separation of allied species has been made in many cases for the reason that there are no intermediate forms between the two and respective forms are distributed limitedly in different water masses'. This means that ecotypes, formae and dependent subspecies can live close to each other without producing intermediates. '... when the question concerns the neritic (inevitably eurythermal and euryhaline) or oceanic but eurythermal species, and when the twins are distributed separately but in adjoining different water masses, it is very likely that the twins are only different generations of the same single species, developed respectively under different environmental conditions. The different features are ecotypic ... in such cases the barrier between the adjoining water masses is not structured so sharply that it is very difficult to pass across it. The barrier will be strengthened by such vertical currents as divergence or convergence, and with the increase of the steadiness and distinctness of these currents the interbreeding between the adjoining populations will become more difficult and the different features will become genotypic. This might be the case for those, in which the distributional areas of two forms are far separated by another intervening form.'

This approach of the problem by Tokioka shows that formae may participate in speciation without clear geographic isolation, and it shows also that distribution of morphologically different but closely related populations has to be treated with care.

The subspecies is the taxon in which all kinds of infraspecific variation can be brought under a valid taxonomic name. Mayr (1971) gives a definition restricting the subspecies to an acceptable well-limited number of infraspecific variants, but he adds at the same moment so many different groups which can be considered subspecies that the subspecies sensu Mayr (1971) is of no value in taxonomy and only remains as a practical unit for nomenclature.

Huxley (1940, 1942) considered two kinds of subspecies: the dependent subspecies and the independent subspecies. Pierrot-Bults (1976) recently considered these two subspecies in relation to holoplankton and concluded that the dependent subspecies is identical with the concept forma (sensu Van der Spoel, 1971). The semispecies, forming the components of a superspecies (sensu Fleminger, 1972), seem partly to cover the concept 'independent subspecies'. There must be yet another infraspecific kind of taxon not covered by the concepts forma and semispecies. This group is exactly what Mayr (1971) considered a subspecies as can be concluded from his definition 'A subspecies is an aggregate of phenotypically similar populations of a species inhabiting a geographic subdivision of the range of the species and differing horizontally from other populations of the species.' There is no sharp difference between the forma, real subspecies and semispecies, all three are connected by a complete spectrum of intermediates. But what phenomena in biology are not connected by all possible intermediates? In biology it is not the question whether a concept, a group or a phenomenon is sufficiently separated from other such concepts, groups or phenomena, but one should ask if the concept, group or phenomenon is a useful unit.

The practical value of the forma and semispecies has already been demonstrated, so that the restricted 'subspecies', according to the above definition, should be made available to cover the other groups. The problems around subspecies are induced by an incorrect interpretation of the concept 'cline' which was discussed above (p. 39).

The superspecies is well studied for some holoplanktonic taxa. Fleminger (1972) gave a good description of holoplanktonic superspecies among calanoid Copepoda. All constituent species in seven of the studied superspecies were assignable to the same neritic or eutrophic sectors or to two or more of the four eutrophic sectors. One other superspecies occupied both the offshore neritic and two eutrophic oceanic habitats. Fleminger (1972) found 'pronounced conformity in hydrographic sectors occupied by constituent populations of monophyletic lineage', which supports the circulation hypothesis of Damas.

The species constituting a superspecies are exactly the mirror picture of the formae discussed above. The constituent formae of an epiplanktonic polytypic species always occur in different oceanic sections and they will remain formae due to the gene-flow caused by instability of the populations and specimens which proves that the circulation hypothesis of Damas does not hold for these polytypic species. The distributions of a superspecies and a polytypic species will therefore be different. This fits in rather well with the opinion of Mayr (1971) that superspecies might be more common in animals with low dispersal facilities.

A superspecies is composed of allopatric species of monophyletic origin and with distinct morphological differences. As the species are not only allopatric but also live in comparable oceanic sectors, they will also be distributed discontinuously, which is excluded among formae.

Variation in holoplanktonic groups leads to different degrees of speciation. In Copepoda several superspecies and sibling species have been described; in Pteropoda formae and subspecies are more commonly found. Though environmental conditions which influenced both groups are to be considered equal, their effects on speciation have been different. This can be explained by the typical differences in reproduction of the groups; pteropods are hermaphroditic and have simply structured copulatory organs, while copepods are of separate sex and have complicated copulatory organs and secondary sexual characters. This special phenomenon may, in copepods, constitute a strong precopular barrier in absence of strong geographic barriers.

Fleminger & Hulsemann (1974) gave a carefully studied example of sibling species in the copepod genus *Pontellina*. An attempt will be made to compare these siblings with the formae in *Cavolinia tridentata*. *Pontellina* is a circumglobal warm-water genus composed of four species. This monophyletic genus shows three basic derivates, *plumata*, *platychela* and a third that underwent more recently speciation into *morii* and *sobrina*. *P. plumata* (Dana, 1849) and *P. platychela* Fleminger & Hulsemann, 1974 form a pair of siblings, while *P. morii* Fleminger & Hulsemann, 1974 and *sobrina* Fleminger & Hulsemann, 1974 form the other pair of siblings. In these species, except for *P. plumata*, panmixis proves to prevail, the Atlantic and Indo-Pacific populations of this species prove to differ morphologically. These differences are, like those among the species, subtle and restricted to secondary sexual structures. There are three phenomena which make this group more complicated. The first complication is the polymorphism due to the occurrence of four morphs of the female fifth leg. The difference between the Atlantic and Indo-Pacific populations of *P. plumata* is due to differences in the percentual occurrence of the four morphs in this species.

The second phenomenon is that the competition between the siblings *P. plumata* and *P. platychela* causes character displacement, resulting in the varying occurrence of the four morphs in Atlantic and Indo-Pacific.

The third typical phenomenon is the isolated position of the East Pacific populations of *P. plumata* and *P. morii*. Fleminger & Hulsemann (1974) paid no attention to this phenomenon but it is known that there are numerous parallels in other holoplanktonic groups. The separation of East and West Pacific populations will be of a more recent age than the speciation giving rise to the four siblings.

P. plumata shows the greatest north-south range (42°N-30°S) of the four species, it is circumglobal in distribution and it is most abundant in Subtropical waters and downstream of upwellings; in typically Equatorial waters this species is less abundant. The other three siblings are found predominantly between 20°N and 20°S and they are typical of Tropical, Equatorial waters.

Hypothetically one can consider *P. plumata* to have avoided the Tropics in warmer geological periods. During such a warmer period *P. platychela* may have developed in the Tropical Atlantic, *P. morii* in the Tropical Indian Ocean and *P. sobrina* in the Tropical Pacific. In recent times *P. plumata* migrated to the Tropics and *P. morii* invaded the Pacific which resulted in the present-day sympatric distributions. This hypothesis explains recent distributions and probably the two pairs of siblings. However, *P. morii* can also have developed as an ally of *P. platychela* and split off the E. Pacific *P. sobrina* (fig. 184).

A comparison of this speciation and distribution with those found in *Cavolinia tridentata* (Forskål, 1774) makes two things clear at once, viz. the north-south range of the species is much wider than in *Pontellina* and only two of the nine taxa occur in the Tropics. The two groups have in common that in the East Pacific special taxa developed and that there is sympatry of the taxa occurring in the Indonesian Archipelago and in the West Pacific.

That *Cavolinia tridentata* avoids the Tropical waters proves that speciation will have been different from that of *Pontellina*. According to the Frost & Fleminger rule (p. 56), longitudinal diversity in *Pontellina* will be greater than in *Cavolinia tridentata*. The connections between populations are in the latter greater than in *Pontellina*, as they extend south of the continents. The populations reach higher latitudes as they are transported by cyclical currents around the oceans.

The speciation giving rise to *P. sobrina* will have been in all probability identical with the speciation giving rise to *Cavolinia tridentata* forma *occidentalis* and forma *affinis*. *P. sobrina*, however, is considered (Fleminger & Hulsemann, 1974) a sibling of *P. morii*; and *occidentalis* and *affinis* are considered formae of *Cavolinia tridentata* (Van der Spoel, 1974). Probably the hermaphroditic sexual behaviour of *Cavolinia* is responsible for the low degree of speciation compared with *P. sobrina*.

Effects of homeostasis in sibling species seems to be very important. Some genes (suppressors) can prevent other genes from influencing the phenotype and likewise modifiers may change the expression of other genes. Goldberger (1974) introduced the term autogenous regulation, which comprises the phenomena in which the product of a gene suppresses the gene in the formation of the same product. This autogenous regulation can buffer the effect of mutated genes and in this way mask all effects of speciation, giving rise to siblings.

Recent taxa at the species-level or below can frequently be explained as having developed during the Ice-Ages or during the drifting of the continents. This does not exclude that some taxa developed more recently. Probably the double taxa are examples of a more recent speciation. Double taxa are two, or sometimes more, taxa which are closely related to each other and clearly less closely related to other taxa of the same group. Such double taxa are not to be confused with sibling species, as the former are morphologically different, and they are not vicariating species either as they do not replace one another in the ecosystem. There are three phenomena typical of double taxa, viz.: 1) different pairs of double taxa in a group usually show the same type of morphological difference; 2) double taxa are largely sympatric; 3) double taxa usually occur in special taxonomic groups, being absent in other such groups though these groups may be closely related.

Typical examples of double taxa are found among

Cavoliniidae (Pteropoda) and Atlantidae (Heteropoda). In the other families of Pteropoda and Heteropoda no double taxa are found. The double taxa in both groups are, like *Atlanta helicinoides* forma *helicinoides* and *Atlanta helicinoides* forma '*A*', largely sympatric (Van der Spoel, 1976). The difference between the composing taxa of all double groups in Cavoliniidae is chiefly restricted to the difference in the length/width ratio and shape of the lateral spines; in the double groups of Atlantidae it is the diameter increase of the successive whorls which forms the difference between the taxa.

Competition among closely related taxa in small areas is always stronger than in larger areas. The double taxa under consideration are closely related, but they live in a very large area, which makes their occurrence acceptable. Probably the advantage of the difference for one of the components of the double taxon is too slight to have any value in competition. The most typical thing which double taxa have in common is their warm-water distribution. A hypothesis could be that in some animal groups a special phenomenon split species of the warm-water belt into double taxa.

The development of double taxa, without a period of isolation, is only possible starting from polymorphic species, in which the morphs become the 'taxa' of which the double taxon consists. In this situation, the genetic basis for the double taxon is not difficult to explain. The nature of the double taxa, however, does not resemble polymorphism, as usually geographic variation is found in the composing taxa and as it is common for one of the taxa of a double taxon to show a slightly wider range than the other.

Even when double taxa are composed of species, one may be dealing with a special kind of sibling species in which a restricted number of characters influences the phenotype. As the phenotype is partly stable, geographic variation is only realised in the characters not suppressed by homeostasis, therefore, the characters in which the two species of the double taxon differ. By this geographic variation the extreme values of the variable characters in which the two species differ may tend to smooth out real constant differences between them.

Very small differences in genome will exist between the constituting members of double taxa, when no homeostasis is active, and this small difference has then to be either of great selective profit or of great influence on heredity as it has to be the only reason for intersterility. A great selective profit has never been demonstrated so that the only explanation left is the presence of different polyploids in double taxa, in which small differences of the phenotype are due to polyploidity, in combination with intersterility only.

Normal geographic speciation has not yet been considered as it does not provide the most plausible explanation for the phenomenon under scrutiny. As the double taxa are restricted to the warm-water belt, and as some are endemic to the Indo Pacific, one can think of an influence of the Ice-Ages. During the glacial periods, the Atlantic warm-water belt has been separated from the other oceans, and the Indian-Pacific warm-water belt was split into two sections. The lowering of the sea-level caused the closure of the Indonesian passages between the two oceans except for a passage 10 km wide and deeper than 200 m between the Timor and Alor islands. This was a passage smaller than the Strait of Bab el Mandab and comparable with the recent Strait of Gibraltar.

So far, speciation and the taxonomic subdivision of plankton have been discussed on the basis of zoogeographic features. However, the zoogeography has also been influenced by special types of speciation. Speciation in plankton is slow, recent speciation only gave rise to variation and polytypic species, and this implies that many species could either develop large ranges or could maintain large ranges. Ecosystems are usually of a simple structure, barriers and boundaries in the ocean are rare so that there are few possibilities for geographic speciation, but great possibilities for dispersal.

A barrier forms an insurmountable obstruction for an organism; a boundary is the actual place where the distributional range of organisms comes to an end. The most illustrative example of a boundary is the limit of the range of an organism in an area where all the environmental conditions change gradually but where the species is unable to follow these changes. In this example there is no question of a barrier which cannot be crossed.

Especially the 'ecological boundary', commonly used in zoogeography, has to be defined as this term is rather obscure for holoplanktonic animals with little ecological specialisation and without formation of clear biocenoses in the large planktonic biotope. Instead of biocenoses the concept 'community' holds better for planktonic animals and an ecological boundary then limits the area where the maximum abundance of the species belonging to a community occurs.

The biogeographical boundary is the border of the distribution of a taxon. As holoplanktonic organisms are passively transported and consequently extremely mobile, the biogeographical boundary is multi-interpretable. To distinguish the total distribution (the entire area where the taxon is recorded) and the 'range' (the area where the taxon lives and reproduces without being supplied with specimens from elsewhere), though useful for terrestrial and lacustrial animals, seems for the present group inadequate. Vertically and horizontally the distribution of a taxon changes in time: daily, seasonally or yearly, but active and passive movements, and on the other hand the supply with specimens from elsewhere, determine the population at a certain place. Therefore it is proposed not to use the concept 'range' for holoplanktonic organisms, but to speak only of their total distribution, limited by the biogeographical boundary. When in a particular season or year the distribution of a taxon is smaller than its total distribution by limitation of transport or local temporarily unfavourable conditions, the taxon occurs as a rule within its ecological boundary.

Heinrich (1969) found that ecological boundaries in the western Pacific coïncide with biogeographical boundaries. This implies that both types of boundaries are in essence the same but along the line which limits the optimal ecological environment for one species, run the limits of maximal distribution of another species.

The borders of distribution of different species usually do not coïncide exactly, as they do not exactly coïncide with a particular hydrographical boundary. The ecological or biogeographical boundary is in reality a more or less broad zone in which conditions change more abruptly than outside this zone. The individual tolerance of each species determines its exact penetration into the boundary zone.

When the boundary zone is broad, certain taxa can adapt to the gradient of conditions in the zone, making it look like a special biotope, usually called a transitional zone.

Plankton communities and patchiness

Communities are described on the basis of species composition and individual species abundances. Differences between communities are therefore based on presence and absence of species. Presence and absence can be proven for a sample, but only presence can be proven for the community. So in plankton it is very difficult to study the community.

The concept 'community' can be treated also in different ways, in general, one can distinguish between the ecological and the geographic community concept. The community in the ecological concept is the unit of ecologically interdependent taxa, bound to specific environmental conditions, which develops and persists independently of other such communities. The geographic concept of the community concerns limited parts of faunas related to environmental conditions. Beklemishev (1971) uses the concept 'cyclical communities' for subsurface waters which are well-delimited and of which only 10% of the population is lost every year into adjacent gyres. Between pairs of gyres there are belts of mixed water in which 'mixed', 'secondary' or 'transitional' communities are found. The 'terminal' communities, according to Beklemishev, desintegrate when the water which carried them diverges, approaching the coasts; this type is deprived of recirculation. In the near-coastal area he recognizes the 'distant-neritic' community.

Beklemishev's communities are thus zoogeographic units. The ecological community concept is not frequently used in holoplanktonic studies and moreover, real plankton communities do not exist, as in the ecological interaction nekton, benthos and frequently birds form essential links. The ecological community is sometimes confused with patches. Is patchiness then a better concept?

As already mentioned, plankton is not homogeneously dispersed in the ocean, in the horizontal as well as in the vertical planes abundances strongly vary. These local high abundances of organisms influence the dispersal of the organism itself and that of ecologically related taxa and, as such, a difference in abundance is a zoogeographic phenomenon.

To describe the local high abundance, the terms clusters, aggregates, patches, swarms and shoals are used; but each of these concepts has a slightly different meaning. The use of these terms in the literature, however, is far from consistent and some authors even use one term for all the concepts (Steele, 1976). Depending on the mechanism responsible for the higher abundance, three different phenomena can be distinguished.

A swarm (or shoal, in nekton) is the total of individuals which stay close to each other due to genetically determined behaviour. Such a swarm can be considered identical with a local population.

A patch is the total of individuals brought together by hydrographical conditions.

A cluster or aggregate is a mathematically described locally occurring high frequency of specimens.

Studying the irregularities in abundance and the influence responsible for it, one has to know: 1) the importance of change; 2) the stability in time and place of the abundances detected; 3) the relation between the hydrography and the organism; 4) the ecological relations of the organism and 5) the behaviour of the organism.

When an organism is found in abundances which cannot be explained by mere chance, in an otherwise homogeneously distributed organism, there must be other influences at work. This idea is only valid if a homogeneous distribution is accepted to occur in the ocean normally; but there is no proof at all that there is no heterogeneous distribution of specimens. In most publications it is a priori accepted that the chance factor has nearly no influence on irregularities in abundance, and the research focuses on biotic and abiotic factors. The first problem then met directly with is that the interaction between the biotic and abiotic factors is so complex that it is usually impossible to discriminate properly between them. For the present discussion a further problem is that we found in the literature either discussions on the ecology of abundance differences or on the zoogeographic implication of them without any link between the two.

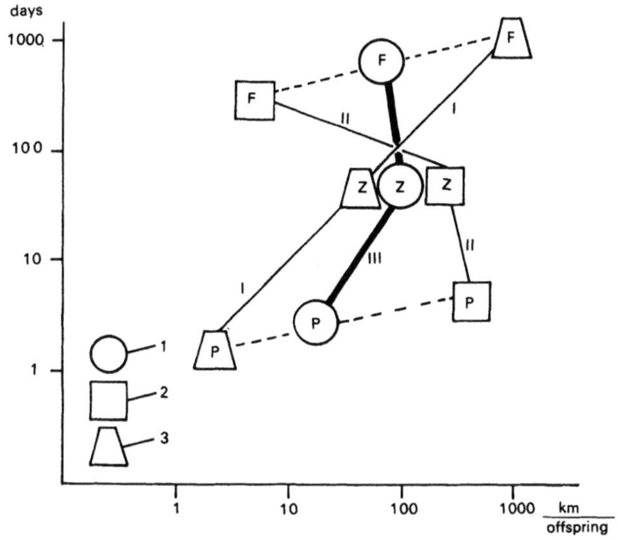

FIG. 193 Graph of the position of the pelagic organisms in their trophic level in space and time. 1 & III = population structure, 2 & II = reproductive structure, 3 & I = specimens in space and time. F = Fish, P = Phytoplankton, Z = Zooplankton.

The relation between ecology and dispersal is still easily visible in the three trophic levels as explained by Steele (1976). The primary producers, phytoplankton and bacteria, have a short life-span and cannot move quickly or over long distances. The last step in the food chain, the third trophic level, represented by e.g. fish, comprises fast and far swimming taxa with a long life-span. In between is the zooplankton, as also shown in fig. 193 (line I). The reproduction, however, contributes just in an opposite way to dispersal of the organisms (line II in fig. 193) so that the final population structure is determined by line III in fig. 193.

In relation to zoogeography the high abundances are only interesting when of larger scale in time and space, and as seen in fig. 193 only the third and second trophic levels then have to be considered. The interaction of the different trophic levels on each other remains, however, important.

In diagram IV after Haury et al. (1977) a more detailed account is given of the different scales of patterns, and they are subdivided into categories. The fine and micro-scale patterns do not persist long enough to contribute to dispersal (Platt & Denman, 1975) and macro- and mega-patterns are zoogeographic patterns. Therefore, the coarse and meso-scale patterns are most important for zoogeographic studies, and these are the categories in which all influences, hydrographic, ecological and behavioural may be active.

Diagram IV

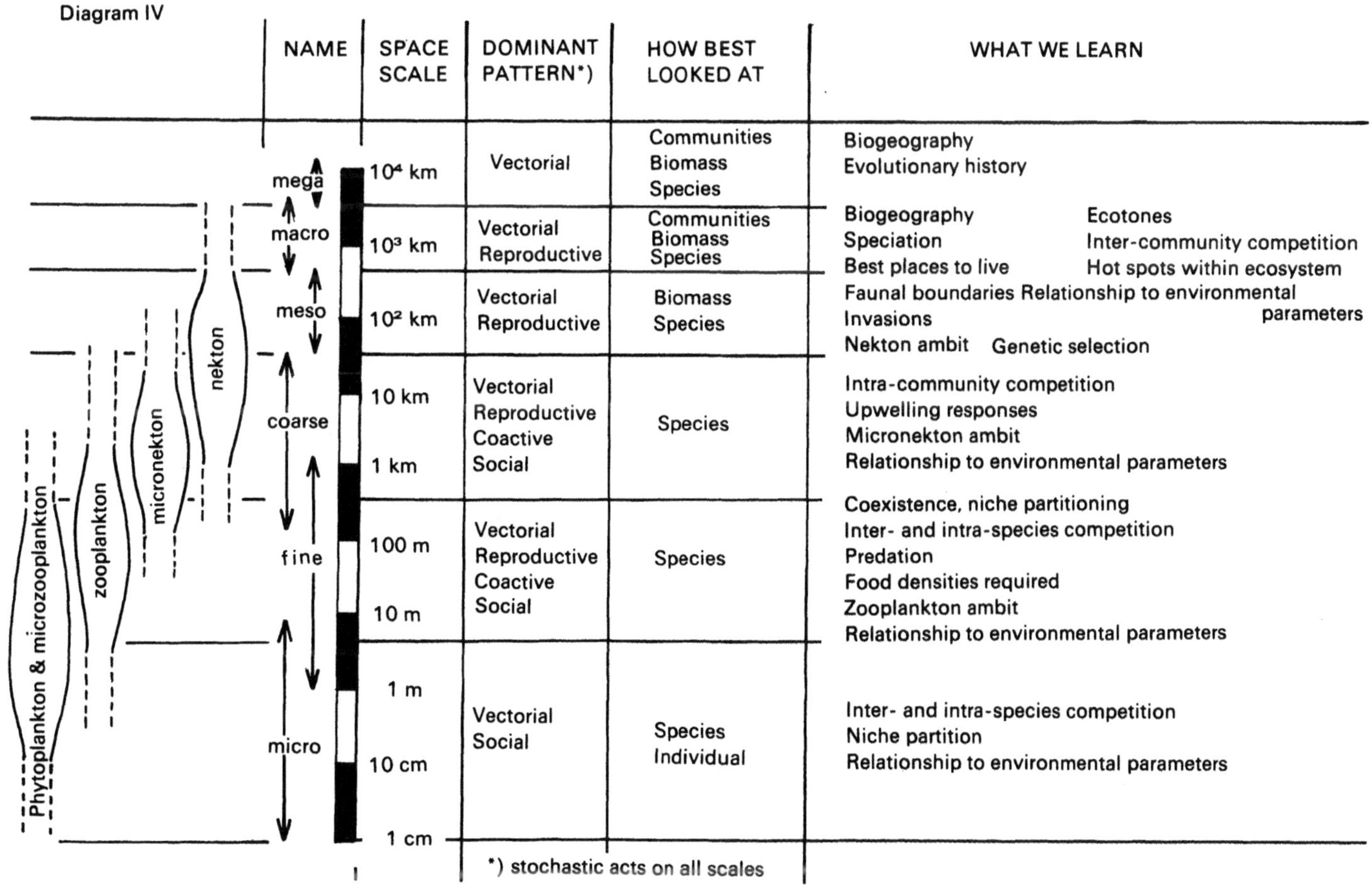

All discussions in the literature concerning coarse- and meso-scale patterns focus on the distribution of single species. The fine-scale patches studied by Haury & Wiebe (1982) usually appear to be multispecies concentrations. In addition, Haury & Wiebe (op. cit.) could not indicate a direct causal relation between distribution and physical conditions. For the fine-scale patterns they consider, as a consequence, the biological interaction of primary importance. Are the meso-scale and coarse-patterns physically determined or not? When upwelling and eddies are the major influence, physical conditions are of course important, but also biological factors seem of great significance on this scale. It is questionable whether in the meso- and coarse-scale patterns single species patches occur. Usually, zoogeographic studies describe single-species patches, due to the fact that the samples are studied in view of one selected group, or of the distribution of one species over larger areas, while ecological and multi-species studies are not available.

The general trends in plankton zoogeography have not yet sufficiently developed to provide a basis for the present study. When ecology and zoogeography become more interwoven and more data become available on vertical and seasonal patterns, the above discussion can probably provide a starting point.

6 Dualistic postscript

This atlas cannot be rounded off whith firm conclusions or definitions. The authors, however, are not disenchanted by this outcome, as the organisms studied and the environment considered are intrinsically dualistic.

Firstly, plankton itself is passively floating and actively moving at the same time. Plankton is defined as the passively floating organisms, transported by currents they are unable to withstand, as their locomotive organs are too weakly developed. Yet, vertical migration is executed by many representatives and this migration is brought forward in this book as a behaviour warranting the geographic stability of populations. Consequently, when explaining phenomena in plankton, one should always consider both the ability, and the inability of a specimen to maintain a stable geographic position.

For the vertical migration itself a second dualism is found. Geographic locality, hydrographical conditions and specific behaviour may lead to completely different results of shift in different populations. In some cases the vertical migration may add to the horizontal stability, in other cases it does just the reverse: stimulate the horizontal displacement of populations. Several types of vertical migration have been distinguished, but are they all that well-defined? Especially ontogenetic and seasonal migrations are contrasting concepts. Only in a few cases was it possible to prove either the seasons or ontogeny to form the basis for the phenomenon, and in many species it is rather by intuition that a name is fixed to the migration found.

As a third dualism the approach to the vertical distribution can be mentioned. A clear subdivision into epi-, meso- and bathypelagic biotopes is applied, as these realms clearly differ in ecological and hydrological respects. Yet, even if one leaves the latitudinal transformation of these layers out of consideration, for the moment, it is hard even then to discriminate between epi- and mesopelagic patterns.

In the ecology of plankton, another series of dualisms is found, as the approach to plankton biotopes, communities and patches is, for example, determined by the acceptance of either stability or instability of the components, as explained above.

When discussing speciation in marine organisms, and in plankton specifically, two conflicting viewpoints have to be considered before reaching a conclusion. One viewpoint is that speciation processes are currently impossible, due to absence of geographical isolation, and that only variation can develop; the other theory is that speciation is possible without geographic isolation, the more so, once variation has developed.

Related to this problem are the contradicting opinions on faunal centres. If speciation is possible and if successive regressions and expansions of ranges do occur, faunal centres can be accepted as realities. If, however, no speciation is found, and if regression and expansion due to climate changes are, in plankton, replaced by chiefly latitudinal displacements of ranges, no faunal centres can be accepted. And besides this dualism, there is even a second one, viz. that of the geographic instability of the geographically fixed faunal centres. The faunal centres are specific geographic areas where speciation took place, or to which, during regression, the populations withdrew; but as the faunal centres, with their specific conditions, can shift with changes in climate and hydrography, their localities are not geographically fixed.

In the characteristics of the marine environment also a great amount of dualism is found. For example, even the sea-water with its gradually changing temperature, salinity and dynamics, should it be considered one continuous body, or a structure comprising hundreds of water masses and current systems, in other words, should one think in terms of gradual changes or of discontinuities? To solve problems in one group, the accent should be laid on the continuities; in another group, the discontinuities in the ocean should be stresses; and in an ecological approach, the separate water masses are more important, while historical zoogeography should rather deal with a continuous ocean in most cases. This also implies, that one and the same phenomenon may act as a barrier to one species, while having no influence on another. Frequently, barriers are formed by currents and in these cases the current prevents further dispersal, but in that same barrier other species may find a vector for transport determining their whole distribution pattern.

Only a small selection of contradictory ideas and phenomena are put forward for this illustration. The correct scientific approach should be the elimination of all contradictions from the theories. Still, this does not seem to be an adequate solution for plankton zoogeography; the future progression of research is only guaranteed when, based on more data than available at present, a theory becomes established which takes into account that nature is too complex for it to be caught in one single reasoning.

A maximum of conformity in the treatment of pelagic geography is desirable, also if theories remain dualistic. The first step in this direction is to introduce conformity in the use of terms and concepts. As explained in the introduction, the interpretation of terms and concepts has frequently not been agreed upon. This atlas tries to demonstrate the inadequacy of some old terms and proposes new terms for distribution patterns.

As pelagic geography is still in its infancy, we hope that some of the terms will prove to stand, and we give the maps as a collection to compare with new patterns to be drawn.

7 Regional index to the distribution maps

SPECIES:	RANGE TYPE:	FIG.:	1 SUBANTARCTIC	2 ANTARCTIC	3 ARCTIC	4 N.W. ATLANTIC	5 N.E. ATLANTIC	6 TROPICAL N. ATLANTIC	7 TROPICAL S. ATLANTIC	8 S.W. ATLANTIC	9 S.E. ATLANTIC	10 S.W. INDIAN	11 S.E. INDIAN	12 TROPICAL N. INDIAN	13 TROPICAL S. INDIAN	14 TROPICAL N. PACIFIC	15 TROPICAL S. PACIFIC	16 TROPICAL E. PACIFIC	17 S.E. PACIFIC	18 S.W. PACIFIC	19 N.W. PACIFIC	20 N.E. PACIFIC	21 RED SEA	22 MEDITERRANEAN
		(these numbers are given in fig. 194)	REGIONS: (for the regions see fig. 194)																					
C. chilensis	CALANUS CHILENSOIDES	79	●	.	.	.	.	.	.	.	.	.	.	.	.	.	.	.	●	.	.	.	.	.
L. r. australis	EUPHAUSIA TRIACANTHOIDES	119	●	.	.	.	.	.	.	●	.	●	●	.	.	.	.	.	.	.	.	.	.	.
D. bulleri	SAGITTA ATLANTICOIDES	140	●	.	.	.	.	.	.	.	.	.	.	.	.	.	.	.	●	●	.	.	.	.
C. brachiatus	CENTROPAGUS BRACHIATOIDES	131	●	.	.	.	.	.	.	●	.	.	.	.	.	.	.	●	●	.	.	.	.	.
P. fusca	THALIA LONGICAUDOIDES	140	●	.	.	.	.	.	.	●	●	●	●	.	.	.	.	.	.	.	.	.	.	.
C. p. antarctica	EUPHAUSIA LUCENSOIDES	57	●	.	.	.	.	.	.	●	●	●	●	.	.	.	.	.	●	●	.	.	.	.
E. lucens	EUPHAUSIA LUCENSOIDES	84	●	.	.	.	.	.	.	●	●	●	●	.	.	.	.	.	●	●	.	.	.	.
L. h. rangi	EUPHAUSIA LUCENSOIDES	82	●	.	.	.	.	.	.	●	●	●	●	.	.	.	.	.	●	●	.	.	.	.
L. peroni	EUPHAUSIA LUCENSOIDES	142	●	.	.	.	.	.	.	●	●	●	●	.	.	.	.	.	●	●	.	.	.	.
P. palpebrata	EUPHAUSIA LUCENSOIDES	139	●	.	.	.	.	.	.	●	●	●	●	.	.	.	.	.	●	●	.	.	.	.
C. australis	CALANUS AUSTRALOIDES	79	●	.	.	.	.	.	.	●	●	●	●	.	.	.	.	.	●	●	.	.	.	.
E. similis	EUPHAUSIA SIMILOIDES	85	●	.	.	.	.	.	.	●	●	●	●	●	●	●	.	.	●	●	●	.	.	.
S. tasmanica	SAGITTA TASMANICOIDES	117	●	.	.	●	●	.	.	●	●	●	●	.	.	.	.	.	●	●	.	.	.	.
B. riisei	-----	61	●	.	.	●	●	●	●	●	●	●	●	.	.	.	●	.	●	●	.	.	.	●
P. sedentaria	PHRONIMA SEDENTARIOIDES	49	●	.	.	●	●	●	●	●	●	●	●	●	●	●	●	●	●	●	●	●	●	●
C. pseudopallida	STYLOCHEIRON MAXIMOIDES	149	●	.	.	●	●	●	●	●	●	●	●	●	●	●	●	.	●	●	●	●	.	.
S. p. zetesios	SAGITTA ZETESIOIDES	71	●	.	.	●	●	●	●	●	●	●	●	●	●	●	●	●	●	●	●	●	.	.
S. maximum	STYLOCHEIRON MAXIMOIDES	72	●	.	.	●	●	●	●	●	●	●	●	●	●	●	●	●	●	●	●	●	.	●
C. p. sulcata	EUPHAUSIA TRIACANTHOIDES	57	●	●	.	.	.	.	.	.	.	.	.	.	.	.	.	.	.	.	.	.	.	.
T. macrura	THYSANOESSA MACRUROIDES	87	●	●	.	.	.	.	.	.	.	.	.	.	.	.	.	.	.	.	.	.	.	.
L. h. antarctica	THYSANOESSA MACRUROIDES	82	●	●	.	.	.	.	.	.	.	.	.	.	.	.	.	.	.	.	.	.	.	.
S. gracilis	EUPHAUSIA TRIACANTHOIDES	146	●	●	.	.	.	.	.	.	.	.	.	.	.	.	.	.	.	.	.	.	.	.
S. gerlachei	SALPA GERLACHEOIDES	51	●	●	.	.	.	.	.	.	.	.	.	.	.	.	.	.	.	.	.	.	.	.
C. l. antarctica	THYSANOESSA MACRUROIDES	134	●	●	.	.	.	.	.	.	.	●	.	.	.	.	.	.	.	.	.	.	.	.
M. noveanglia	-----	141	●	●	.	●	●	.	.	.	.	.	.	.	.	.	.	●	.	.	●	●	.	.
S. australis	SPONGIOBRANCHAEA AUSTRALOIDES	120	●	●	.	.	.	.	.	●	●	●	.	.	.	.	.	.	●	●	.	.	.	.
E. triacantha	EUPHAUSIA TRIACANTHOIDES	86	●	●	.	.	.	.	.	.	●	●	●	.	.	.	.	.	.	.	.	.	.	.
C. brunnea	-----	121	●	●	.	.	.	●	●	●	●	●	●	.	●	.	.	●	.	●	.	.	.	.
S. thompsoni	EUPHAUSIA TRIACANTHOIDES	51	●	●	.	.	.	.	.	●	●	●	●	.	.	.	.	.	●	●	.	.	.	.
C. atraria	CYCLOTHONE ATRARIOIDES	150	●	●	.	●	●	●	●	●	●	●	●	.	.	.	.	.	●	●	.	.	.	.
B. abyssicola	LIMACINA HELICOIDOIDES	154	●	●	.	●	●	●	●	.	●	●	.	●	.	.	.	●	.	.	.	.	.	.
N. cylindrica	NITZSCHIA CYLINDRICOIDES	81	●	●	●	●	●	.	.	.	.	.	.	.	.	.	.	.	.	.	.	.	.	.

			1	2	3	4	5	6	7	8	9	10	11	12	13	14	15	16	17	18	19	20	21	22
G. pachyderma	GLOBOQUADRINA PACHYDERMOIDES	80	●	●	●	●	●	●	●	●	●	●	●	●	●	·	·	·	●	●	●	●	·	●
R. alata	RHIZOSOLENIA ALATOIDES	70	●	●	●	●	●	●	●	●	●	●	●	●	●	●	●	●	●	●	●	●	●	●
E. crystallorophias	EUPHAUSIA CRYSTALLOROPHIOIDES	135	·	●	·	·	·	·	·	·	·	·	·	·	·	·	·	·	·	·	·	·	·	·
G. armata	–––––	61	·	●	·	●	●	●	●	·	●	●	·	·	·	·	·	·	·	·	●	●	·	●
T. sagittatus	PERACLIS BISPINOSOIDES	155	·	·	●	·	●	●	●	·	●	●	·	·	●	·	·	·	·	·	·	·	·	●
L. h. helicina	CALANUS HELGOLANDICOIDES	82	·	·	●	●	●	·	·	·	·	·	·	·	·	·	·	·	·	·	·	·	·	·
G. homari	GAMMARELLUS HOMAROIDES	136	·	·	●	●	●	·	·	·	·	·	·	·	·	·	·	·	·	·	·	·	·	·
C. finmarchicus	CALANUS HELGOLANDICOIDES	78	·	·	●	●	●	·	·	·	·	·	·	·	·	·	·	·	·	·	·	·	·	●
C. helgolandicus	CALANUS HELGOLANDICOIDES	79	·	·	●	●	●	·	·	·	·	·	·	·	·	·	·	·	·	·	·	·	·	●
S. e. elegans	SAGITTA ELEGANSOIDES	124	·	·	●	●	●	·	·	·	·	·	·	·	·	·	·	·	·	·	●	●	·	·
C. glacialis	CALANUS GLACIALOIDES	78	·	·	●	●	●	·	·	·	·	·	·	·	·	·	·	·	·	·	●	●	·	·
O. banksi	STYLOCHEIRON CARINATOIDES	74	·	·	●	●	●	●	●	●	●	●	·	●	●	●	●	●	●	●	●	●	·	●
S. abbreviatum	STYLOCHEIRON CARINATOIDES	64	·	·	·	●	●	●	●	●	●	●	●	●	●	●	●	●	●	●	●	●	●	●
Pterotrachea	PTEROTRACHOIDES	126	·	·	·	●	●	●	●	·	●	●	●	●	●	●	●	●	·	●	●	●	●	●
E. sewelli	EUCALANUS SEWELLOIDES	54	·	·	·	●	●	●	●	●	●	·	·	●	●	●	●	·	●	·	●	●	·	·
S. carinata	STYLOCHEIRON CARINATOIDES	63	·	·	·	●	●	●	●	●	●	●	●	●	●	●	●	●	●	●	●	●	●	·
S. longicorne 'Long form'	STYLOCHEIRON CARINATOIDES	66	·	·	·	●	●	●	●	●	●	●	●	●	●	●	●	●	●	●	●	●	·	●
S. fusiformis	SALPA FUSIFORMOIDES	51	·	·	·	●	●	●	●	●	●	●	●	●	●	●	●	●	●	●	●	●	·	●
C. p. lanceolata	SALPA FUSIFORMOIDES	57	·	·	·	●	●	●	●	●	●	●	●	●	●	●	●	●	●	●	●	●	·	●
C. cuspidata	STYLOCHEIRON CARINATOIDES	48	·	·	·	●	●	●	●	●	●	●	●	●	●	●	●	●	●	●	●	●	·	●
L. lesueuri	STYLOCHEIRON CARINATOIDES	47	·	·	·	●	●	●	●	●	●	●	●	●	●	●	●	●	●	●	●	●	·	·
P. obliquiloculata	PULLENIATINA OBLIQUILOCULATOIDES	76	·	·	·	●	●	●	●	●	●	●	●	●	●	●	●	●	●	●	●	●	·	·
E. subtenuis	EUCALANUS SUBTENUOIDES	52	·	·	·	●	●	●	●	●	●	●	●	●	●	●	●	●	●	●	●	●	·	·
S. p. planctonis	–––––	71	·	·	·	●	●	●	●	●	●	●	●	●	●	●	●	●	●	●	●	●	·	·
C. paupulus	CLAUSOCALANUS PAUPULOIDES	56, 73	·	·	·	●	●	●	●	●	●	●	●	●	●	●	●	●	●	●	●	●	·	·
S. elongatum	CLAUSOCALANUS PAUPULOIDES	65	·	·	·	●	●	●	●	●	●	●	●	●	●	●	●	●	●	●	●	●	·	·
T. obtusifrons	THYSANOPODA OBTUSIFRONSOIDES	93	·	·	·	●	●	●	●	●	●	●	●	·	●	●	●	·	●	●	●	●	·	·
N. atlantica	–––––	108	·	·	·	●	●	●	●	●	●	●	●	·	●	●	●	·	●	●	●	●	·	●
S. suhmi	STYLOCHEIRON SUHMOIDES	90	·	·	·	●	●	●	●	●	●	●	●	·	●	●	●	·	●	●	●	●	·	●
D. danae	STYLOCHEIRON CARINATOIDES	123	·	·	·	●	●	●	●	●	●	●	·	●	●	●	●	●	·	●	·	●	●	●
O. carribaea	–––––	61	·	·	·	●	●	●	●	●	●	●	·	●	●	●	●	●	·	●	●	●	·	●
L. p. pelagica	–––––	153	·	·	·	●	●	●	●	·	●	●	·	●	●	·	·	·	·	·	·	·	·	·
E. gibboides	EUPHAUSIA GIBBOIDOIDES	107	·	·	·	●	●	●	●	●	●	·	·	·	·	·	·	●	·	·	●	●	·	·
L. helicoides	LIMACINA HELICOIDOIDES	152	·	·	·	●	●	●	●	●	●	·	·	·	·	·	·	·	·	●	·	·	·	·
C. arcuicornis	STYLOCHEIRON SUHMOIDES	56	·	·	·	●	●	●	●	●	●	●	●	●	●	·	·	·	●	●	●	●	·	·
E. hyalinus	STYLOCHEIRON SUHMOIDES	58	·	·	·	●	●	●	●	●	●	●	●	·	·	●	·	·	●	●	●	●	·	·
D. t. trispinosa	DIACRIA TRISPINOSOIDES	95	·	·	·	●	●	●	●	●	●	●	●	●	●	●	●	·	·	●	●	●	·	·
P. plumata	PONTELLINA PLUMATOIDES	115	·	·	·	●	●	●	●	●	●	●	●	●	●	●	●	●	●	●	●	●	·	·
G. hirsuta	GLOBOROTALIA HIRSUTOIDES	38	·	·	·	●	●	●	●	●	●	●	●	●	·	·	·	·	●	●	·	·	·	·
A. profundorum	–––––	153	·	·	·	●	●	●	●	·	●	●	●	●	●	●	●	·	·	●	●	·	·	·
P. solitaria	PHRONIMA SOLITARIOIDES	110	·	·	·	●	●	●	●	·	●	●	●	●	●	●	●	·	·	·	●	·	●	·
D. papilio	PHRONIMA SOLITARIOIDES	44, 111	·	·	·	●	●	●	●	·	●	●	●	●	●	●	·	·	·	·	·	·	●	●
P. tornata	–––––	153	·	·	·	●	●	●	●	·	●	●	·	●	●	·	·	·	·	·	·	·	·	·
C. placenta	CARDIAPODA PLACENTOIDES	128	·	·	·	●	●	●	●	·	·	●	·	●	●	●	●	●	·	●	·	·	·	·
P. triacantha	PERACLIS BISPINOSOIDES	60	·	·	·	●	●	●	●	·	●	·	·	·	·	·	·	·	·	·	·	·	·	·
P. apicifulva	PERACLIS APICIFULVOIDES	59	·	·	·	●	●	●	●	·	●	●	·	●	●	●	●	●	·	●	·	●	·	●
P. reticulata	PERACLIS RETICULATOIDES	59	·	·	·	●	●	●	●	·	●	●	·	●	●	●	●	●	·	●	●	●	·	●
P. p. phocoena	SAGITTA FRIDERICOIDES	144	·	·	·	●	●	●	●	·	·	●	·	·	·	·	·	●	●	●	·	·	·	●
P. bispinosa	PERACLIS BISPINOSOIDES	60	·	·	·	●	●	●	●	·	●	·	·	·	·	·	·	·	·	·	·	·	·	●
P. a. atlanticum	PNEUMODERMA ATLANTICOIDES	88	·	·	·	●	●	●	●	·	·	·	·	●	●	●	●	●	●	●	·	·	·	●
S. euphrosyne 'Atlantic form'	PERACLIS BISPINOSOIDES	142	·	·	·	●	●	●	●	·	●	·	·	·	·	·	·	·	·	·	·	·	·	●
S. serratodentata	SAGITTA SERRATODENTATOIDES	118	·	·	·	●	●	●	●	●	●	●	·	·	·	·	·	·	·	·	·	·	·	●
S. spirula	PERACLIS APICIFULVOIDES	155	·	·	·	●	●	●	·	·	·	●	·	●	●	●	●	·	·	●	·	·	·	·

			1	2	3	4	5	6	7	8	9	10	11	12	13	14	15	16	17	18	19	20	21	22
E. americana	EUPHAUSIA AMERICANOIDES	109	·	·	·	●	●	●	●	·	·	·	·	·	·	·	·	·	·	·	·	·	·	·
P. moluccensis	PERACLIS RETICULATOIDES	59	·	·	·	●	●	●	●	·	·	●	·	●	●	●	●	·	·	●	·	·	·	·
C. jobei	CLAUSOCALANUS JOBEOIDES	55	·	·	·	●	●	●	·	·	●	●	●	·	·	●	·	●	●	●	·	●	·	·
S. brevibarbatus	EUPHAUSIA AMERICANOIDES	147	·	·	·	●	●	●	·	·	·	·	·	·	·	·	·	·	·	·	·	·	·	·
E. tenera	EUPHAUSIA TENEROIDES	75	·	·	·	●	·	●	●	·	●	●	·	●	●	●	●	●	·	·	●	●	·	·
S. affinis	EUPHAUSIA PSEUDOGIBBOIDES	146	·	·	·	●	·	●	●	·	·	·	·	●	●	●	●	●	·	●	●	·	●	·
E. pseudogibba	EUPHAUSIA PSEUDOGIBBOIDES	77	·	·	·	●	·	●	●	·	·	·	·	●	·	●	●	●	·	·	●	·	·	·
S. edwardsi	SERGESTES EDWARDSOIDES	101	·	·	·	●	·	●	●	·	·	·	·	·	·	·	·	·	·	·	·	·	·	·
A. a. carolinae	MELICERTES AZTECOIDES	103	·	·	·	●	·	●	·	·	·	·	·	·	·	·	·	·	·	·	·	·	·	·
C. t. bermudensis	CAVOLINIA BERMUDENSOIDES	53	·	·	·	●	·	●	·	·	·	·	·	·	·	·	·	·	·	·	·	·	·	·
E. hemigibba	EUPHAUSIA HEMIGIBBOIDES	97	·	·	·	●	●	·	●	●	●	●	●	·	●	·	·	·	·	·	●	●	·	●
E. brevis	EUPHAUSIA BREVOIDES	89	·	·	·	●	●	·	·	●	●	●	●	●	●	●	●	●	●	●	●	●	·	●
T. gregaria	THYSANOESSA GREGARIOIDES	83	·	·	·	●	●	·	·	●	●	●	●	·	●	·	·	·	●	●	●	●	·	●
D. major	DIACRIA MAJOROIDES	96	·	·	·	●	●	·	·	·	·	●	·	·	●	·	·	·	●	●	●	●	·	·
H. bonelli	GLOBOROTALIA HIRSUTOIDES	50	·	·	·	●	●	·	·	·	●	●	●	·	·	·	·	·	·	·	·	·	·	●
T. raschii	THYSANOESSA RASCHIOIDES	134	·	·	·	●	●	·	·	·	·	·	·	·	·	·	·	·	·	·	●	●	·	·
C. rufobrunnea	-----	121	·	·	·	●	●	·	·	·	·	·	·	·	·	·	·	·	·	·	●	●	·	·
C. pelagicus	-----	39	·	·	·	●	●	·	·	·	·	·	·	·	·	·	·	·	·	·	●	●	·	·
L. r. retroversa	LIMACINA RETROVERSOIDES	119	·	·	·	●	●	·	·	·	·	·	·	·	·	·	·	·	·	·	·	·	·	·
L. r. balea	LIMACINA BALEOIDES	119	·	·	·	●	●	·	·	·	·	·	·	·	·	·	·	·	·	·	·	·	·	·
L. acutus	DIACRIA ATLANTICOIDES	142	·	·	·	●	●	·	·	·	·	·	·	·	·	·	·	·	·	·	·	·	·	·
S. b. ferox	DIACRIA ATLANTICOIDES	146	·	·	·	●	●	·	·	·	·	·	·	·	·	·	·	·	·	·	·	·	·	·
C. p. pyramidata	DIACRIA ATLANTICOIDES	57	·	·	·	●	●	·	·	·	·	·	·	·	·	·	·	·	·	·	·	·	·	·
D. t. atlantica	DIACRIA ATLANTICOIDES	95	·	·	·	●	●	·	·	·	·	·	·	·	·	·	·	·	·	·	·	·	·	●
P. inermis	PONTEGENIA INERMOIDES	137	·	·	·	●	●	·	·	·	·	·	·	·	·	·	·	·	·	·	●	●	·	·
S. tenuis	SAGITTA TENUOIDES	129	·	·	·	●	·	●	●	·	·	·	·	·	·	·	·	●	·	·	·	·	·	·
C. spectabilis	COROLLA SPECTABILOIDES	106	·	·	·	●	·	●	●	·	·	·	·	·	·	·	·	·	·	·	·	●	·	·
M. aztecus	MELICERTES AZTECOIDES	131	·	·	·	●	·	·	·	·	·	·	·	·	·	·	·	·	·	·	·	·	·	·
T. m. latirostris	MELICERTES AZTECOIDES	145	·	·	·	●	·	·	·	·	·	·	·	·	·	·	·	·	·	·	·	·	·	·
T. longicornis	TEMORA LONGICORNOIDES	132	·	·	·	●	·	·	·	·	·	·	·	·	·	·	·	·	·	·	·	·	·	·
P. valdiviae	PERACLIS VALDIVIAOIDES	60	·	·	·	·	●	●	●	●	●	●	●	●	●	●	·	·	·	●	·	·	·	·
L. p. hyperborea	PHORMOSPYRIS STABILOIDES	114	·	·	·	·	●	●	●	●	●	●	·	●	●	●	●	●	●	·	●	●	·	·
S. friderici	SAGITTA FRIDERICOIDES	129	·	·	·	·	●	●	●	·	·	·	·	·	·	·	·	●	·	·	·	·	·	●
S. lampropeltis	EUPHAUSIA HANSENOIDES	147	·	·	·	·	●	●	●	·	·	·	·	·	·	·	·	·	·	·	·	·	·	·
P. depressa	PERACLIS BISPINOSOIDES	60	·	·	·	·	●	●	●	·	●	·	·	·	·	·	·	·	·	·	·	·	·	·
C. alba	-----	121	·	·	·	·	●	●	●	·	·	·	·	·	·	·	·	·	·	·	●	·	·	·
C. t. dakarensis	CAVOLINIA DAKARENSOIDES	53	·	·	·	·	●	●	·	·	·	·	·	·	·	·	·	·	·	·	·	·	·	·
E. eximia	EUPHAUSIA EXIMIOIDES	62	·	·	·	·	●	·	·	·	·	·	·	·	·	●	●	●	●	·	·	●	·	●
S. b. boa	-----	146	·	·	·	·	●	·	·	●	●	●	●	·	·	·	·	·	●	●	·	·	·	●
S. setosa	SAGITTA SETOSOIDES	124	·	·	·	·	●	·	·	·	·	·	·	·	·	·	·	·	·	·	·	·	·	●
F. zetezios	FOWLERINA ZETEZIOIDES	120	·	·	·	·	●	·	·	·	·	·	·	·	·	·	·	·	·	·	·	·	·	●
C. norvegica	-----	121	·	·	·	·	●	·	·	·	·	·	·	·	·	·	·	·	·	·	·	·	·	·
D. rampali	EUPHAUSIA PSEUDOGIBBOIDES	96	·	·	·	·	·	●	●	●	●	·	·	·	●	●	●	·	·	·	·	·	·	·
C. obscura	EUPHAUSIA PSEUDOGIBBOIDES	151	·	·	·	·	·	●	●	●	·	·	·	●	●	●	●	●	·	·	·	·	·	·
A. a. americana	LITOPENAEUS SETIFEROIDES	103	·	·	·	·	·	●	●	●	·	·	·	·	·	·	·	·	·	·	·	·	·	·
W. telescopa	*cf.*: EUPHAUSIA RECURVOIDES	100	·	·	·	·	·	●	·	●	●	●	●	·	●	●	●	·	●	●	●	●	·	·
E. rissoi	ELECTRONA RISSOOIDES	127	·	·	·	·	·	●	●	·	●	●	●	●	●	●	●	●	●	●	·	●	·	·
P. s. scaphipes	PHORMOSPYRIS STABILOIDES	114	·	·	·	·	·	●	●	·	●	●	·	●	●	·	●	●	·	·	●	·	·	·
E. hanseni	EUPHAUSIA HANSENOIDES	120	·	·	·	·	·	●	●	·	●	·	·	·	·	·	·	·	·	·	·	·	·	·
C. chaptali	EUPHAUSIA PSEUDOGIBBOIDES	152	·	·	·	·	·	●	●	·	·	·	·	●	●	●	●	●	·	·	·	·	·	·
T. senegalensis	EUPHAUSIA HANSENOIDES	145	·	·	·	·	·	●	●	·	·	·	·	·	·	·	·	·	·	·	·	·	·	·
P. platychela	SERGESTES EDWARDSOIDES	106	·	·	·	·	·	●	●	·	·	·	·	·	·	·	·	·	·	·	·	·	·	·
L. setiferus	LITOPENAEUS SETIFEROIDES	131	·	·	·	·	·	●	●	●	·	·	·	·	·	·	·	·	·	·	·	·	·	·
T. m. manatus	LITOPENAEUS SETIFEROIDES	145	·	·	·	·	·	●	·	·	·	·	·	·	·	·	·	·	·	·	·	·	·	·
A. marinus	ACETES MARINOIDES	105	·	·	·	·	·	●	·	·	·	·	·	·	·	·	·	·	·	·	·	·	·	·

			1	2	3	4	5	6	7	8	9	10	11	12	13	14	15	16	17	18	19	20	21	22
Hypothetical Central-Water range	-----	92	.	.	.	.	.	.	•	•	•	•	•	.	•	.	•	.	•	•	•	•	.	.
C.t.atlantica	CAVOLINIA ATLANTICOIDES	53	.	.	.	.	.	.	•	•	•	.	.	.	.	.	.	.	.	.	.	.	.	.
S.longicorne 'Short form'	STYLOCHEIRON ROBUSTOIDES	66	.	.	.	.	.	.	.	•	•	•	•	.	•	•	•	•	•	•	•	•	.	.
E.recurva	EUPHAUSIA RECURVOIDES	99	.	.	.	.	.	.	.	•	•	•	•	.	.	.	.	.	•	•	•	•	.	.
C.t.australis	EUPHAUSIA LUCENSOIDES	53	.	.	.	.	.	.	.	•	•	•	•	.	.	.	.	.	•	•	.	.	.	.
E.paucirastra	ELECTRONA PAUCIRASTROIDES	148	.	.	.	.	.	.	.	•	•	•	•	.	.	.	.	.	.	•	.	.	.	.
T.longicauda	THALIA LONGICAUDOIDES	122	.	.	.	.	.	.	.	•	•	•	•	.	.	.	.	.	.	•	.	.	.	.
M.paulensis	MELICERTES PAULENSOIDES	132	.	.	.	.	.	.	.	•	.	.	.	.	.	.	.	.	.	.	.	.	.	.
S.robustum	STYLOCHEIRON ROBUSTOIDES	94	.	.	.	.	.	.	.	.	•	•	•	.	•	•	•	.	•	•	•	•	.	.
E.attenuatus	CLAUSOCALANUS FARRANOIDES	54	.	.	.	.	.	.	.	.	•	•	.	•	•	•	•	•	.	.	•	•	.	.
E.langae	EUCALANUS LANGAOIDES	54	.	.	.	.	.	.	.	.	•	•	•	.	.	.	.	.	•	•	.	.	.	.
S.orientalis	CAVOLINIA GLOBULOSOIDES	101	.	.	.	.	.	.	.	.	•	•	.	•	•	•	•	.	.	.	•	•	.	.
S.pacifica	SAGITTA PACIFICOIDES	117	.	.	.	.	.	.	.	.	.	•	•	•	•	•	•	•	•	•	•	•	•	.
C.farrani	CLAUSOCALANUS FARRANOIDES	56	.	.	.	.	.	.	.	.	.	•	•	•	•	•	•	•	•	•	•	•	.	.
S.affine 'Central form'	STYLOCHEIRON AFFINOIDES	91	.	.	.	.	.	.	.	.	.	•	•	.	•	•	•	.	•	•	•	•	.	.
D.gardineri	DESMOPTERUS GARDINEROIDES	111	.	.	.	.	.	.	.	.	.	•	•	•	•	.	.	.	.	.	.	.	.	.
P.morii	CAVOLINIA GLOBULOSOIDES	98	.	.	.	.	.	.	.	.	.	•	.	•	•	•	•	•	.	•	•	.	.	.
S.bedoti	SAGITTA BEDOTOIDES	129	.	.	.	.	.	.	.	.	.	•	•	•	•	•	•	•	.	•	•	.	.	.
C.convexa	CLIO CONVEXOIDES	57	.	.	.	.	.	.	.	.	.	•	.	•	•	•	•	•	.	•	.	.	•	.
S.microphthalma	CAVOLINIA GLOBULOSOIDES	65	.	.	.	.	.	.	.	.	.	•	.	•	•	•	•	•	.	.	•	.	.	.
S.onalaniensis	CAVOLINIA GLOBULOSOIDES	50	.	.	.	.	.	.	.	.	.	•	.	•	•	•	•	•	.	.	•	.	.	.
D.dugong	ACETES JAPONICOIDES	145	.	.	.	.	.	.	.	.	.	•	.	•	•	•	•	.	.	.	•	.	•	.
S.affine 'W.Equatorial form'	HYDROMYLES GLOBULOSOIDES	67	.	.	.	.	.	.	.	.	.	•	.	•	•	•	•	•	.	.	•	.	.	.
D.quadridentata	HYDROMYLES GLOBULOSOIDES	123	.	.	.	.	.	.	.	.	.	•	.	.	•	•	•	•	.	.	.	.	.	.
M.spinifer	-----	116	.	.	.	.	.	.	.	.	.	•	.	•	•	•	•	•	.	.	•	.	.	.
C.globulosa	CAVOLINIA GLOBULOSOIDES	98	.	.	.	.	.	.	.	.	.	•	.	•	•	•	•	•	.	•	•	•	.	.
L.penicillifer	LUCIFER HANSENOIDES	131	.	.	.	.	.	.	.	.	.	•	.	•	.	•	•	.	.	.	•	.	.	.
A.natalensis	ACETES NATALENSOIDES	105	.	.	.	.	.	.	.	.	.	•	.	.	.	.	.	.	.	.	.	.	.	.
C.minor	CLAUSOCALANUS MINOROIDES	55	.	.	.	.	.	.	.	.	.	.	.	•	•	•	•	•	.	.	•	.	.	.
H.globulosa	HYDROMYLES GLOBULOSOIDES	112	.	.	.	.	.	.	.	.	.	.	.	•	•	•	•	•	.	•	•	•	.	.
C.cristata	CARINARIA CRISTATOIDES	113	.	.	.	.	.	.	.	.	.	.	.	•	•	•	•	•	.	•	•	•	.	.
L.hanseni	LUCIFER HANSENOIDES	130	.	.	.	.	.	.	.	.	.	.	.	•	•	•	.	.	.	•	.	.	•	.
S.nebulosus	CAVOLINIA GLOBULOSOIDES	147	.	.	.	.	.	.	.	.	.	.	.	•	•	•	•	.	.	.	•	.	.	.
S.affine 'Indo-Aust. form'	cf.: ACETES JAPONICOIDES	67	.	.	.	.	.	.	.	.	.	.	.	•	•	•	•	.	.	.	•	.	.	.
D.costata	HYDROMYLES GLOBULOSOIDES	123	.	.	.	.	.	.	.	.	.	.	.	•	•	•	•	.	.	.	.	.	.	.
A.s.sibogae	-----	104	.	.	.	.	.	.	.	.	.	.	.	•	.	•	•	.	.	.	.	.	.	.
S.insulare	THALIA SIBOGAOIDES	63	.	.	.	.	.	.	.	.	.	.	.	•	•	•	•	.	.	.	.	.	.	.
A.indicus	ACETES JAPONICOIDES	103	.	.	.	.	.	.	.	.	.	.	.	•	.	•	.	.	.	.	.	.	.	.
A.erythraeus	LUCIFER HANSENOIDES	105	.	.	.	.	.	.	.	.	.	.	.	•	•	•	•	.	.	.	.	.	.	.
S.euphrosyne 'Tropical form'	HYDROMYLES GLOBULOSOIDES	142	.	.	.	.	.	.	.	.	.	.	.	•	•	•	.	.	.	.	.	.	.	.
A.japonicus	ACETES JAPONICOIDES	103	.	.	.	.	.	.	.	.	.	.	.	•	.	•	.	.	.	.	•	.	.	.
E.elongatus	EUCALANUS ELONGATOIDES	58	.	.	.	.	.	.	.	.	.	.	.	•	•	•	•	.	.	.	•	.	.	.
S.indicus	-----	66	.	.	.	.	.	.	.	.	.	.	.	•	.	•	•	.	.	.	.	.	.	.
D.erythra	DIACRIA ERYTHROIDES	123	.	.	.	.	.	.	.	.	.	.	.	•	•	.	.	.	.	.	.	.	•	.
C.t.danae	DESMOPTERUS GARDINEROIDES	53	.	.	.	.	.	.	.	.	.	.	.	•	•	.	.	.	.	.	.	.	.	.
A.s.sibogalis	ACETES JOHNOIDES	104	.	.	.	.	.	.	.	.	.	.	.	•	.	•	•	.	.	.	.	.	.	.
A.johni	ACETES JOHNOIDES	105	.	.	.	.	.	.	.	.	.	.	.	•	.	.	.	.	.	.	.	.	.	.
S.semissis	STYLOCHEIRON LONGICORNOIDES	102	.	.	.	.	.	.	.	.	.	.	.	•	.	.	.	.	.	.	.	.	.	.
S.longicorne 'N.Indian form'	STYLOCHEIRON LONGICORNOIDES	66	.	.	.	.	.	.	.	.	.	.	.	•	.	.	.	.	.	.	.	.	.	.
A.vulgaris	cf.: ACETES JAPONICOIDES	105	.	.	.	.	.	.	.	.	.	.	.	.	•	•	•	.	.	.	.	.	.	.
A.serrulatus	ACETES SERRULATOIDES	104	.	.	.	.	.	.	.	.	.	.	.	.	•	.	•	.	.	.	.	.	.	.

			1	2	3	4	5	6	7	8	9	10	11	12	13	14	15	16	17	18	19	20	21	22
D.nigripes	CALANUS PACIFICOIDES	139	·	·	·	·	·	·	·	·	·	·	·	·	·	●	·	·	·	·	●	●	·	·
D.immutabilis	SAGITTA PSEUDOSERRATODENTATOIDES	140	·	·	·	·	·	·	·	·	·	·	·	·	·	●	·	·	·	·	●	●	·	·
S.pseudoserratodentata	SAGITTA PSEUDOSERRATODENTATOIDES	118	·	·	·	·	·	·	·	·	·	·	·	·	·	●	·	·	·	·	●	●	·	·
C.t.teschi	CAVOLINIA TESCHOIDES	53	·	·	·	·	·	·	·	·	·	·	·	·	·	●	●	●	·	·	●	·	·	·
S.gibbilobatus	SERGESTES GIBBILOBATOIDES	102	·	·	·	·	·	·	·	·	·	·	·	·	·	●	●	●	·	·	·	·	·	·
S.tantillus	SERGESTES TANTILLOIDES	101	·	·	·	·	·	·	·	·	·	·	·	·	·	●	●	●	·	·	·	·	·	·
S.euphrosyne 'Pacific form'	SERGESTES TANTILLOIDES	142	·	·	·	·	·	·	·	·	·	·	·	·	·	●	·	●	●	·	·	·	·	·
S.microps 'White belly form'	SERGESTES BREVISPINATOIDES	143	·	·	·	·	·	·	·	·	·	·	·	·	·	●	·	●	·	·	·	·	·	·
N.lobata	-----	108	·	·	·	·	·	·	·	·	·	·	·	·	·	●	·	·	·	·	·	·	·	·
D.s.occidentalis	ACETES JAPONICOIDES	123	·	·	·	·	·	·	·	·	·	·	·	·	·	●	·	·	·	·	·	·	·	·
A.intermedius	ACETES JAPONICOIDES	105	·	·	·	·	·	·	·	·	·	·	·	·	·	●	·	·	·	·	·	·	·	·
T.sibogae	THALIA SIBOGAOIDES	122	·	·	·	·	·	·	·	·	·	·	·	·	·	●	●	·	·	·	·	·	·	·
P.meseres	POEOBIUS MESEROIDES	125	·	·	·	·	·	·	·	·	·	·	·	·	·	·	·	●	·	·	●	●	·	·
S.columbrinus	CAVOLINIA AFFINOIDES	147	·	·	·	·	·	·	·	·	·	·	·	·	·	·	·	●	●	·	·	·	·	·
C.t.affinis	CAVOLINIA AFFINOIDES	53	·	·	·	·	·	·	·	·	·	·	·	·	·	·	·	●	●	·	·	·	·	·
T.discaudata	TEMORA DISCAUDATOIDES	131	·	·	·	·	·	·	·	·	·	·	·	·	·	·	·	●	·	·	·	·	·	·
S.affine 'E.equatorial form'	EUCALANUS INERMOIDES	67	·	·	·	·	·	·	·	·	·	·	·	·	·	·	·	●	·	·	·	·	·	·
S.atriventer	SAGITTA BIERIOIDES	146	·	·	·	·	·	·	·	·	·	·	·	·	·	·	·	●	·	·	·	·	·	·
S.microps 'Eastern form'	LABIDOCERA LUBBOCKOIDES	143	·	·	·	·	·	·	·	·	·	·	·	·	·	·	·	●	·	·	·	·	·	·
S.geminus	SERGESTES BREVISPINATOIDES	102	·	·	·	·	·	·	·	·	·	·	·	·	·	·	·	●	·	·	·	·	·	·
S.brevispinatus	SERGESTES BREVISPINATOIDES	101	·	·	·	·	·	·	·	·	·	·	·	·	·	·	·	●	·	·	·	·	·	·
S.bierii	SAGITTA BIERIOIDES	117	·	·	·	·	·	·	·	·	·	·	·	·	·	·	·	●	·	·	·	·	·	·
P.perla	-----	153	·	·	·	·	·	·	·	·	·	·	·	·	·	·	·	●	·	·	·	·	·	·
P.sobrina	EUCALANUS INERMOIDES	106	·	·	·	·	·	·	·	·	·	·	·	·	·	·	·	●	·	·	·	·	·	·
L.p.dalli	-----	153	·	·	·	·	·	·	·	·	·	·	·	·	·	·	·	●	·	·	·	·	·	·
L.lubbocki	LABIDOCERA LUBBOCKOIDES	132	·	·	·	·	·	·	·	·	·	·	·	·	·	·	·	●	·	·	·	·	·	·
E.inermis	EUCALANUS INERMOIDES	58	·	·	·	·	·	·	·	·	·	·	·	·	·	·	·	●	·	·	·	·	·	·
L.diandra	-----	132	·	·	·	·	·	·	·	·	·	·	·	·	·	·	·	●	·	·	·	·	·	·
A.californica	-----	153	·	·	·	·	·	·	·	·	·	·	·	·	·	·	·	●	·	·	·	·	·	·
A.binghami	LABIDOCERA LUBBOCKOIDES	103	·	·	·	·	·	·	·	·	·	·	·	·	·	·	·	●	·	·	·	·	·	·
D.s.schmidti	EUCALANUS INERMOIDES	123	·	·	·	·	·	·	·	·	·	·	·	·	·	·	·	●	·	·	·	·	·	·
D.irrorata	SERGESTES BREVISPINATOIDES	140	·	·	·	·	·	·	·	·	·	·	·	·	·	·	·	●	·	·	·	·	·	·
Symbolopharus spec.	-----	148	·	·	·	·	·	·	·	·	·	·	·	·	·	·	·	·	●	·	·	·	·	·
E.gibba	EUPHAUSIA GIBBOIDES	97	·	·	·	·	·	·	·	·	·	·	·	·	·	·	·	·	●	●	·	·	·	·
S.s.atlantica	SAGITTA ATLANTICOIDES	118	·	·	·	·	·	·	·	·	·	·	·	·	·	·	·	·	●	●	·	·	·	·
A.s.australis	ACETES AUSTRALOIDES	104	·	·	·	·	·	·	·	·	·	·	·	·	·	·	·	·	·	●	·	·	·	·
D.albatrus	CALANUS PACIFICOIDES	139	·	·	·	·	·	·	·	·	·	·	·	·	·	·	·	·	·	·	●	●	·	·
C.microdon	CALANUS PACIFICOIDES	150	·	·	·	·	·	·	·	·	·	·	·	·	·	·	·	·	·	·	●	●	·	·
O.bulbosus	LIMACINA ACUTOIDES	151	·	·	·	·	·	·	·	·	·	·	·	·	·	·	·	·	·	·	●	●	·	·
C.plumchrus 'Northern form'	CALANUS PLUMCHROIDES pars	136	·	·	·	·	·	·	·	·	·	·	·	·	·	·	·	·	·	·	●	●	·	·
C.pacificus	CALANUS PACIFICOIDES	79	·	·	·	·	·	·	·	·	·	·	·	·	·	·	·	·	·	·	●	●	·	·
E.parki	EUCALANUS PARKOIDES	54	·	·	·	·	·	·	·	·	·	·	·	·	·	·	·	·	·	·	●	●	·	·
I.antrostomia	IDIACANTHUS ANTROSTOMOIDES	133	·	·	·	·	·	·	·	·	·	·	·	·	·	·	·	·	·	·	●	●	·	·
L.obliquidens	CALANUS PACIFICOIDES	142	·	·	·	·	·	·	·	·	·	·	·	·	·	·	·	·	·	·	●	●	·	·
L.h.acuta	LIMACINA ACUTOIDES	82	·	·	·	·	·	·	·	·	·	·	·	·	·	·	·	·	·	·	●	●	·	·
L.h.pacifica	LIMACINA PACIFICOIDES	82	·	·	·	·	·	·	·	·	·	·	·	·	·	·	·	·	·	·	●	●	·	·
M.robusta	CALANUS PACIFICOIDES	155	·	·	·	·	·	·	·	·	·	·	·	·	·	·	·	·	·	·	●	●	·	·
P.a.pacificum	EUCALANUS PARKOIDES	88	·	·	·	·	·	·	·	·	·	·	·	·	·	·	·	·	·	·	●	●	·	·
E.b.californicus	EUCALANUS PARKOIDES	58	·	·	·	·	·	·	·	·	·	·	·	·	·	·	·	·	·	·	●	●	·	·

			1	2	3	4	5	6	7	8	9	10	11	12	13	14	15	16	17	18	19	20	21	22
E.b.bungi	EUCALANUS BUNGOIDES	58	.	.	.	.	.	.	.	.	.	.	.	.	.	.	.	.	.	.	●	●	.	.
S.consobrinus 'Central form'	SERGESTES CONSOBRINOIDES	102	.	.	.	.	.	.	.	.	.	.	.	.	.	.	.	.	.	.	●	●	.	.
S.microps 'Hawaiian form'	STENELLA MICROPSOIDES	143	.	.	.	.	.	.	.	.	.	.	.	.	.	.	.	.	.	.	●	●	.	.
C.plumchrus 'Western form'	CALANUS PLUMCHROIDES pars	136	.	.	.	.	.	.	.	.	.	.	.	.	.	.	.	.	.	.	●	.	.	.
C.sinicus	-----	79	.	.	.	.	.	.	.	.	.	.	.	.	.	.	.	.	.	.	●	.	.	.
R.gigas	CALANUS PLUMCHROIDES	145	.	.	.	.	.	.	.	.	.	.	.	.	.	.	.	.	.	.	●	.	.	.
P.p.vomerina	PHOCOENA VOMERINOIDES	144	.	.	.	.	.	.	.	.	.	.	.	.	.	.	.	.	.	.	●	.	.	.
A.chinensis	ACETES CHINENSOIDES	103	.	.	.	.	.	.	.	.	.	.	.	.	.	.	.	.	.	.	●	.	.	.
D.pacificus	DESMOPTERUS PACIFICOIDES	111	.	.	.	.	.	.	.	.	.	.	.	.	.	.	.	.	.	.	.	●	.	.
O.acanthias	DESMOPTERUS PACIFICOIDES	151	.	.	.	.	.	.	.	.	.	.	.	.	.	.	.	.	.	.	.	●	.	.
C.pedunculata	-----	121	.	.	.	.	.	.	.	.	.	.	.	.	.	.	.	.	.	.	.	●	.	.
C.t.occidentalis	DESMOPTERUS PACIFICOIDES	53	.	.	.	.	.	.	.	.	.	.	.	.	.	.	.	.	.	.	.	●	.	.
C.plumchrus 'Eastern form'	DESMOPTERUS PACIFICOIDES	136	.	.	.	.	.	.	.	.	.	.	.	.	.	.	.	.	.	.	.	●	.	.
C.marshallae	CALANUS MARSHALLOIDES	78	.	.	.	.	.	.	.	.	.	.	.	.	.	.	.	.	.	.	.	●	.	.
L.jollae	LABIDOCERA JOLLAOIDES	132	.	.	.	.	.	.	.	.	.	.	.	.	.	.	.	.	.	.	.	●	.	.
L.kolpos	LABIDOCERA KOLPOSOIDES	---	.	.	.	.	.	.	.	.	.	.	.	.	.	.	.	.	.	.	.	●	.	.
S.consobrinus 'Calif. C. form'	DESMOPTERUS PACIFICOIDES	102	.	.	.	.	.	.	.	.	.	.	.	.	.	.	.	.	.	.	.	●	.	.
S.affine 'Calif. C. form'	DESMOPTERUS PACIFICOIDES	67	.	.	.	.	.	.	.	.	.	.	.	.	.	.	.	.	.	.	.	●	.	.
C.t.tridentata	CAVOLINIA TRIDENTATOIDES	53	.	.	.	.	.	.	.	.	.	.	.	.	.	.	.	.	.	.	.	.	.	●
S.euximia	SAGITTA EUXIMIOIDES	124	.	.	.	.	.	.	.	.	.	.	.	.	.	.	.	.	.	.	.	.	.	.
P.p.relicta	SAGITTA EUXIMIOIDES	144	BLACK SEA					.	.	.	.	.	.	.	.	.	.	.	.	.	.	.	.	.
S.e.balthica	SAGITTA BALTHICOIDES	124	BALTHIC SEA						.	.	.	.	.	.	.	.	.	.	.	.	.	.	.	.
A.paraguayensis	-----	105	-----fresh-water species-----																					
T.inunguis	-----	145	-----fresh-water species-----																					

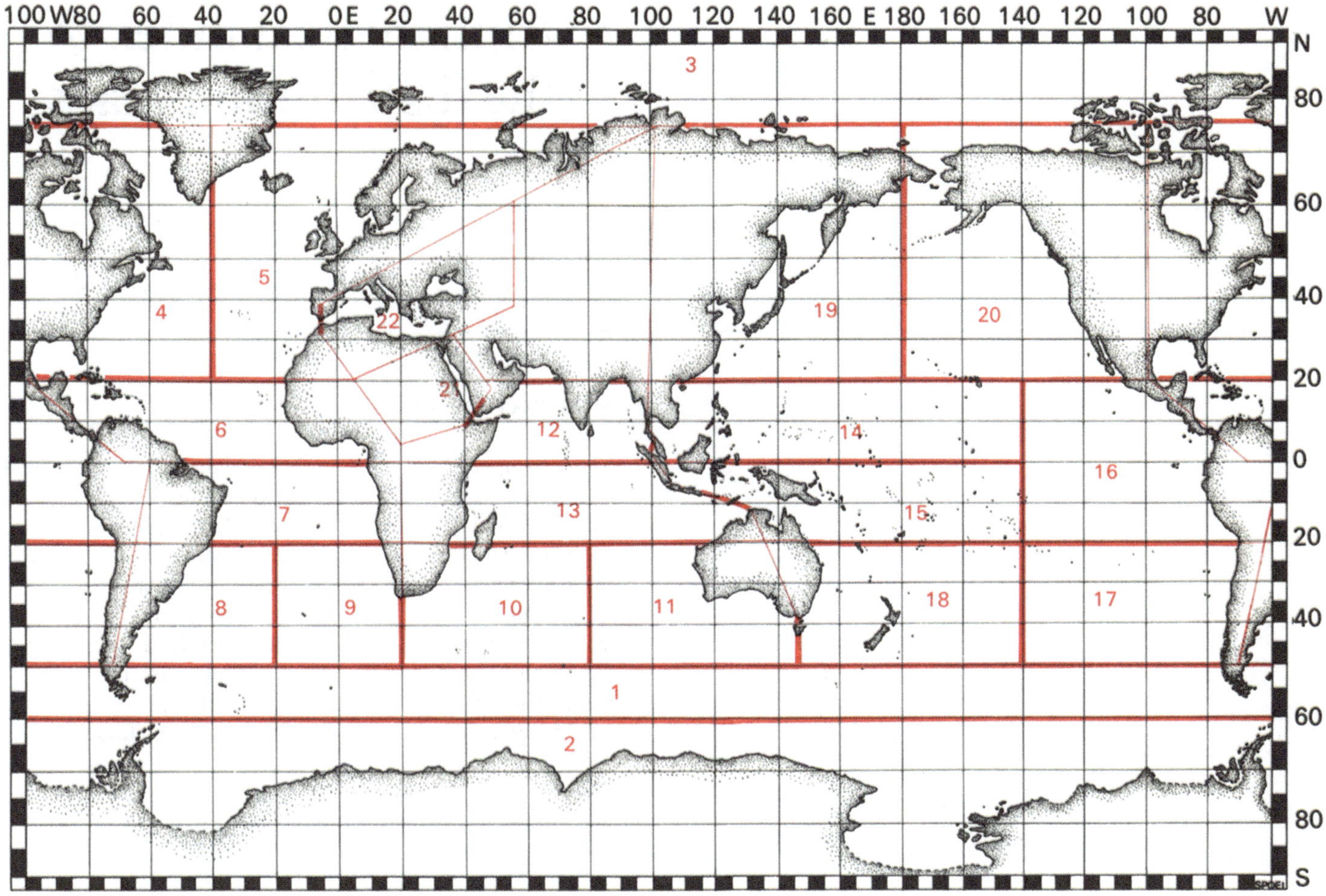

FIG. 194 The areal division of the oceans into the 22 regions used to compile the Regional Index to the Distribution Maps.

8 Alphabetical index to the distribution maps

ABBREVIATIONS USED:

AM	Amphipoda	MA	Mammalia
AV	Aves	MO	Mollusca
CE	Cephalopoda	MS	Mastigophorea
CH	Chaetognatha	PI	Pisces
CO	Copepoda	PO	Polychaeta
DE	Decapoda	PT	Pteropoda
DI	Diatomacea	RA	Radiolaria
EU	Euphausiacea	SI	Siphonophora
FO	Foraminifera	TH	Thaliacea
HE	Heteropoda	TR	Trachymedusae

Eucalanus attenuatus (Dana, 1849)	CO	Fleminger, 1973	54	42
Eucalanus bungi bungi Giesbrecht, 1892	CO	Fleminger & Hulsemann, 1973	58	94
Eucalanus bungi californicus Johnson, 1938	CO	Fleminger & Hulsemann, 1973	58	94
Eucalanus elongatus (Dana, 1849)	CO	Fleminger & Hulsemann, 1973	58	86
Eucalanus hyalinus (Claus, 1866)	CO	Fleminger & Hulsemann, 1973	58	46
Eucalanus inermis Giesbrecht, 1892	CO	Fleminger & Hulsemann, 1973	58	96
Eucalanus langae Fleminger, 1973	CO	Fleminger, 1973	54	94
Eucalanus parki Fleminger, 1973	CO	Fleminger, 1973	54	94
Eucalanus sewelli Fleminger, 1973	CO	Fleminger, 1973	54	95
Eucalanus subtenuis Giesbrecht, 1888	CO	Fleminger, 1973	52	40, 60
Eukrohnia hamata (Möbius, 1875)	CH	-----	171	139
Euphausia americana Hansen, 1911	EU	Mauchline & Fisher, 1969	109	85
Euphausia brevis Hansen, 1905	EU	Brinton, 1975	89	70
Euphausia crystallorophias Holt & Tattersall, 1906	EU	Johnes, 1936	135	108
Euphausia eximia Hansen, 1911	EU	Mauchline & Fisher, 1969	62	48
Euphausia gibba G.O. Sars, 1883	EU	Brinton, 1975	97	76
Euphausia gibboides Ortmann, 1893	EU	Brinton, 1975 (partly)	107	82
Euphausia hanseni Zimmer, 1915	EU	Mauchline & Fisher, 1969	120	96
Euphausia hemigibba Hansen, 1910	EU	Brinton, 1975	97	76
Euphausia lucens Hansen, 1905	EU	Mauchline & Fisher, 1969	84	66
Euphausia pseudogibba Ortmann, 1893	EU	Brinton, 1975	77	62
Euphausia recurva Hansen, 1905	EU	Brinton, 1975	99	78
Euphausia similis G.O. Sars, 1883	EU	Brinton, 1975	85	66
Euphausia tenera Hansen, 1905	EU	Brinton, 1975	75	60, 84
Euphausia triacantha Holt & Tattersall, 1906	EU	Mauchline & Fisher, 1969	86	68
Expatriation in diatoms	DI	Semina, 1979	37	30
Faunal barriers and boundaries	--	-----	68	54
Faunal centres	--	Van der Spoel, 1976	16	16
Faunal zones	--	Beklemishev, 1971	4	6
Fertility	--	-----	22	20
Foraminiferan zonation	FO	Bé & Tolderlund, 1971	8	8
Fowlerina zetezios Pelseneer, 1906	PT	Van der Spoel, 1976	120	96
Galiteuthis armata Joubin, 1898	CE	Clarke, 1966	61	48
Gammarellus homari (Fabricius, 1779)	AM	Dunbar, 1954	136	110
Gennadas elegans Smith, 1886	DE	Moore, 1958	160	130
Globorotalia hirsuta (D'Orbigny, 1839)	FO	Bé, 1977; Bé & Tolderlund, 1971	38	30, 85
Globoquadrina pachyderma (Ehrenberg, 1861)	FO	Bé, 1977	80	64
Histioteuthis bonelli Férussac, 1835	CE	Clarke, 1966	50	38
Hydromyles globulosa (Rang, 1825)	PT	Van der Spoel, 1976	112	88
Hypothetical range of Central-Water taxa	--	-----	92	72
Ichtyoplankton zones	PI	Beklemishev et al., 1977	11	12
Idiacanthus antrostomia Gilbert, 1890	PI	Beklemishev et al., 1977	133	108
Interocean connections	--	-----	---	78
Isotherms of the sea-surface	--	-----	18	18
Labidocera diandra Fleminger, 1967	CO	Fleminger, 1975	132	106
Labidocera jollae Esterly, 1906	CO	Fleminger, 1975	132	106
Labidocera kolpos Fleminger, 1967	CO	Fleminger, 1975	---	106
Labidocera lubbocki	CO	Fleminger, 1975	132	106
Laginorhynchus acutus (Gray, 1828)	MA	Mörzer Bruyns, 1971	142	115
Laginorhynchus obliquidens Gill, 1865	MA	Mörzer Bruyns, 1971	142	115
Latitudinal physical zonation of oceans	--	-----	2	4
Lensia campanella (Moser, 1925)	SI	Musaeva, 1976	162	133
Lensia subtiloides (Lens & Van Riemsdijk, 1908)	SI	Musaeva, 1976	162	132
Light penetration	--	Moore, 1958	157	128
Limacina helicina forma *acuta* Van der Spoel, 1967	PT	Van der Spoel, 1967	82	64, 92
Limacina helicina forma *antarctica* Woodward, 1854	PT	Van der Spoel, 1967	82	64
Limacina helicina forma *helicina* (Phipps, 1774)	PT	Van der Spoel, 1967	82	64
Limacina helicina forma *pacifica* Dall, 1871	PT	Van der Spoel, 1967	82	64, 92
Limacina helicina forma *rangi* (D'Orbigny, 1836)	PT	Van der Spoel, 1967	82	64
Limacina helicoides Jeffreys, 1877	PT	Van der Spoel, 1967	152	122
Limacina lesueuri (D'Orbigny, 1836)	PT	Bé & Gimer, 1977	47	36
Limacina retroversa forma *australis* (Eydoux & Souleyet, 1840)	PT	Van der Spoel, 1967	119	94

Rhizosolenia alata Brightwell	DI	Beklemishev et al., 1977	70	57
Sagitta bedoti Beraneck, 1895	CH	Tokioka, 1974	129	102
Sagitta bierii Alvariño, 1961	CH	Pierrot-Bults, 1974	117	96
Sagitta elegans balthica (Ritter-Zahony, 1911)	CH	-----	124	100
Sagitta elegans elegans Verrill, 1873	CH	-----	124	66
Sagitta euxima Moltschanoff, 1909	CH	-----	124	99
Sagitta friderici Ritter-Zahony, 1911	CH	Tokioka, 1979	129	102
Sagitta pacifica Tokioka, 1940	CH	Pierrot-Bults, 1974	117	92
Sagitta planctonis forma *planctonis* Steinhaus, 1896	CH	Pierrot-Bults, 1975	172	58, 140
Sagitta planctonis forma *zetesios* Fowler, 1905	CH	Pierrot-Bults, 1975	71, 172	58, 140
Sagitta pseudoserratodentata Tokioka, 1939	CH	Pierrot-Bults, 1974	118	92
Sagitta serratodentata atlantica Thomson, 1947	CH	Pierrot-Bults, 1974	118	98
Sagitta serratodentata serratodentata Krohn, 1853	CH	Pierrot-Bults, 1974	118	85
Sagitta setosa Müller, 1847	CH	-----	124	108
Sagitta tasmanica Thomson, 1947	CH	Pierrot-Bults, 1974	117	92
Sagitta tenuis Conant, 1896	CH	Tokioka, 1979	129	102
Salinity profile of the Atlantic Ocean	--	Deacon, 1962	28	24
Salinity profile of the Indian Ocean	--	-----	31	27
Salinity profile of the Pacific Ocean	--	-----	34	29
Salinity of the sea-surface	--	Sverdrup et al., 1942	21	20
Salpa fusiformis Cuvier, 1804	TH	Van Soest, 1974	51	40, 57
Salpa gerlachei Foxton, 1961	TH	Van Soest, 1974	51	98
Salpa thompsoni Foxton, 1961	TH	Van Soest, 1974	51	41
Sergestes brevispinatus Judkins, 1978	DE	Judkins, 1978	101	78, 96
Sergestes consobrinus Milne, 1968 'California Curr. form'	DE	Judkins, 1978	102	80, 97
Sergestes consobrinus Milne, 1968 'Central form'	DE	Judkins, 1978	102	80, 97
Sergestes edwardsi Krøyer, 1855	DE	Judkins, 1978	101	78, 86
Sergestes geminus Judkins, 1978	DE	Judkins, 1978	102	78
Sergestes gibbilobatus Judkins, 1978	DE	Judkins, 1978	102	78, 97
Sergestes orientalis Hansen, 1919	DE	Judkins, 1978	101	78, 116
Sergestes semissis Burkenroad, 1940	DE	Judkins, 1978	102	52, 80
Sergestes tantillus Burkenroad, 1940	DE	Judkins, 1978	101	78, 96
Spirula spirula (Linnaeus, 1758)	CE	Clarke, 1966	155	124
Spongiobranchaea australis D'Orbigny, 1836	PT	Van der Spoel, 1976	120	95
Stenella euphrosyne Flower, 1885 'Atlantic form'	MA	Mörzer Bruyns, 1971	142	114
Stenella euphrosyne Flower, 1885 'Pacific form'	MA	Mörzer Bruyns, 1971	142	114
Stenella euphrosyne Flower, 1885 'Tropical form'	MA	Mörzer Bruyns, 1971	142	114
Stenella microps (Gray, 1846) 'Eastern form'	MA	Mörzer Bruyns, 1971; David et al., 1979	143	116
Stenella microps (Gray, 1846) 'Hawaiian form'	MA	Mörzer Bruyns, 1971; David et al., 1979	143	116
Stenella microps (Gray, 1846) 'White Belly form'	MA	Mörzer Bruyns, 1971; David et al., 1979	143	116
Stomias affinis Günther, 1887	PI	Gibbs, 1969	146	118
Stomias atriventer Garman, 1899	PI	Gibbs, 1969	146	118
Stomias boa boa (Risso, 1810)	PI	Gibbs, 1969	146	118
Stomias boa ferox Reinhardt, 1843	PI	Gibbs, 1969	146	118
Stomias brevibarbatus Ege, 1933	PI	Gibbs, 1969	147	118
Stomias columbrinus Garman, 1899	PI	Gibbs, 1969	147	118
Stomias gracilis Garman, 1899	PI	Gibbs, 1969	146	118
Stomias lampropeltis Gibbs, 1969	PI	Gibbs, 1969	147	118
Stomias nebulosus Alcock, 1889	PI	Gibbs, 1969	147	118
Stylocheiron abbreviatum G.O. Sars, 1883	EU	Brinton, 1975	64	50
Stylocheiron affine Hansen, 1910 'California Curr. form'	EU	Brinton, 1975	67	50
Stylocheiron affine Hansen, 1910 'Central form'	EU	Brinton, 1975	91	50, 72
Stylocheiron affine Hansen, 1910 'E. Equatorial form'	EU	Brinton, 1975	67	50
Stylocheiron affine Hansen, 1910 'Indo-Australian form'	EU	Brinton, 1975	67	50
Stylocheiron affine Hansen, 1910 'W. Equatorial form'	EU	Brinton, 1975	67	50
Stylocheiron carinatum G.O. Sars, 1883	EU	Brinton, 1975	63	50, 60, 84
Stylocheiron elongatum G.O. Sars, 1883	EU	Brinton, 1975	65	52
Stylocheiron indicus Silas & Mathew, 1967	EU	Brinton, 1975	66	50
Stylocheiron insulare Hansen, 1910	EU	Brinton, 1975	63	50
Stylocheiron longicorne G.O. Sars, 1883 'Long form'	EU	Brinton, 1975	66	50, 98
Stylocheiron longicorne G.O. Sars, 1883 'N. Indian form'	EU	Brinton, 1975	66	50, 98
Stylocheiron longicorne G.O. Sars, 1883 'Short form'	EU	Brinton, 1975	66	50, 98

Stylocheiron maximum Hansen, 1908	EU	Brinton, 1975; Mauchline & Fisher, 1969	72	58, 84
Stylocheiron microphthalma Hansen, 1910	EU	Brinton, 1975	65	50, 52
Stylocheiron robustum Brinton, 1962	EU	Brinton, 1975	94	50, 74,
Stylocheiron suhmi G.O. Sars, 1883	EU	Brinton, 1975	90	50, 70,
Symbolopharus spec.	PI	McGinnis, 1974	148	120
Symplectoteuthis onalaniensis (Lesson, 1830)	CE	Clarke, 1966	50	38
Temora discaudata (Thompson & Scott, 1898)	CO	Fleminger, 1975	131	105
Temora longicornis (O. F. Müller, 1785)	CO	Fleminger, 1975	132	107
Temperature profile of Atlantic Ocean	--	Deacon, 1962	25	23
Temperature profile of the Indian Ocean	--	-----	30	27
Temperature profile of the N. Atlantic Ocean	--	Richardson et al., 1981	27	23
Temperature profile of the N. Atlantic Ocean	--	Van der Spoel, 1981	26	23
Temperature profile of the Pacific Ocean	--	Deacon, 1962	33	29
Temperature fluctuations	--	-----	19	18
Temperature of sea-surface	--	Sverdrup et al., 1942	18	18
Temperature of sea-surface, 18000 BP	--	CLIMAP, 1976	20	18
Temperature at 400 m	--	-----	24	22
Thalia longicauda (Quoy & Gaimard, 1824)	TH	Van Soest, 1973	122	97
Thalia sibogae Van Soest, 1973	TH	Van Soest, 1973	122	84, 97
Thaliacea zonation	TH	Van Soest, 1973	9	10
Thysanopoda obtusifrons G. O. Sars, 1883	EU	Brinton, 1975	93	72
Thysanoessa gregaria (G. O. Sars, 1885)	EU	Mauchline & Fisher, 1969	83	66
Thysanoessa macrura G. O. Sars, 1885	EU	Mauchline & Fisher, 1969	87	68
Thysanoessa raschi (M. Sars, 1864)	EU	Mauchline & Fisher, 1969	134	108
Todarodes sagittatus (Lamarck, 1799)	CE	Clarke, 1966	155	124
Trichechus inunguis (Natterer, 1883)	MA	Mohr, 1957	145	118
Trichechus manatus latirostris (Harlan, 1824)	MA	Mohr, 1957	145	118
Trichechus manatus manatus Linnaeus, 1758	MA	Mohr, 1957	145	118
Trichechus senegalensis Link, 1795	MA	Mohr, 1957	145	118
Vertical migration types	--	Vinogradov, 1970 (partly)	156	127
Water mass profile of the Atlantic Ocean	--	-----	29	24
Water mass profile of the Indian Ocean	--	-----	32	27
Water mass profile of the Pacific Ocean	--	-----	35	29
Winteria telescopa Brauer, 1901	PI	Beklemishev et al., 1977	100	78
Zooplankton patterns	--	Beklemishev et al., 1977	6	8

9 Literature

ALLEN, J.A. & R.S.SCHELTEMA, 1972. The functional morphology and geographical distribution of Planktomya henseni, a supposed neotonous pelagic bivalve.—*J. mar. biol. Ass. UK. 52*: 19-31.

ANGEL, M.V. & A. DE C.BAKER, 1982. Vertical distribution of the standing crop of plankton and micronekton at three stations in the Northeast Atlantic.—*Biol. Oceanogr. 2* (1): 1-30.

ANGEL, M.V. & M.J.R.FASHAM, 1975. Analysis of the vertical and geographic distribution of the abundant species of planktonic ostracods in the North-East Atlantic.—*J. mar. biol. Ass. UK. 55*: 709-737.

ANGEL, M.V., P.HARGREAVES, P.KIRKPATRICK & P.DOMANSKI, 1982. Low variability in planktonic and micronektonic populations at 1,000 m depth in the vicinity of 42°N 17°W; evidence against diel migratory behaviour in the majority of species.—*Biol. Oceanogr. 1* (3): 287-319.

BACKUS, R.H. & J.E.CRADDOCK, 1977. Pelagic faunal provinces and sound-scattering levels in the Atlantic Ocean.—In: Oceanic Sound Scattering prediction (N.R.Anderson & B.J.Zahuranec Eds) Plenum Press, New York, London: 529-547.

BÉ, A.W.H., 1977. An ecological, zoogeographic and taxonomic review of recent planktonic Foraminifera.—In: Ocean Micropaleontology I (A.T.S.Ramsay Ed.) Acad. press, London, N.York, S.Francisco: 1-100.

BÉ, A.W.H. & R.W.GILMER, 1977. A zoogeographic and taxonomic review of Euthecosomatous Pteropoda.—In: Ocean Micropaleontology I (A.T.S.Ramsay Ed.) Acad press, London, N.York, S.Francisco: 733-808.

BÉ, A.W.H. & D.S.TOLDERLUND, 1971. Distribution and ecology of living planktonic Foraminifera in surface waters of the Atlantic and Indian Oceans.—In: The micropaleontology of oceans (B.M.Funnel & W.R.Riedel Eds) Cambridge Univ. Press, Cambridge: 105-149.

BEKLEMISHEV, C.W., 1960. Southern atmospheric cyclones and the whale feeding grounds in Antarctica.—*Nature, 187*: 530-531.

BEKLEMISHEV, C.W., 1961. On the spacial structure of plankton communities in dependence of type of oceanic circulation. Boundaries of ranges of oceanic plankton animals in the North Pacific.—*Okeanologia, 6*: 1059-1072 (in Russian).

BEKLEMISHEV, C.W., 1971. Distribution of plankton as related to micropaleontology.—In: The micropaleontology of oceans (B.M.Funnel & W.R.Riedel Eds) Cambridge Univ. Press, Cambridge: 75-87.

BEKLEMISHEV, C.W., N.B.PARIN & G.N.SEMINA, 1977.

Pelagial.—In: Biogeographical structure of the ocean (M. Vinogradov Ed) (Ocean Biogeography I) Moscow: 219-261 (in Russian).
BODEN, P.B. & E.BRINTON, 1957. The euphausiid crustaceans Thysanopoda aequalis Hansen and Thysanopoda subaequalis Boden, their taxonomy and distribution in the Pacific.—*Limnol. Oceanogr. 2*: 337-341.
BOGDANOV, D.W., 1963. Map of the natural zones of the Ocean.—*Deep Sea Res. 10*: 520-523.
BOGOROV, V.G., 1974. Plankton of the world ocean.—Akad. press CCCP, Moscow: 1-319 (in Russian).
BOWMAN, T.E., 1973. Pelagic amphipods of the genus Hyperia and closely related genera (Hyperiidae).—Smithson. Contr. Zool. 136: 1-76.
BRINTON, E., 1975. Euphausiids of southeastern Asian waters.—*Naga Rep. 4* (5): 1-287.
BUZAS, M.A. & S.J.CULVER, 1980. Foraminifera: distribution of provinces in the western north Atlantic.—*Science 209*: 687-689.
CASEY, R.E., 1971. Distribution of polycystine Radiolaria in the oceans in relation to physical and chemical conditions.—In: The micropaleontology of oceans (B.M. Funnell & W.R. Riedel Eds) Cambridge Univ. Press, Cambridge: 331-341.
CONOVER, R.J., 1979. Secondary production as an ecological phenomenon.—In: Zoogeography and diversity of plankton (S. van der Spoel & A.C. Pierrot-Bults Eds) Bunge, Utrecht: 50-86.
CLARKE, M.R., 1966. A review of the systematics and ecology of oceanic squids.—*Adv. mar. Biol. 4*: 91-300.
CLIMAP project members, 1976. The surface of the Ice-Age earth.—*Science, 191*: 1131-1137.
DAHL, F., 1923. Ökologische Tiergeographie II. Fischer Verl., Jena: 1-122.
DAVID, W.K.A., W.L.PERRYMAN & W.F.PERRIN, 1979. Dolphin distribution and the relationship to environmental features in the Eastern Tropical Pacific.—*S. Fish. Centre Adm. Rep. L.J.*: 79-143.
DEACON, G.E.R., 1962. Oceans. Paul Hamlyn, London, 297 pp.
DUNBAR, M.J., 1954. The amphipod Crustacea of Ungave Bay, Canadian Eastern Arctic.—*J. Fish. Res. Bd Canada 11* (6): 709-798.
DUNBAR, M.J., 1979. The relation between oceans.—In: Zoogeography and diversity of plankton (S. van der Spoel & A.C. Pierrot-Bults Eds) Bunge, Utrecht: 112-125.
DUPONT, L., 1979. Note on variation in Diacria Gray, 1847, with description of a species new to science, Diacria rampali nov. spec. and a forma new to science Diacria trispinosa forma atlantica nov. forma.—*Malacologia 18*: 37-52.
EMILIANI, C., S.GARTNER, B.LIDZ, K.ELDRIDGE, D.K.ELVEY, T.C.HUANG, J.J.STIP & M.F.SWANSON, 1975. Paleoclimatological analysis of the late Quarternary cores from the Northeastern Gulf of Mexico.—*Science, 189*: 1083-1088.
ENDLER, J.A., 1977. Geographic variation, speciation, and clines.—Monogr. pop. Biol. 10. Princeton, New Jersey, 246 pp.
ERICSON, D.B. & G.WOLLIN, 1970. Pleistocene climates in the Atlantic and Pacific Oceans: a comparison based on deepsea sediments.—*Science, 167*: 1483-1485.
FLEMINGER, A., 1972. Habitat patterns among epiplanktonic calanoid copepods.—*Trans. Amer. Microsc. Soc. 91* (1): 86-87.
FLEMINGER, A., 1973. Pattern, number, variability and taxonomic significance of integumental organs (sensilla and glandular pores) in the genus Eucalanus (Calanoida, Crustacea).—*Fish. Bull. 71* (4): 965-1010.
FLEMINGER, A., 1975. Geographical distribution and morphological divergence in American coastalzone planktonic copepods of the genus Labidocera.—In: Estuarine research (L.E. Cronin Ed.). Acad. Press., N. York: 392-419.
FLEMINGER, A. & K.HULSEMANN, 1973. Relationship of Indian Ocean epiplanktonic calanoids to the World Ocean.—In: The biology of the Indian Ocean (B. Zeitzschel & S.A. Gerlach Eds) Springer Verl., Berlin, Heidelberg, New York: 339-348.
FLEMINGER, A. & K.HULSEMANN, 1974. Systematics and distribution of the four sibling species comprising the genus Pontellina Dana (Copepoda, Calanoida).—*Fish. Bull. 72* (1): 63-120.
FLEMINGER, A. & K.HULSEMANN, 1977. Geographic range and taxonomic divergence in North Atlantic Calanus (C. helgolandicus, C. finmarchicus, C. glacialis).—*Mar. Biol. 40*: 233-248.
FOXTON, P., 1961. Salpa fusiformis Cuvier and related species.—*Disc. Rep. 32*: 1-32.
FROST, B.W. & A.FLEMINGER, 1968. A revision of the genus Clausocalanus (Copepoda, Calanoida) with remarks on distributional patterns in diagnostic characters.—*Bull. Scripp Inst. Oceanogr. 12*: 1-235.
GARDINER, L.F., 1975. The systematics, postmarsupial development and ecology of the deep-sea family Neotanaidae (Crustacea; Tanaidacea).—*Smithson. Contr. Zool. 170*: 1-265.
GEITZENAUR, K.R., M.B.ROCHE & A.MCINTYRE, 1977. Coccolith biogeography from North Atlantic and Pacific surface sediments.—In: Ocean Micropaleontology II (A.T.S. Ramsay Ed.) Acad. Press, London, N. York, S. Francisco: 973-1008.
GIBBS, R.H., 1969. Taxonomy, dimorphism, vertical distribution and evolutionary zoogeography of the bathypelagic fish genus Stomias.—*Smithson. Contr. Zool. 31*: 1-25.
GOLDBERGER, R.F., 1974. Autogenous regulation of gene expression.—*Science, 183*: 810-816.
GOLL, R.H., 1976. Morphological intergradation between modern populations of Lophospyris and Phormospyris (Trissocyclidae, Radiolaria).—*Micropal. 22* (4): 379-418.
GRAVES, J.E. & G.N.SOMERO, 1982. Electrophoretic and functional enzymic evolution in four species of eastern Pacific Barracudas from different thermal environments.—*Evolution 36* (1): 97-106.
HAMMOND, A.L., 1974. Undersea storms; experiment in the Atlantic.—*Science, 185*: 244-247.
HANSEN, B., 1967. The taxonomy and zoogeography of the deep-sea Holothurians in their evolutionary aspects.—*Stud. trop. Oceanogr. 5*: 480-501.
HARDY, A.C., 1935. The plankton community, the whale fisheries, and the hypothesis of animal exclusion.—*Disc. Rep. 11*: 273-370.
HASLE, G.R., 1976. The biogeography of some marine planktonic diatoms.—*Deep Sea Res. 23*: 319-338.
HAURY, L.R., J.A.MCGOWAN & P.H.WIEBE, 1977. Patterns and processes in the time-space scales of plankton

distributions.—In: Spatial pattern in plankton communities (J.H. Steele Ed.) Plenum Press, N. York, London: 277-327.
HAURY, L.R. & P.H. WIEBE, 1982. Fine-scale multi-species aggregation of oceanic plankton.—*Deep Sea Res. 29* (7A): 915-921.
HERMAN, Y., 1979. Plankton distribution in the past.—In: Zoogeography and diversity of plankton (S. van der Spoel & A.C. Pierrot-Bults Eds) Bunge, Utrecht: 29-49.
HESSE, R., 1924. Tiergeographie. Fischer Verl., Jena, 613 pp.
HEINRICH, A.K., 1969. On the tropical plankton communities in the Western Pacific.—*J. Cons. intern. Expl. Mer. 33* (1): 45-52.
HUXLEY, J.S., 1940. Towards the new systematics.—In: The new systematics (J.S. Huxley Ed.) Oxford Univ. press, Oxford: 1-46.
HUXLEY, J.S., 1942. Evolution, the modern synthesis. George Allen & Unwin, London, 645 pp.
JUDKINS, D.C., 1978. Pelagic shrimps of the Sergestes edwardsii species group (Crustacea, Decapoda, Sergestidae).—*Smithson. Contr. Zool. 256*: 1-34.
KNUDSEN, J., 1970. The systematic and biology of abyssal and hadal bivalvia.—*Galathea Rep. 11*: 1-241.
KOSLOW, J.A. & A. OTA, 1981. The ecology and vertical migration in three common zooplankters in the La Jolla Bight, April-August 1967.—*Biol. Oceanogr. 1* (2): 107-134.
KRAMP, P.L., 1959. The Hydromedusae of the Atlantic Ocean and adjacent waters.—*Dana Rep. 46*: 1-283.
KRAMP, P.L., 1968. The Hydromedusae of the Pacific and Indian Oceans, sections II and III.—*Dana Rep. 72*: 1-200.
LAWSON, T.J., 1977. Community interaction and zoogeography of the Indian Ocean Candaciidae (Copepoda, Calanoida).—*Mar. Biol. 42*: 71-92.
LOMAKINA, N.B., 1978. Euphausiids (Euphausiacea) of the World Ocean. Nauka, Leningrad, 222 pp. (in Russian).
LONGHURST, A.R., 1976. Vertical migration.—In: Ecology of the seas (D.H. Cushing & J.J. Walsh Eds) Blackwell Sc. Publ., Oxford, London: 116-137.
MAUCHLINE, J. & L.R. FISHER, 1969. The biology of Euphausiids.—*Adv. mar. Biol. 7*: 1-454.
MAYR, E., 1963. Animal species and evolution. Belknap press, Harvard Univ. press, Cambridge, 797 pp.
MAYR, E., 1971. Populations, species and evolution. Belknap press, Harvard Univ. press, Cambridge, 453 pp.
MCGINNIS, R.F., 1974. Counterclock wise circulation in the Pacific subantarctic sector of the Southern Ocean.—*Science 186*: 736-738.
MCGOWAN, J.A., 1954. The relationship of the distribution of the planktonic worm Poeobius meseres Heath, to the water masses of the North Pacific.—*Deep Sea Res. 6*: 125-139.
MCGOWAN, J.A., 1967. Distribution Atlas of pelagic molluscs in the California current region.—*Calcofi Atlas 6*: 1-218.
MCGOWAN, J.A., 1971. Ocean biogeography of the Pacific.—In: The micropaleontology of oceans (B.M. Funnel & W.R. Riedel Eds) Cambridge Univ. press, London: 3-74.
MCINTYRE, A., A.W.H. BÉ & M.B. ROCHE, 1970. Modern Pacific Coccolithophorida.—*Trans. N. York Acad. Sci.* (II) *32* (6): 720-731.
MEISENHEIMER, J., 1905. Pteropoda.—*Wiss. Ergebn. dt. Tiefsee Exp. 'Valdivia' 1898-1899. 9* (1): 1-134.
MIKKELSE, P.M., 1982. Studies on euphausiacean crustaceans from the Indian river region of Florida. I. Systematics of the Stylocheiron longicorne species group with emphasis on reproductive morphology.—*Proc. biol. Soc. Wash. 94* (4): 1174-1202.
MILLER, C.B., 1970. Some environmental consequences of vertical migration in marine zooplankton.—*Limnol. Oceanogr. 15*: 727-741.
MOHR, E., 1957. Sirenen oder Seekühe. Ziemsen Verl., Wittenberg, 61 pp.
MOORE, H.B., 1958. Marine Ecology. Wiley & Son, N. York, 493 pp.
MÖRZER-BRUYNS, W.F.J., 1971. Field guide of whales and dolphins. Tor, Amsterdam, 258 pp.
MURINA, G.V.V., 1977. Marine Sipunculid worms of the Arctic and Boreal waters of Eurasia. Nauka, Leningrad, 281 pp (in Russian).
MUSAEVA, E.I., 1976. Distribution of Siphonophores in the eastern part of the Indian Ocean.—Ecol. Biogeogr. Plankton. *Transact. P.P. Shirshov Inst. Oceanology 105*: 171-197.
OMORI, M., 1975. The systematics, biogeography and fishery of epipelagic shrimps of the genus Acetes (Crustacea, Decapoda, Sergestidae).—*Bull. Ocean. Res. Inst. Univ. Tokio. 7*: 1-91.
OMORI, M., 1977. Distribution of warm water epiplanktonic shrimps of the genera Lucifer and Acetes (Macrura, Penaeidae, Sergestidae).—*Proc. Symp. Warm Water Zoopl.* (suppl. publ. UNESCO/NIO): 1-12.
PAFORT-VAN IERSEL, T., 1981. Variation in Diacria trispinosa (De Blainville, 1821) and Clio pyramidata Linnaeus, 1767 (Pteropoda, Mollusca) of the Amsterdam Mid North Atlantic Plankton Expedition 1980.—*Beaufortia 32* (6): 97-115.
PIERROT-BULTS, A.C., 1974. Taxonomy and zoogeography of certain members of the Sagitta serratodentata group (Chaetognatha).—*Bijdr. Dierk. 44* (2): 215-234.
PIERROT-BULTS, A.C., 1975. Taxonomy and zoogeography of Sagitta planctonis Steinhaus, 1896 (Chaetognatha) in the Atlantic Ocean.—*Beaufortia 23* (297): 27-51.
PIERROT-BULTS, A.C., 1976. Zoogeographic patterns in Chaetognaths and some other planktonic organisms.—*Bull. zool. Mus. Univ. Amsterdam 5* (8): 59-72.
PIERROT-BULTS, A.C. & S. VAN DER SPOEL, 1979. Speciation in macrozooplankton.—In: Zoogeography and diversity of plankton (S. van der Spoel & A.C. Pierrot-Bults Eds) Bunge, Utrecht: 144-167.
PLATT, T. & K.L. DENMAN, 1975. Spectral analysis in ecology.—*A. Rev. Ecol. Syst. 6*: 189-210.
ROPER, C.F.E. & R.E. YOUNG, 1975. Vertical distribution of pelagic cephalopods. *Smithson. Contr. Zool. 209*: 1-51.
ROTTMAN, M.L.G., 1977. Euthecosomatous pteropods and planktonic Foraminifera in Southeast Asian marine waters.—Ph. D. Thesis Univ. Colorado dept. Geology: 1-235.
SAKTHIVEL, M., 1972. Studies on Desmopterus Chun, 1889 species in the Indian Ocean.—*'Meteor' Forsch. Ergebn.* (D) *10*: 46-57.
SCHALK, P.H., M.S. Sound scattering patterns in the Mid North Atlantic.
SCHELTEMA, R.S., 1971a. Larval dispersal as a means of genetic exchange between geographically separated populations of shallow water benthic marine gastropods.—*Biol. Bull. Woods Hole 140* (2): 284-322.
SCHELTEMA, R.S., 1971b. The dispersal of larvae of shoal-water benthic invertebrate species over long distances by ocean currents.—In: Proc. fourth Europ. Marine Biol. Symp. Cambridge Univ. press, London: 7-28.

SEMINA, H.J., 1979. The geography of plankton diatoms of the Southern Ocean.—*Nova Hedwigia* (beih.) *64*: 341-356.
SHIH, C.-T., 1969. The systematics and biology of the family Phronimidae (Crustacea, Amphipoda).—*Dana Rep. 74*: 1-100.
SHIH, C.-T., 1979. East-west diversity.—In: Zoogeography and diversity of plankton (S. van der Spoel & A.C. Pierrot-Bults Eds) Bunge, Utrecht: 87-102.
STEELE, J.H., 1976. Patchiness.—In: The ecology of the Seas (D.H. Cushing & J.J. Walsh Eds) Blackwell Sc. Publ., Oxford, London, Edinburgh, Melbourne: 98-115.
STUBBINGS, H.G., 1938. Pteropoda.—*John Murray Rep. 5* (2): 1-33.
SVERDRUP, H.U., M.W. JOHNSON & R.H. FLEMINGER, 1942. The oceans, their physics, chemistry and general biology. Prentice-Hall. Inc., N. York?, 1087 pp.
TOKIOKA, T., 1974. On the specific validity in species pairs or trios of plankton animals, distributed respectively in different but adjoining water masses as seen in Chaetognaths.—*Publs. Seto mar. biol. Lab. 21* (5/6): 393-408.
TOKIOKA, T., 1979. Neritic and oceanic plankton.—In: Zoogeography and diversity of plankton (S. van der Spoel & A.C. Pierrot-Bults Eds) Bunge, Utrecht: 126-143.
VAN DER SPOEL, S., 1967. Euthecosomata, a group with remarkable developmental stages (Gasteropoda, Pteropoda) Noorduyn & Zn, Gorinchem, 375 pp.
VAN DER SPOEL, S., 1971. Some problems in infraspecific classification of holoplanktonic animals.—*Z. zool. Syst. Evol. Forsch. 9*: 107-138.
VAN DER SPOEL, S., 1974. Geographic variation in Cavolinia tridentata (Mollusca, Pteropoda).—*Bijdr. Dierk. 44* (1): 100-112.
VAN DER SPOEL, S., 1976. Pseudothecosomata, Gymnosomata and Heteropoda. Bohn, Scheltema & Holkema, Utrecht, 484 pp.
VAN DER SPOEL, S., 1981. List of discrete depth samples and open net hauls of the Amsterdam Mid North Atlantic Plankton Expedition 1980 (Project 101A).—*Bull. zool. Mus. Univ. Amsterdam 8* (1): 1-10.
VAN DER SPOEL, S., 1983. Patterns in plankton distribution and their relation to speciation. The dawn of pelagic biogeography.—In: Evolution, time and space: the emergence of the biosphere (R.W. Sims, J.H. Price & P.E.S. Whalley Eds) Acad. press, London.
VAN DER SPOEL, S. & A.C. PIERROT-BULTS, 1979.—Zoogeography of the Pacific Ocean.—In: Zoogeography and diversity of plankton (S. van der Spoel & A.C. Pierrot-Bults Eds) Bunge, Utrecht: 291-327.
VAN LEYEN, A. & S. VAN DER SPOEL, 1982. A new taxonomic and zoogeographic interpretation of the Diacria quadridentata group (Mollusca, Pteropoda).—*Bull. zool. Mus. Univ. Amsterdam 8* (13): 101-118.
VAN SOEST, R.W.M., 1973. The genus Thalia Blumenbach, 1798 (Tunicata, Thaliacea) with description of two new species.—*Beaufortia 20* (271): 193-212.
VAN SOEST, R.W.M., 1974. A revision of the genera Salpa Forskål, 1775, Pegea Savigny, 1861 and Ritteriella Metcalf, 1919 (Tunicata, Thaliacea).—Beaufortia 22 (293): 153-191.
VAN SOEST, R.W.M., 1975. Zoogeography and speciation in the Salpidae.—*Beaufortia 23* (307): 181-215.
VAN SOEST, R.W.M., 1979. North-south diversity.—In: Zoogeography and diversity of plankton (S. van der Spoel & A.C. Pierrot-Bults Eds) Bunge, Utrecht: 103-111.
VINOGRADOV, M.E., 1970. Vertical distribution of the oceanic zooplankton (transl. from Russian) U.S. dept. Inter., Nat. Sc. Found Washington (Israel progr. sc. transl.), 339 pp.
VORONINA, N.M., 1978. Variability of ecosystems.—In: Advances in Oceanography (H. Charnock & G. Deacon Eds) Plenum, N. York, London: 221-244.
WORTHINGTON, L.V., 1976. On the North Atlantic circulation.—*John Hopkins Oceanogr. Stud. 6*: 1-110 (John Hopkins Univ. press, Baltimore, London).
ZVEREVA, Z.A., 1975. Geographical and seasonal variability of marine plankton.—*Explor. Mar. Fauna 12* (20) Akad. Nauk. CCCP Zool. Inst. (Israel progr. sc. transl.).

GPSR Compliance

The European Union's (EU) General Product Safety Regulation (GPSR) is a set of rules that requires consumer products to be safe and our obligations to ensure this.

If you have any concerns about our products, you can contact us on ProductSafety@springernature.com

In case Publisher is established outside the EU, the EU authorized representative is:

Springer Nature Customer Service Center GmbH
Europaplatz 3
69115 Heidelberg, Germany

Batch number: 10615097

Printed by Printforce, the Netherlands